Life in Extreme Environments

Insights in Biological Capability

From deep ocean trenches and the geographical poles to outer space, organisms can be found living in remarkably extreme conditions. This book provides a captivating account of these systems and their extraordinary inhabitants, 'extremophiles'. A diverse, multidisciplinary group of experts discuss responses and adaptations to change; biodiversity, bioenergetic processes, and biotic and abiotic interactions; polar environments; and life and habitability, including searching for biosignatures in the extraterrestrial environment. The editors emphasise that understanding these systems is important for increasing our knowledge and utilising their potential, but this remains an understudied area. Given the threat to these environments and their biota caused by climate change and human impact, this timely book also addresses the urgency to document these systems. It will help graduate students and researchers in conservation, marine biology, evolutionary biology, environmental change and astrobiology better understand how life exists in these environments and its susceptibility or resilience to change.

GUIDO DI PRISCO was Professor of Biochemistry and CNR (National Research Council) Research Associate, Institute of Biosciences and Bioresources, Naples, Italy. He was the CNR Research Director up until his retirement in 2004. He took part in numerous expeditions in both the Antarctic and Arctic. On 29 September 2019, Guido passed away after a serious illness.

HOWELL G.M. EDWARDS is Emeritus Professor of Molecular Spectroscopy, Faculty of Life Sciences, University of Bradford, UK. He is a member of the International Science Team on the RLS Raman instrument for the ExoMars 2020 mission. He has published over 1300 papers on Raman spectroscopy and its applications.

JOSEF ELSTER is Professor at the Centre for Polar Ecology, Faculty of Science, University of South Bohemia, České Budějovice, and senior scientist, Phycology Centre, Institute of Botany, Academy of Science of the Czech Republic, Třeboň, Czech Republic. His expertise is in the field and laboratory study of polar cyanobacteria and microalgae. He has led or been a member of many polar research expeditions and was founder of the Czech Arctic Research Infrastructure 'Josef Svoboda Station', Svalbard.

AD H.L. HUISKES is a guest scientist at the Royal Netherlands Institute for Sea Research (NIOZ), Yerseke, Netherlands. He led the Unit of Polar Ecology at the Netherlands Institute of Ecology, Yerseke, was acting Director of the Centre of Estuarine and Marine Ecology (now a division of NIOZ), and subsequently acting Director of the Yerseke branch of NIOZ. In addition, he was Vice President of the Scientific Committee for Antarctic Research (SCAR, 2008–2012) and Lecturer of Polar Ecology at the University of Groningen.

Ecological Reviews

Ecological Reviews publishes books at the cutting edge of modern ecology, providing a forum for volumes that discuss topics that are focal points of current activity and likely long-term importance to the progress of the field. The series is an invaluable source of ideas and inspiration for ecologists at all levels from graduate students to more-established researchers and professionals. The series has been developed jointly by the British Ecological Society and Cambridge University Press and encompasses the Society's Symposia as appropriate.

Life in Extreme Environments

Insights in Biological Capability

Edited by

GUIDO DI PRISCO
National Research Council of Italy

HOWELL G.M. EDWARDS
University of Bradford

JOSEF ELSTER
University of South Bohemia

AD H.L. HUISKES
Royal Netherlands Institute for Sea Research

CAMBRIDGE
UNIVERSITY PRESS

CAMBRIDGE
UNIVERSITY PRESS

University Printing House, Cambridge CB2 8BS, United Kingdom

One Liberty Plaza, 20th Floor, New York, NY 10006, USA

477 Williamstown Road, Port Melbourne, VIC 3207, Australia

314–321, 3rd Floor, Plot 3, Splendor Forum, Jasola District Centre, New Delhi – 110025, India

79 Anson Road, #06–04/06, Singapore 079906

Cambridge University Press is part of the University of Cambridge.

It furthers the University's mission by disseminating knowledge in the pursuit of education, learning, and research at the highest international levels of excellence.

www.cambridge.org
Information on this title: www.cambridge.org/9781108498562
DOI: 10.1017/9781108683319

First published 2021

Printed in the United Kingdom by TJ International Ltd, Padstow Cornwall

A catalogue record for this publication is available from the British Library.

Library of Congress Cataloging-in-Publication Data
Names: Di Prisco, Guido, 1937– editor. | Edwards, Howell G. M., 1943–
editor. | Elster, J. (Josef), editor. | Huiskes, A. H. L., editor.
Title: Life in extreme environments : insights in biological capability / edited by Guido di Prisco, National Research Council of Italy, Howell G.M. Edwards, University of Bradford, Josef Elster, University of South Bohemia, Czech Republic, Ad H.L. Huiskes, Royal Netherlands Institute of Sea Research.
Description: Cambridge, UK ; New York, NY : Cambridge University Press, 2021. | Series: Ecological reviews | Includes bibliographical references and index
Identifiers: LCCN 2020014069 (print) | LCCN 2020014070 (ebook) | ISBN 9781108498562 (hardback) | ISBN 9781108683319 (ebook)
Subjects: LCSH: Extreme environments – Microbiology. | Biodiversity – Climatic factors. | Microbial ecology. | Exobiology. | Adaptation (Physiology) | Acclimatization. | Climatic changes.
Classification: LCC QR100.9 .L54 2021 (print) | LCC QR100.9 (ebook) | DDC 578.75/8–dc23
LC record available at https://lccn.loc.gov/2020014069
LC ebook record available at https://lccn.loc.gov/2020014070

ISBN 978-1-108-49856-2 Hardback
ISBN 978-1-108-72420-3 Paperback

Cambridge University Press has no responsibility for the persistence or accuracy of URLs for external or third-party internet websites referred to in this publication and does not guarantee that any content on such websites is, or will remain, accurate or appropriate.

The EU FP7 Coordination Action for Research on Life in Extreme
Environments (CAREX, 2008–2010) brought together scientists
from all over Europe and beyond to discuss the various challenges
for organisms living in extreme environments such as polar,
desert, deep sea, anoxic, mine spoil and even extraterrestrial
environments. Guido di Prisco was one of the initiators of the
project and the promoter of the publication of a book dealing with
the topics discussed during the various workshops of the CAREX
programme. It is with deep regret that we had to finish the
production of this book without him, as he passed away in
September 2019.

We dedicate this book to him.

Contents

Colour plates can be found between pages 142 and 143.

Contributors

AMANDA M. ACHBERGER
Department of Oceanography
Texas A&M University
College Station
USA
aachberger@tamu.edu

NICOLETTA ADEMOLLO
Institute of Polar Science
National Research Council
Venezia, Italy
ademollo@irsa.cnr.it

STEFANIA ANCORA
Department of Physics
Earth and Environmental Sciences
University of Siena
Siena
Italy
stefania.ancora@unisi.it

CARLOS ANGULO-PRECKLER
Department of Evolutionary Biology
Ecology and Environmental Sciences
& Biodiversity Research Institute
Universitat de Barcelona
Barcelona
Catalonia, Spain
carlospreckler@hotmail.com

CONXITA AVILA
Department of Evolutionary Biology

Ecology and Environmental Sciences
& Biodiversity Research Institute
Universitat de Barcelona
Barcelona
Catalonia, Spain

JIŘÍ BÁRTA
Department of Ecosystem Biology
University of South Bohemia
České Budějovice
Czech Republic
jiri.barta@prf.jcu.cz

JIŘÍ ČERNÝ
Institute of Parasitology, Biology
Centre of the Czech Academy of
Sciences
České Budějovice
Czech Republic
Faculty of Tropical AgriSciences
Czech University of Life Sciences in
Prague
Prague
Czech Republic
Centre for Polar Ecology
University of South Bohemia
České Budějovice
Czech Republic
cerny@paru.cas.cz

STEVEN L. CHOWN
School of Biological Sciences

Monash University
Victoria
Australia
steven.chown@monash.edu

JØRGEN S. CHRISTIANSEN
Department of Arctic and Marine
Biology
UiT The Arctic University of Norway
Tromsø
Norway
Environmental and Marine Biology
Åbo Akademi University
Turku
Finland
jorgen.s.christiansen@uit.no

MELODY S. CLARK
Natural Environment Research
Council
British Antarctic Survey
Cambridge
UK
mscl@bas.ac.uk

DANIELA COPPOLA
National Research Council (CNR) –
Institute of Bioscience and
BioResources (IBBR)
Naples
Italy
daniela.coppola@ibbr.cnr.it

CINZIA CORINALDESI
Department of Life and
Environmental Sciences
Polytechnic University of Marche
Ancona
Italy
c.corinaldesi@univpm.it

SIMONETTA CORSOLINI
Department of Physics, Earth and
Environmental Sciences
University of Siena

Siena
Italy
corsolini@unisi.it

ROBERTO DANOVARO
Department of Life and
Environmental Sciences
Polytechnic University of Marche
Ancona
Italy
Zoological Station Anton Dorhn
Naples
Italy
r.danovaro@staff.univpm.it

PAULA DE CASTRO-
FERNANDEZ
Department of Evolutionary Biology
Ecology and Environmental Sciences
and Biodiversity Research Institute
Universitat de Barcelona
Barcelona
Catalonia, Spain
pauladecastrofer@gmail.com

ANTONIO DELL'ANNO
Department of Life and
Environmental Sciences
Polytechnic University of Marche
Ancona
Italy
a.dellanno@univpm.it

GERHARD DIECKMANN
Helmholtz Centre for Polar and
Marine Research
Alfred Wegener Institute
Bremerhaven
Germany
gerhard.dieckmann@awi.de

GUIDO DI PRISCO
National Research Council (CNR) –
Institute of Bioscience and
BioResources (IBBR)

Naples
Italy
guido.diprisco@ibbr.cnr.it

HOWELL G.M. EDWARDS
School of Chemistry and Biosciences
University of Bradford
Bradford
UK
H.G.M.Edwards@bradford.ac.uk

JOSEF ELSTER
Centre for Polar Ecology, Faculty of
Science
University of South Bohemia
České Budějovice
Czech Republic
Centre for Phycology, Institute of
Botany
Academy of Sciences of the Czech
Republic
Třeboň
Czech Republic
jelster@prf.jcu.cz, Josef.Elster@ibot
.cas.cz

JANA ELSTEROVÁ
Institute of Parasitology
Biology Centre of the Czech
Academy of Sciences
České Budějovice
Czech Republic
University of South Bohemia
České Budějovice
Czech Republic
Veterinary Research Institute
Brno
Czech Republic
elsterova@paru.cas.cz

SARA FERRANDO
DISTAV, University of Genoa
Genoa

Italy
sara.ferrando@unige.it

BLANCA FIGUEROLA
Smithsonian Tropical Research
Institute
Balboa
Republic of Panama
bfiguerola@gmail.com

SILVIA FINESCHI
National Research Council (CNR) –
Institute for Plant Protection
Sesto Fiorentino (Florence)
Italy
s.fineschi@ipp.cnr.it

CRISTINA GAMBI
Department of Life and
Environmental Sciences
Polytechnic University of Marche
Ancona
Italy
c.gambi@staff.univpm.it

LAURA GHIGLIOTTI
IAS, National Research Council (CNR)
Genoa
Italy
laura.ghigliotti@ge.ismar.cnr.it

DANIELA GIORDANO
National Research Council (CNR) –
Institute of Bioscience and
BioResources (IBBR)
Naples
Italy
daniela.giordano@ibbr.cnr.it

LIBOR GRUBHOFFER
Institute of Parasitology
Biology Centre of the
Czech Academy of Sciences
České Budějovice

Czech Republic
Centre for Polar Ecology
University of South Bohemia
České Budějovice
Czech Republic
libor.grubhoffer@bc.cas.cz

JULIAN GUTT
Helmholtz Centre for
Polar and Marine Research
Alfred Wegener Institute
Bremerhaven
Germany
julian.gutt@awi.de

FATEMEH HEIDARI
Faculty of Life Sciences and
Biotechnology
Shahid Beheshti University
Tehran
Iran
Fateme.hidary@yahoo.com

EVA HEJDUKOVÁ
Department of Ecology,
Faculty of Science
Charles University in Prague
Prague
Czech Republic
eva.hejdukova@natur.cuni.cz

STÉPHANE HOURDEZ
Laboratoire d'écogéochimie
des environnements benthiques
(LECOB)
Observatoire Océanologique de
Banyuls
Banyuls-sur-Mer
France
hourdez@obs-banyuls.fr

AD H.L. HUISKES
Royal Netherlands Institute
of Sea Research
Yerseke

The Netherlands
ad.huiskes@nioz.nl

JAN JEHLICKA
Institute of Geochemistry,
Mineralogy and Mineral Resources,
Faculty of Science
Charles University in Prague
Prague
Czech Republic
jehlicka@natur.cuni.cz

DIDIER JOLLIVET
Adaptation et Diversité en Milieu
Marin (AD2 M)
Station Biologique de Roscoff
France
jollivet@sb-roscoff.fr

JANA KVÍDEROVÁ
Centre for Polar Ecology, Faculty of
Science
University of South Bohemia
České Budějovice
Czech Republic
jana.kviderova@objektivem.net

FRANCESCO LORETO
Department of Biology, Agriculture
and Food Sciences (Roma)
Italy
francesco.loreto@cnr.it

ARVE LYNGHAMMAR
The Norwegian College of Fishery
Science
UiT The Arctic University
of Norway
Tromsø
Norway
arve.lynghammar@uit.no

RAFAEL MARTÍN-MARTÍN
Department of Evolutionary Biology,
Ecology and Environmental

Sciences, and Biodiversity Research
Institute
Universitat de Barcelona
Barcelona
Catalonia, Spain
ginkopsida@gmail.com

ALEXANDER B. MICHAUD
Center for Geomicrobiology,
Department of Bioscience
Aarhus University
Aarhus
Denmark
a.b.michaud@gmail.com

LINDA NEDBALOVÁ
Department of Ecology
Charles University
Prague
Czech Republic
Centre for Phycology, Institute of
Botany
The Czech Academy of Sciences
Třeboň
Czech Republic
lindane@natur.cuni.cz

JULIUS NIELSEN
Marine Biological Section,
Department of Biology
University of Copenhagen
Copenhagen
Denmark
Greenland Institute of Natural
Resources
Nuuk
Greenland
juliusnielsen88@gmail.com

LLOYD S. PECK
Natural Environment Research
Council
British Antarctic Survey
Cambridge

UK
lspe@bas.ac.uk

MARTINA PICHRTOVÁ
Department of Botany
Charles University
Prague
Czech Republic
martina.pichrtova@natur.cuni.cz

EVA PISANO
IAS, National Research Council (CNR)
Genoa
Italy
pisano@ge.ismar.cnr.it

JOHN C. PRISCU
Department of Land Resources and
Environmental Science
Montana State University
Bozeman
USA
jpriscu@montana.edu

ANTONIO PUSCEDDU
Department of Life and
Environmental Sciences
University of Cagliari
Cagliari
Italy
apusceddu@unica.it

HOSSEIN RIAHI
Faculty of Life Sciences and
Biotechnology
Shahid Beheshti University
Tehran
Iran
H-Riahi@sbu.ac.ir

ROBERTA RUSSO
National Research Council
(CNR) – Institute of Biosciences
and BioResources (IBBR)
Naples

Italy
roberta.russo@ibbr.cnr.it

DANIEL RŮŽEK
Institute of Parasitology
Biology Centre of the
Czech Academy of
Sciences
České Budějovice
Czech Republic
Veterinary Research Institute
Brno
Czech Republic
ruzekd@paru.cas.cz

ZEINAB SHARIATMADARI
Faculty of Life Sciences and
Biotechnology
Shahid Beheshti University
Tehran
Iran
z_shariat@sbu.ac.ir

JOHN F. STEFFENSEN
Marine Biological Section,
Department of Biology

University of Copenhagen
Copenhagen
Denmark
jfsteffensen@bio.ku.dk

MICHAEL TANGHERLINI
Zoological Station Anton Dorhn
Naples
Italy
michael.tangherlini@szn.it

CINZIA VERDE
National Research Council (CNR) –
Institute of Bioscience and
BioResources (IBBR)
Naples
Italy
cinzia.verde@ibbr.cnr.it;
c.verde@ibp.cnr.it

TRISTA J. VICK-MAJORS
Flathead Lake Biological Station
University of Montana
Polson
USA
tristyv@gmail.com

Introduction

GUIDO DI PRISCO,
AD H.L. HUISKES,
JOSEF ELSTER
and
HOWELL G.M. EDWARDS

A few years ago, extreme environments were defined as having one or more environmental parameters showing values permanently or periodically close to the limits known for life in its various forms; only specialised organisms are able to cope with such extreme environments. Such environments can be considered as end-members of a continuum of environmental conditions that constitute limits for life as we know it. Extreme environments have been identified in marine and terrestrial biomes across the globe (deep sea, hydrothermal vents, continental margins, polar regions, hot springs, high altitude and glaciers, hot arid regions and deserts, acidic and alkaline environments, continental and seafloor subsurfaces, intertidal coastal areas, hypersaline environments, atmosphere). Outer space is also an extreme environment, comprising planetary bodies, space vessels and space itself.

Extreme environments show enormous diversity and can be either stable or unstable. In the former (e.g. polar oceans, deep biosphere), organisms *constantly* live near the limits of their physiological potential or absolutely require such conditions and can persist for long periods of time with development and selection of genotypes. Such habitats are occupied by their own well-adapted organisms (e.g. psychrophiles, hyperthermophiles and acidophiles), with usually a restricted ecological amplitude. In unstable environments, organisms live *intermittently* near the limits of their physiological potential, and must implement diverse strategies to survive. Physicochemical parameters can change in a predictable (e.g. tidal and seasonal) or chaotic (e.g. change in hydrothermal-vent regime) manner.

Cyclical environmental patterns may require a number of different adaptations. Where irregular catastrophic events occur (e.g. a volcanic eruption), the capacity of survival may be exceeded and may lead to massive mortality and extinction of local populations. Recovery to a stable ecosystem state then involves pre-adapted organisms adaptation at the level of populations and communities that drive their resilience. When studying life in extreme environments (LEXEN), the overarching question is: *How is life (meaning the forms of life caught in ensuing extreme environments to which they had to adapt) limited by, and how has it adapted to, extreme external biotic and abiotic factors?* While some biotas require harsh conditions to thrive and evolve, the stresses imposed on organisms and communities living in extreme environments have led to unique adaptation and survival mechanisms.

Several fundamental questions about life in extreme environments (LEXEN) are relevant to any life forms from microbes to vertebrates:

- What limits LEXEN (e.g. changes in temperature, salinity, pressure, pH, chemical composition of the environment or radiation, light/darkness, seasonality) and how do extremophilic organisms (living predominantly in extreme environments) differ from conventional organisms? (Of course they differ in the unique adaptations they have been forced to develop.)
- What limits life on Earth in general?
- How do species colonise new environments or confront new stressors (e.g. high concentration of metals in soils and water, salinisation of soils, urban environment pollution)? This requires studying how large is the molecular, ecological and physiological amplitude, and how much these overlap, and requires studies on all stages in the life cycle; also the life cycle as a whole needs to be investigated.
- What limits life with respect to certain groups or processes?
- What is the capacity of species to face variations in their environmental conditions? We need a definition of the limits of adaptive mechanisms, as a key to predicting future patterns of biodiversity loss or increase or ecosystem change. This question is of relevance when considering that global change leads some environments to become hostile for their organisms (e.g. ocean acidification, enhanced UV radiation in polar regions, extension of deserts, pollution, reduction in sea ice and warming of the ocean).
- What limits individual species and to what extent can they usually acclimatise or adapt?
- What limits communities or ecosystems evolving in extreme environments? Little is known about interactions among species, e.g. multitrophic relationships involved in LEXEN. A multidisciplinary effort is needed to characterise ecosystems in extreme environments.
- Which parameters shape an ecosystem, and will any change in these parameters drastically alter that ecosystem?
- What are the molecular and biochemical bases of response to conditions involving factors and stresses not yet experienced? A comprehensive understanding of the survival strategies of species through comparative genomics and proteomics would provide biological insights. These insights would not only involve new aspects of physiology and evolution that may be unique, but could also uncover unknown adaptations required for life in stressful

environments that may be shared across all domains of life. A special focus should be given to adaptations to absence of gravity, low temperature, low pressure, pollution and unusual radiation. Studying these adaptations brings valuable new insights to our knowledge of biological processes, from molecular biology to physiology, ecology and evolution.
- What happens to species facing a new environment (environment being a new set of biotic and/or abiotic parameters)?

When studying LEXEN, answers to these detailed, fundamental questions are within reach, and are opportune as the biosphere is changing as a result of atmospheric changes (climate change, ozone depletion) and of the increasing human population. The study of LEXEN may also provide answers to questions such as:

- What is the role of a particular extreme environment for regional/global ecology and element cycles?
- What is the potential societal and economic value of extreme adaptation?
- In what way are certain extreme environments analogues for ancient ter-restrial and extraterrestrial environments?

New perspectives gained through the investigation of LEXEN allow refine-ment of theories and development of new ideas, as well as considerable opportunities in terms of biotechnological, industrial and medical applications.

Whilst investigations have been undertaken for many decades on some extremes, such as hot deserts or high-altitude ecosystems, the true breadth of extreme environments in which life has been found has emerged only comparatively recently. As examples, mid-ocean ridge hydrothermal-vent biota was only discovered in the late 1970s, the extent and diversity of deep subsurface ecosystems began to be recognised in the late 1980s and the first clues for life in subglacial lakes (Antarctica) were only reported in the past decades.

In this still maturing field, the issue of exploration, inventory, character-isation and mapping of extreme environments and their associated biota can be seen as an overarching priority. What is the geographical extent of life? Are there still unknown extreme environments? New knowledge will not only be created through investigation of known ecosystems, but it is also strongly dependent on exploration and discovery of new extreme environments.

In recent decades, the Earth has experienced some major environmental changes which are having (and will have) an impact on the biosphere. This is particularly true in regions where the changes are amplified (e.g. polar regions are warming more rapidly than any other area on the planet). Life adapted to specific extreme environments will be particularly sensitive to change and so may provide a useful indicator, but is also threatened by rapid changes.

Research on LEXEN represents a major challenge at the beginning of the twenty-first century, needing to be multidisciplinary in content and interdisciplinary in execution. Its success depends upon the close coordination of diverse scientific disciplines and programmes and upon technological developments. The timing is pertinent, as the past decade has seen remarkable advances in technology for accessing extreme environments (e.g. autonomous vehicles AUV and ROV, *in situ* instrumentation) and in molecular and systems biology. Omics (genomics, proteomics, transcriptomics, metabolomics, etc.), in particular, give us the ability to recognise and identify elusive organisms and gain new insights into the metabolic capabilities and functionality of organisms from extreme environments.

It thus appears that the study of LEXEN is bound to address a very high number of questions, also in view of the current impact of current climate change, particularly rapid in some extreme environments, e.g. in polar regions. Based on these factors, the EU FP7 Coordination Action for Research on Life in Extreme Environments (CAREX, 2008–2010) had the features typical of an appropriate tool to meet the challenges outlined above. Four research themes, relevant to all extreme environments and to all types of organisms evolving within these environments, were identified and discussed by the over 220 participants from 28 countries. The themes have strong connections and relevance to important contemporary challenges, whether scientific (e.g. biodiversity, evolutionary biology, biogeochemical cycles, astrobiology), societal (e.g. food production, remediation of polluted areas, medical and biotechnological applications) or both (e.g. impacts of climate change). Each theme comprises several sub-themes. Some of these can be addressed in relatively short time scales (3 to 5 years), but many will need a decade or more to achieve significant progress. International collaboration and new technology will be necessary. LEXEN research is likely to contribute to fundamental and applied science.

This volume brings together the current LEXEN knowledge. The volume is structured in four parts. Each part pertains to one of the CAREX research themes.

Part I – Extreme environments – responses and adaptation to change:
- how do organisms respond to the stresses of extreme environments?
- are there unique/common paths for responses to stresses?
- how have proteins and genomes evolved under extreme conditions?
- how do organisms in extreme environments respond to change?

Part II – Biodiversity, bioenergetic processes, and biotic and abiotic interactions in extreme environments:

- what characterises biodiversity in extreme environments?
- how diverse are bioenergetic processes in extreme environments?
- what characterises the nature and extent of biotic and abiotic interactions in extreme environments?

Part III – Life in extreme environments and the responses to change; the example of polar environments:

- how do organisms function in an extreme environment such as the polar environment?
- how resilient is life in extreme environments to environmental change? (this question also refers to Part I)
- how do communities change as a result of environmental change?

Part IV – Life and habitability:

- what are the physicochemical boundary conditions for habitability?
- where are the terrestrial analogues for putative extraterrestrial habitats?
- what bio-signatures facilitate life detection?
- what is the role of technological developments for accessing life in extreme environments?

PART I Extreme environments: responses and adaptation to change

GUIDO DI PRISCO

Introduction

Environmental conditions have changed considerably over the history of the Earth, finally leading to the development of a large variety of harsh ecological niches (in terms of temperature, pressure, oxidative stress, desiccation, salt, pH, seasonality and radiation, etc). Populating these environments required adaptations which include the genome and protein level, and we need to understand the mechanisms that allow genes and proteins to function under these conditions.

Organisms living in extreme environments have to cope with and adapt to harsh conditions. Some are well adapted, others become dependent and get stressed by environmental variations. Any response to stress starts with stress perception, followed by activation of defence and repair mechanisms that may result in hardiness to the encountered stress. Survival is clearly the first objective when dealing with life in extreme environments, but the capacity to grow, develop and reproduce is also relevant to ensure life under these conditions.

Stress conditions can have great impact on community structure and function. The community response could be short-term, e.g. changes in net primary productivity, temporary migration of species; or long-term, e.g. permanent changes in community types. Extreme environments allow relatively fast changes in the genome of the species because the selection pressures and the interspecific competition are higher than in non-extreme environments. In some cases the community can better cope with the stress as a result of symbiosis among different organisms (e.g. lichens in cold environments).

In studying stressful situations, the spatial dimension and the level of organisation also need to be considered. Acquiring data on the genome, gene expression, protein structure and function (e.g. by crystallography, 3D molecular modelling and functional dynamics modelling) will be a precondition to studying how ecosystems are responding to climate change. The molecular properties of relevant genes and proteins need to be interpreted in a whole species/community context and then in terms of biodiversity. The focus should span from cell to ecosystem level, involving investigations of intra- and interspecific adaptation and interaction mechanisms. A deep knowledge of the genome of the studied organism would be required. The use of micro- array or proteomic approaches would facilitate understanding of the genome response to extreme conditions; gene analyses (sequence and expression) would be essential to quantify the response.

Due to the realities of acclimation, adaptation and evolution, organismal responses to extreme environments have to be investigated at both short- and long-term periods. The phenotypic and physiological plasticity is likely to be fundamental in allowing species to adjust to the current and future condi- tions, but plastic responses have limits and are unlikely to provide long-term solutions. Genetic modification is needed, via gene flow between populations and/or genetic adaptation. The time needed for genetic adaptation may not be sufficient for some multicellular organisms such as invertebrates and fish to respond effectively to changing conditions.

Some organisms develop avoidance mechanisms to escape extreme con- ditions. Besides migration, the most widespread adaptation to environmen- tal stress is dormancy. Dormancy stages are cysts, spores, seeds and desiccation of leaves for plants, desiccation for lichens, cysts or desiccated forms for animals, spores; and there are as yet unknown forms for prokar- yotes. These are more resistant stages, commonly seen as overwintering stages. Dormancy can be subdivided into *diapause* (controlled endogenously: connected with external stressors, but not directly induced by them) and *quiescence* (decrease of metabolic activity under exogenous control; a transformation to a resistant state without visible cell morphological differentiation). Starvation and entering the stationary phase can induce changes in ultrastructure (e.g. thicker cell walls, thicker cuticles for plants, optimisation of the surface:volume ratio) and biochemistry (e.g. sucrose and trehalose accumulation, change in composition of fatty acids, secretion of extracellular polysaccharides) of stressed cells. Some quiescent organisms are capable of reducing their metabolic activities to an undetectable level (cryptobiosis). They can persist for even centuries in an inactive state and resume their metabolic activity when they are brought back to growing conditions. Regular cycles of favourable and unfavourable conditions could lead to development of complex life cycles that include vegetative as well as dormant stages. The strategy of dormancy, i.e. changing the rate of physio- logical processes, is combined with additional protective mechanisms against various stress factors. Hibernation in selected invertebrates can be considered as a synonym of dormancy.

The identification of possible common mechanisms of response to extreme conditions among different species and different environments is a key issue to be investigated. Common responses being observed in a wide range of organisms should imply that these mechanisms evolved over a long time period and have been preserved for millennia. One example is oxidative stress, a common process found in several extreme situations. The mechanisms adopted by different organisms to cope with this stress are often similar. On the other hand, specific and unique mechanisms adopted by small populations in a specific environment may be indicative of their adaptation.

CHAPTER ONE

Physiological traits of the Greenland shark *Somniosus microcephalus* obtained during the TUNU-Expeditions to Northeast Greenland

GUIDO DI PRISCO

National Research Council (CNR) – Institute of Bioscience and BioResources (IBBR)

NICOLETTA ADEMOLLO

Water Research Institute, National Research Council (CNR)

STEFANIA ANCORA

University of Siena

JØRGEN S. CHRISTIANSEN

UiT The Arctic University of Norway & Åbo Akademi University, Finland

DANIELA COPPOLA

National Research Council (CNR) – Institute of Bioscience and BioResources (IBBR)

SIMONETTA CORSOLINI

University of Siena

SARA FERRANDO

University of Genoa

LAURA GHIGLIOTTI

IAS, National Research Council (CNR)

DANIELA GIORDANO

National Research Council (CNR) – Institute of Bioscience and BioResources (IBBR)

ARVE LYNGHAMMAR

UiT The Arctic University of Norway

JULIUS NIELSEN

Greenland Institute of Natural Resources

EVA PISANO

IAS, National Research Council (CNR)

ROBERTA RUSSO

National Research Council (CNR) – Institute of Bioscience and BioResources (IBBR)

JOHN F. STEFFENSEN

University of Copenhagen

and

CINZIA VERDE

National Research Council (CNR) – Institute of Bioscience and BioResources (IBBR)

1.1 Introduction

Arctic regions are inhabited by cold-adapted stenothermal or eurythermal species. Unlike in the Antarctic, eurythermal species predominate, because of opportunities for migrations to temperate latitudes. In the Antarctic sea, the modern chondrichthyan genera are scarcely represented. In contrast, in the Arctic, sharks and skates are present with about 8% of the species (Mecklenburg et al., 2011; Lynghammar et al., 2013). The distribution of the Greenland shark *Somniosus microcephalus* is quite wide; in fact, this species typically thrives in deep and extremely cold waters, seasonally covered by sea ice (MacNeil et al., 2012), but is also known to enter more temperate waters in the North Atlantic (Bigelow & Schroeder, 1948; Skomal & Benz, 2004; Campana et al. 2015). Widespread climate changes in the arctic ecosystem have led to increased attention on trophic dynamics and on the role of this apex predator in the structure of arctic marine food webs (MacNeil et al., 2012).

The Greenland shark (Bloch & Schneider, 1801; Campana et al. 2015) (Squaliformes, Somniosidae) is distributed across the North Atlantic and the adjacent Arctic from surface waters to >2900-m depth (Porteiro et al., 2017; Mecklenburg et al., 2018). It is among the largest fish species known with a verified total length (TL) of 375 cm for males and 550 cm for females (Campana et al., 2015). The Greenland shark is considered an scavenger and predator that eats practically anything, including marine mammals (Nielsen et al., 2014). But recent studies also suggest that Greenland shark may hunt actively (Leclerc et al., 2012) with an ontogenetic shift in prey from cephalopods for juveniles to fishes and seals for adults (Nielsen et al., 2019).

The Greenland shark has received increasing scientific attention because its life history, abundance and demography are largely unknown (MacNeil et al., 2012). Climate change likely affects the fishes of the North Atlantic and the Arctic Seas, and among them, the Greenland shark. Moreover, the species often turns up as unaccounted bycatch and is viewed as a nuisance in commercial bottom longline fisheries. The Greenland shark is categorised as 'near threatened' (NT) in the Red List of the International Union for Conservation of Nature (IUCN, www.iucnredlist.org/). In the Norwegian Red List (2015, www.artsdatabanken.no/) the given status is 'data deficient' (DD), flagging the precautionary principle and raising pertinent conservation issues and questions about the species' ecological resilience and physiological adaptation.

The ongoing TUNU-Programme (since 2002) at UiT The Arctic University of Norway primarily investigates the diversity and adaptation of arctic marine fishes and so conducts regular expeditions to the fjords and shelves in Northeast Greenland (Christiansen, 2012). In 2010, TUNU became affiliated with the independent Old & Cold Project on the biology of the Greenland shark (headed by John F. Steffensen at the University of Copenhagen, http://bioold.science.ku.dk/jfsteffensen/OldAndCold/). Within TUNU, the Greenland shark has become the target of joint investigations, including migration and genetic structuring, life history, physiology, orga-notropism and age/gender difference of trace elements and organic pollu-tants, and sensory capability. In 2010, the first three animals from Northeast Greenland were sampled. The following TUNU-Expeditions in 2013 and 2015 landed an additional eight specimens – altogether 11 animals (Table 1.1).

The specimen from Peters Bugt (#5) is of particular interest (Table 1.1). The core of the eye lens disclosed an independent radiocarbon marker depicting the onset of the bomb pulse from the atmospheric nuclear tests that took place around year 1960 (Campana et al., 2002; Nielsen et al. 2016). In other words, this particular specimen could be aged to about 50 years at the time of capture in 2013. This finding ignited the Greenland shark chronology from which the age of larger and presumably older animals could be estimated. The Greenland shark attracted worldwide attention in the media and gained iconic status when a Science-paper gave age estimates for the largest animals (TL: c. 500 cm) of at least 272 years – i.e. the oldest among vertebrates (Nielsen et al., 2016).

Recent studies provide new information about the diet and trophic relation-ships of the Greenland shark (Nielsen et al., 2019), and its whereabouts in time

Table 1.1 *Greenland sharks* Somniosus microcephalus *obtained from scientific long-lines during the TUNU-Expeditions and sampled for organs and tissues. N/A indicates that data are not available due to partly cannibalised animals. During TUNU-V, two additional animals were equipped with archival pop-up tags (PSAT) and released*

Animal	Expedition	Date	Location	Latitude N	Longitude W	Depth (m)	Sex	Total length (cm)
1	TUNU-IV	11 Aug 10	Ella Ø	72.47	24.40	442	M	290
2	(2010)	12 Aug 10	Rødsten	73.26	23.48	473	N/A	N/A
3		12 Aug 10	Rødsten	73.26	23.48	473	F	330
4	TUNU-V	13 Aug 13	Kap Herschell	74.05	19.32	320	F	320
5	(2013)	14 Aug 13	Peters Bugt	75.16	20.21	325	F	220
6	TUNU-VI	9 Aug 15	Shelf	74.39	13.55	395	M	291
7	(2015)	9 Aug 15	Shelf	74.39	13.55	395	M	266
8		11 Aug 15	Shelf	75.04	12.42	377	M	271
9		11 Aug 15	Shelf	75.04	12.42	377	F	322
10		11 Aug 15	Shelf	75.04	12.42	377	F	320
11		11 Aug 15	Shelf	75.04	12.42	377	N/A	N/A

and space from satellite tracking (Campana et al., 2015; Hussey et al., 2018; Fisk et al. 2012). A recent review addresses main knowledge gaps, especially reproduction biology such as generation span (age-at-maturity estimates >134 years, Nielsen et al., 2016), Nielsen et al. in press

Physiological studies on the metabolic capacity of the Greenland shark have just started to emerge (Costantini et al., 2016; Herbert et al., 2017; Shadwick et al., 2018), but further physiological insights are imperative both from *in vitro* assays on tissues and from tests on live animals. In light of the extreme life span of the Greenland shark, we sampled a range of organs and tissues from the TUNU animals (Table 1.1), and here address little known physiological traits such as structure and function of haemoglobins (Hbs), sensory capability from anatomical proxies and tissue-specific loads of pollutants. Thus, this contribution mostly deals with oxygen transport: Hbs and ligand-binding properties, ecotoxicology of trace elements and organic contaminants, and sensory capability: the sense of olfaction.

1.2 Oxygen transport: haemoglobins and ligand-binding properties

The evolutionary success of elasmobranchs has raised considerable interest in their respiratory control mechanisms (Butler & Metcalfe, 1988). Structural information on Hbs, unlike in polar teleosts, is scarce. Erythrocytes are larger than those of most vertebrates, which may limit the efficiency of the oxygen-transport system.

Herewith we summarise the structural and functional properties of the Hb system of *S. microcephalus*, with regard to Hb oxygen-binding properties and their modulation by physiological effectors. Hbs have also been structurally characterised by a combination of spectroscopic techniques (UV-Vis and resonance Raman spectroscopy, autoxidation kinetics and CO-rebinding kinetics) to gain information on the heme cavity, heme oxidation and coordination states. This part is described in detail in Russo et al. (2017), in which all experimental details are outlined.

Shark Hb systems see the occurrence of Hb isoforms (Manwell & Baker, 1970; Fyhn & Sullivan, 1975). The erythrocytes of the Greenland shark (Russo et al., 2017) contain three major Hbs, made of two copies of the same α chain combined with two copies of three β chains (β^1, β^2 and β^3), with very similar primary structures (Figure 1.1).

The chain compositions of Hb 1, Hb 2 and Hb 3 are $(\alpha\beta^1)_2$, $(\alpha\beta^2)_2$ and $(\alpha\beta^3)_2$, respectively. The α chain has 141 residues (Figure 1.1a). The β chains have 142 residues (Figure 1.1b). The sequence identity between the β chains approaches 100%. Despite separation by ion-exchange chromatography, β^1 and β^2 appear identical, but the sequences have portions in the CD corner and helix F that could not be sequenced, and some differences may be placed in these portions.

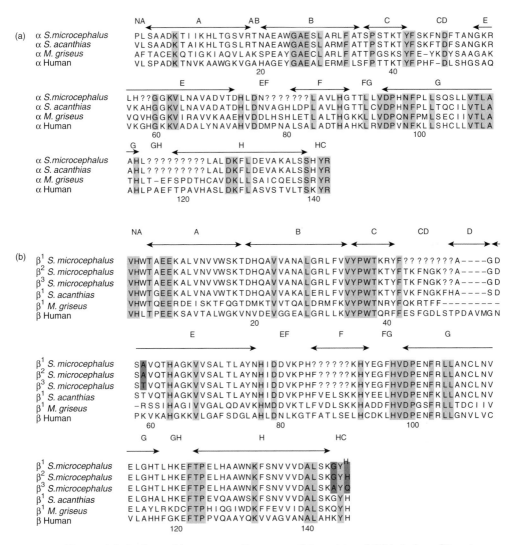

Figure 1.1 Amino acid sequence alignment of the α (a) and β (b) chains of *Somniosus microcephalus* (Greenland shark) Hbs with *Squalus acanthias*, *Mustelus griseus* and *Homo sapiens* α and β chains. Identical residues are in grey; different residues in the β globins of Greenland shark are in dark grey. The chains were aligned using Clustal OMEGA; in the β chains of *S. microcephalus* the position of residues in CD and D have been manually aligned with the sequences of *M. griseus* (Naoi et al., 2001), *S. acanthias* (Aschauer et al., 1985) and HbA (accession numbers: P69905 for α chain and P68871 for β chain). The question mark indicates unsequenced regions; dashes indicate deletions.

The sequence identity of Greenland shark Hbs and human HbA is 50% for the α chain and between 44 and 47% for the β chains, with many replacements of functionally important residues. Compared to other sharks, the globin chains of the *Somniosus microcephalus* (Greenland shark) have much higher identity with those of *S. acanthias* than those of *M. griseus*. The highly conserved C-terminal sequence Tyr-Arg is found in α chains. In all β chains, the four residues of helix D are missing. The lack of β helix D is a feature that distinguishes cartilaginous from teleost Hbs.

The Hbs of several polar skates and sharks (Nash et al., 1976; Fisher et al., 1977; Aschauer et al., 1985; Naoi et al., 2001) and of a temperate skate (Chong et al., 1999) do not display β helix D. Site-directed mutagenesis suggested that this deletion is a neutral modification, neither exerting large functional effect(s) on oxygen binding nor affecting the assembly of cooperative tetramers (Komiyama et al., 1991). The high number of His in Greenland shark Hbs confirms the hypothesis that a high Hb buffer capacity was ancestral in all jawed vertebrates, similar to contemporary elasmobranchs (Berenbrink et al., 2005).

The isoforms display similar Bohr effects (Table 1.2). The Bohr coefficient φ ($\Delta\log P_{50}/\Delta pH$), i.e. the mean number of protons released upon heme oxygenation, is higher in the presence of ATP (around −0.20 without ATP, −0.60 with ATP), with oxygen-linked dissociation of ~0.8 and ~2.0 protons in the absence and presence of ATP, respectively, *per* Hb tetramer. The enhancement by ATP is high in all Greenland shark Hbs, indicating that binding of negatively charged ATP to the low-affinity conformation enhances H^+ uptake.

Greenland shark Hbs show P_{50} values similar to *M. griseus* Hb at pH 7.4 and 6.7 without ATP, under the same conditions (Naoi et al., 2001). P_{50} values of Greenland shark Hbs at 0°C, calculated by extrapolation from vant'Hoff plots (Fago et al., 1997), show the same behaviour as those at 15°C and 25°C, in the absence and presence of ATP, indicating that the oxygen affinity increases with decreasing temperatures up to the physiological conditions of cold waters.

The higher Bohr effect in the presence of ATP found in Greenland shark Hbs in comparison to temperate sharks suggests evolution of molecular adaptations. In Greenland shark Hb 1 and Hb 2, HisβHC3 and LysαC5 are conserved; Hb 3 has Gln at position βHC3, whereas AspβFG1 is replaced by Glu, as in *M. griseus* Hb (Naoi et al., 2001). Valβ1 and Hisβ2, which contribute to the Bohr effect in adult human Hb (HbA) together with HisβHC3, forming salt bridges with Asp94βFG1 and LysαC5 (Perutz, 1998), are also present in the Greenland shark (Figure 1.1b). These residues are present as well in *S. acanthias* and *M. griseus* Hbs, which however show a smaller Bohr effect and weaker cooperativity (Aschauer et al., 1985; Naoi et al., 2001).

The effect of ATP on the oxygenation of Greenland shark Hbs differs from that of the two polar rays *B. eatonii* and *R. hyperborea* (Verde et al., 2005), which show no Bohr effect and ATP sensitivity. The model of the

Table 1.2 *Values of P$_{50}$, n$_H$, ΔH, φ of S. microcephalus Hbs (1 kcal = 4.184 kJ)*

S. microcephalus Hbs	100 mM KCl	2 mM ATP	P$_{50}$ (Torr) pH 7.4	6.7[b]	n$_H$ pH 7.4	6.7[b]	ΔH^a (kcal·mol^{-1}) pH 7.4	6.7[b]	φ pH 7.4–6.7[b]
25°C									
Hb 1	+	−	5.5	8.9	1.2	1.4	−7.9	−10.1	−0.2
	+	+	14.4	46.1	1.8	1.6	−0.7	−5.4	−0.5
Hb 2	+	−	5.7	8.6	1.6	1.9	−8.9	−8.5	−0.2
	+	+	12.5	40.8	1.8	1.6	−2.1	−6.6	−0.6
Hb 3	+	−	7.7	11.6	1.4	1.7	−8.4	−8.3	−0.2
	+	+	13.3	36.2	1.9	1.7	−0.9	−4.8	−0.6
	100 mM KCl	2 mM ATP	pH 7.4	6.8[c]	pH 7.4	6.8[c]	pH		7.4–6.8[b]
15°C									
Hb 1	+	−	2.9	4.1	1.5	1.7			−0.2
	+	+	11.6	28.2	2.1	1.3			−0.4
Hb 2	+	−	2.8	4.4	1.6	2.0			−0.2
	+	+	9.2	23.2	1.9	1.3			−0.4
Hb 3	+	−	3.9	6.0	1.8	1.7			−0.2
	+	+	11.0	22.9	2.0	1.7			−0.4
0°C[†]									
Hb 1	+	−	1.0	1.2					
	+	+	8.1	12.6					
Hb 2	+	−	0.9	1.5					
	+	+	5.6	9.3					
Hb 3	+	−	1.3	2.0					
	+	+	7.3	10.8					

[a] Extrapolated from LogP$_{50}$ (measured at 15–25°C) versus 1/T.
[b] pH 6.08 for Hb 3 in the presence of ATP.
[c] pH 6.01 for Hb 3 in the presence of ATP.
P$_{50}$: O$_2$ partial pressure at 50% saturation; n$_H$, Hill coefficient; ΔH, enthalpy change; Φ, Bohr coefficient
Experimental errors are within 10%

ATP binding site in teleost Hbs (Perutz & Brunori, 1982; Perutz, 1983) indicates that Valβ1, Hisβ2, LysβEF6 and HisβH21, which bind 2,3-diphosphoglycerate (DPG) in human HbA, participate in ATP binding. These residues are conserved in Greenland shark Hbs, with the exception of HisβH21, replaced by Lys, consistent with the allosteric effect of ATP on oxygen affinity and Bohr effect. In M. griseus Hb, although the canonical binding site is not preserved, ATP works as allosteric effector by lowering the oxygen affinity and stabilising the T-state structure (Naoi et al., 2001).

Greenland shark Hbs show cooperativity at all pH values investigated and at both 15°C and 25°C, with n_H ranging from 1.2 to 2.0. Cooperativity increases in the presence of ATP.

Looking at the oxygen affinity between 15 and 25°C (Table 1.2), ΔH (which includes the heat of oxygen solubilisation and the heat of proton and anion dissociation, i.e. processes linked to oxygen binding) shows that the change of oxygenation is constant in the absence of ATP over the whole pH range explored (6.7–7.4), with the exception of Hb 1 for which it is more negative (i.e. oxygenation is more exothermic) at low pH. In the presence of ATP, the heat of oxygenation progressively decreases with decreasing pH, reflecting the endothermic contribution of the heterotropic effectors released upon oxygen binding. The data show that the oxygenation-enthalpy change in Greenland shark Hbs is lower than that of temperate fish Hbs and very similar to that of polar fish Hbs (di Prisco et al., 1991; Verde et al., 2006). The expression of Hbs with reduced ΔH seems a frequent evolutionary strategy of cold-adapted fish, resulting in improved oxygen release to tissues at low temperatures. The low temperature effect on oxygen affinity thus suggests that oxygen delivery may be facilitated by lower heat of oxygenation of Hb, as also reported in polar mammals (Weber & Campbell, 2011).

No important differences were observed in the oxygen-binding parameters in the absence and presence of urea (a very important compound in the physiology of marine elasmobranchs) either in terms of P_{50} or cooperativity. A cooperating effect of urea and ATP on the oxygen affinity has also been found in other elasmobranchs, e.g. the dogfish S. acanthias and the carpet shark Cephaloscyllium isabella (Weber et al., 1983a,1983b; Tetens & Wells, 1984).

The three isoforms of the Greenland shark display identical electronic absorption and resonance Raman spectra (Russo et al., 2017), indicating identical or highly similar heme-pocket structures.

Thus, the Hbs are functionally quite similar, in keeping with the high sequence identity, as seen in other elasmobranchs (Weber et al., 1983a), suggesting microheterogeneity without major functional specialisation of iso-forms associated with migrations within wide latitudinal ranges. Teleosts often exhibit multiplicity of Hbs that display structural and functional differ-ences (Riggs, 1970; Gillen & Riggs, 1973). Multiplicity is usually interpreted as a sign of phylogenetic diversification and molecular adaptation, and generally is the result of gene-related heterogeneity and gene-duplication events (Dettaï et al., 2008; Giordano et al., 2010). However, in cartilaginous fishes, there is no evidence of the functional differentiation often found in teleosts, and no information on the functional consequences of Hb multiplicity is available for sharks. The absence of functional heterogeneity in Greenland shark Hbs is in keeping with the low tolerance of sharks to oxygen-pressure variation, and their low capability to maintain constant uptake rates as oxygen tension falls (Speers-Roesch et al., 2012b). The fact that they have very similar features also suggests that the shark may play regulatory control of oxygen transport at other levels. Alternatively, the isoforms may protect against deleterious muta-tional gene changes, thus providing higher total Hb concentration in the erythrocyte, and increasing the gene expression rate (Speers-Roesch et al., 2012a).

The very similar ligand-binding properties suggest that regulatory control of oxygen transport may be at the cellular level and may involve changes in the cellular concentrations of allosteric effectors and/or variations of other systemic factors. Compared to temperate sharks, the Hbs of the Greenland shark have evolved adaptive decreases in oxygen affinity, similar to that of some antarctic and sub-antarctic teleosts of the suborder Notothenioidei (di Prisco et al., 1991) and of the arctic fish *Arctogadus glacialis* (Verde et al., 2006), thus suggesting that the P_{50} values might be linked to the high oxygen concentration in cold waters, implying some adaptation to cold temperatures (di Prisco et al., 2007). In fact, these observations highlight an important difference from temperate cartilagi-nous Hbs, and, in contrast, some similarity between a polar shark and cold-adapted teleosts thriving in both polar environments.

Although amino acid sequences and ligand binding do not reveal special features associated with cold environmental conditions, the decrease in oxy-gen affinity evolved by the Hb isoforms of the Greenland shark may have adaptive implications. Physiological differences in oxygen transport between polar and temperate sharks, both dispersed across wide latitude and tempera-ture gradients, may be governed at the physiological plasticity level.

1.3 Ecotoxicology of inorganic and organic contaminants

Despite the limited ecotoxicological data available, this species can be considered at high risk of bioaccumulating toxic and persistent organic pollutants (POPs), endocrine-disrupting compounds (EDCs) and trace elements. Most of the organs were investigated for the presence of these contaminants. Studies were focused on assessing the contaminant organotropism and the differences between age classes, gender and sampling areas in Greenland seawaters (NE, E, S and W Greenland) (Corsolini et al., 2014, 2016; Ademollo et al., 2018; Cotronei et al.. 2018a, 2018b). POP levels were determined also in its prey to evaluate the biomagnification and related biomagnification factor (BMF), taking into account that shark prey such as cods and Greenland halibuts, seals and polar bears are all important food resources for the Greenland shark and they play a key ecological role in the POP transfer through the shark food web.

The dioxin-like compound (DLC) and Toxic Equivalent (TEQ) concentrations were also evaluated using the Toxic Equivalency Factors (TEF) method (Van den Berg et al., 1998, 2006) in order to assess the possible risk for the analysed specimens.

The study of this interesting and peculiar species may suggest speculations in the light of global change, being this shark is strictly dependent on deep and cold seawaters. For instance, the stomach content of specimens caught in 2015 in NE Greenland seawaters (74°–78°N, 5–14°W) during the TUNU VI Expedition (Nielsen et al., 2019) included Atlantic cod *Gadus morhua*, a boreal species distributed from temperate to sub-arctic seas (Righton et al., 2010); this cod was reported to move northward to more arctic seawaters following warming (Fossheim et al., 2015; Ingvaldsen et al., 2017). Therefore, the shark might be affected by global change by shifting its diet and migrations.

The longevity, predatory habits and general biological and ecological features make important the ecotoxicological study of the Greenland shark. In the peculiar arctic environment, due to low temperatures and winter darkness, contaminants degrade very slowly; moreover, they can be trapped in snow and ice and eventually be released during summer melting; once in the seawater, they can enter trophic webs and bioaccumulate in the tissues of organisms (Corsolini, 2009).

POPs include a list of a diverse class of synthetic and ubiquitous contaminants. Due to their persistence in the environment, they show long-range transport potency (LRTP) and can bioaccumulate in organisms where they can elicit toxic effects (SC-POPs, 2013). POPs have been used extensively worldwide as pesticides in agriculture and home applications, and in industry. Exposure to POPs can lead to serious health effects among humans and wildlife, and they occur in ecosystems worldwide including the arctic region (Muir & de Wit, 2010; Rigét et al., 2010). For these reasons, the TEQs of most toxic chemicals were calculated by applying WHO1998 TEFs for fish (Van den Berg et al., 1998) and the most recent WHO2005 TEFs for mammals and humans (Van den Berg et al., 2006).

Some POPs are also EDCs, which include a wide range of both natural and synthetic (not only listed as POPs) chemicals that can adversely deregulate the hormone systems with multiple effects, mostly through steroid receptors on peptide/protein hormones. Most EDCs interfere with reproduction acting as either agonists or antagonists of the steroidal sex hormones, estrogens or androgens. In the literature, there are few papers on the distribution and transport of these EDCs in the arctic marine food webs and no such studies have been carried out on the Greenland shark.

Nonylphenols (NPs) are metabolites of nonylphenol ethoxylates (NPEOs), belonging to the group of non-ionic surfactants used as detergents, solubilisers, emulsifiers and dispersants. NPEOs are biodegraded during sewage treatment processes in wastewater treatment plants (WWTPs) and partially in the environment by the loss of the ethoxy-groups, resulting in NPs and other mono- and di-ethoxylate precursors (NP1EO, NP2EO), which are more toxic and oestrogenic than NPEOs (Soares et al., 2008). Once NPs and NPEOs enter aquatic systems, they can elicit an oestrogenic action against the reproductive system of aquatic organisms binding to oestrogen receptors and can block or alter endogenous endocrine functions in various reproductive and developmental stages, also affecting aromatase activity and the function of the aryl hydrocarbon receptor (AhR) (Hong et al., 2004). Bisphenol A (BPA) is an intermediate in the production of epoxy resins and polycarbonate plastics. NPs and BPA have multiple detrimental effects, such as endocrine disorder, developmental abnormalities and reproduction dysfunctions (Zhou et al., 2009). Over time, aquatic organisms are chronically exposed to synergistic effects between these compounds and other xenobiotics in a global change scenario that includes harvesting, habitat degeneration and competitive exclusion. The potential of endocrine disruption, even at low NP concentrations, is likely to be enhanced by additive or synergistic effects due to co-occurrence with other xenoestrogens (Soares et al., 2008) like some POPs are.

Metals, including trace elements, occur naturally and are released into the environment from anthropogenic sources as well. Climate-mediated shift in food web structure may also influence body burdens of primarily dietary-driven contaminants (e.g. mercury, Hg) in arctic food webs.

Despite concern over trace-element toxicity and the potential for biomagnification in arctic wildlife, very limited data are available on the Greenland shark (Dietz et al., 2000; McMeans et al., 2007). A number of essential and non-essential elements were analysed only in liver of Greenland sharks from Cumberland Sound (McMeans et al., 2007). A contribution to fill this data gap was published on the presence of non-essential and essential metals in several organs and tissues of specimens from NE Greenland (Corsolini et al., 2014).

Among POPs, EDCs and trace elements, the following contaminants were studied in the Greenland shark samples: aldrin, chlordanes (CHLs), dichloro diphenyl trichloroethane (DDT), dieldrin, endrin, hexachlorobenzene (HCB), mirex, heptachlor, toxaphenes, hexachlorocyclohexanes (HCHs), pentachlorobenzene (PeCB), polychlorobyphenyls (PCBs) including mono- and non-*ortho* substituted congeners, polychlorinated dibenzo-dioxins (PCDDs) and -furans (PCDFs), the flame retardants polybromodiphenylethers (PBDEs) and hexabromocyclododecane (HBCDs), perfluorooctane sulfonate (PFOS), and perfluorooctanoic acid (PFOA), 4-NP, its mono- (NP1EO) and di-ethoxylate (NP2EO) precursors and bisphenol A (BPA), cadmium (Cd), lead (Pb), mercury (Hg) and selenium (Se).

These chemicals were also investigated in sample tissues of Greenland sharks from E, S and W Greenland seawaters, collected during other expeditions, thanks to projects led by Prof. J.F. Steffensen (University of Copenhagen) and to the collaboration with Dr J. Nielsen (University of Copenhagen).

The main results of investigations on inorganic and organic contaminants in specimens of Greenland shark, with regard to organotropism and age and gender differences, are summarised here. All these results are discussed in detail in Corsolini et al., 2014, 2016; Cotronei et al., 2017, 2018a, 2018b; Ademollo et al., 2018; in which all experimental details are also described. Very briefly, the analyses of POPs were performed by gas-chromatography coupled to mass spectrometry and high-performance liquid chromatography with electrospray ionisation tandem mass spectrometry; the analyses of EDCs implied liquid chromatography coupled to mass spectrometry equipped with a fluorescence detector; graphite furnace atomic absorption spectrometry was used for Cd and Pb determination, cold vapor flow injection for total Hg, and graphite furnace atomic absorption coupled with hydride generation spectrometry for Se. Several tissues (red and white muscle, brain, gonads, adipose tissue, liver, pancreas, spleen, gonad and epidermis) of the shark specimens were analysed, and, prior to residue analyses, their moisture and lipid contents were measured (Corsolini et al., 2014), in order to normalise the inorganic and organic contaminants to dry and lipid weight, respectively; these calculations are useful for comparisons. In Corsolini et al. (2014), the tissue lipid content showed high values (47±6%–24±2%, with brain > liver > gonad > spleen > red muscle > abdominal fat > white muscle), confirming it as an adaptation to cold seawaters in polar fish; moreover, the fatty liver helps buoyancy in the sharks (Malins & Barone, 1970). Interestingly, in the three specimens caught during the TUNU-IV Expedition (August 2010), the lipid content in the fat was lower than in liver, an unusual pattern already reported in the blue shark *Prionace glauca* (Kannan et al., 1996).

POPs used now or in the past were detected in the sharks from NE Greenland with variable concentrations depending on the tissue and contaminant. Results are reported in Corsolini et al. (2014, 2016): pesticides were detected in most tissues and specimens, except for Dieldrin and Aldrin (<0.001 ng/g wet weight (wet wt)); HCB was <0.001 ng/g in spleen and pancreas and HCHs were <0.002 ng/g wet wt in all tissues except brain and white muscle; DDTs (sum of *p,p'*- and *o,p'*-DDD, DDE, DDT isomers) were the most abundant chlorinated pesticides and concentrations ranged from 0.02 to 0.59 ng/g wet wt in pancreas and white muscle, respectively. Shark specimens collected in E, S and W Greenland seawaters showed 1095 ± 818 ng/g lipid wt in muscle and 761 ± 416 ng/g lipid wt in liver samples (Cotronei et al., 2018a,b). These findings confirm that Greenland sharks bioaccumulate DDTs available in the environment because of their long-range global transport from countries where DDT-based insecticides for controlling the malaria disease vector, the *Anopheles* sp. mosquitoes, have been or are used (SC-POPs, 2013). Alternatively, being this shark alive before POP global production and use started in 1940s, the detection of these chemicals may be due to low biotransformation activity of the parent compound *p,p'*-DDT to *p,p'*-DDE.

PCB concentrations in the shark tissues (2.01–65.6 ± 32.8 ng/g wet wt, Corsolini et al., 2014) were of the same order of magnitude as those reported by Molde et al. (2013) in the plasma of the Greenland shark from Svalbard (Norway), and lower than those detected in sharks from the NE Atlantic (Strid et al., 2007). Tetra-, penta- and hexa-CBs made up 65–75% of the total PCB residue and the class of isomer profile in the tissues was similar to that of Aroclor 1248, a technical mixture used in hydraulic fluids, lubricants, plasticisers and adhesives (IARC, 1979).

Brominated flame retardants were also studied in the Greenland sharks: HBCDs and PBDEs were detected in all tissues (Corsolini et al., 2016; Cotronei et al., 2017, 2018b). The γ-HBCD is the major component in the commercial mixtures, but α-HBCD made up 95% of the residue, likely due to its slow metabolism or to a possible bioisomerisation of all isomers mainly to α-HBCD (Janák et al., 2005). PBDEs were 426 pg/g wet wt in NE

Greenland specimens (Corsolini et al., 2016) and 20 ± 25 ng/g lipid wt in specimens from W, S and E Greenland (Cotronei et al., 2018b). BDE47 was the most abundant congener in sharks from all sampling areas and BDE99 was below 5%, although it is a major constituent of the technical Bromkal mixtures (Corsolini et al., 2016). These findings are in agreement with other authors (Strid et al., 2013) and suggest a debromination process may occur in this shark species, following the pattern BDE154→BDE99→BDE47 (Ikonomou et al., 2002). Levels differed significantly between muscle samples of sharks from NW and SE Greenland ($p < 0.05$); notwithstanding Greenland sharks migrate (Nielsen, 2018), the difference may be related to environmental levels as W and E Greenland are influenced by different water currents (Pedersen et al., 2004).

The biomagnification factors were calculated for most contaminants using predator-stomach content and predator–prey relationships and values ranged from 33 to 3892 for DDTs and PCBs and from 0.01 to 239 for flame retardants, confirming this shark is able to highly biomagnify POPs, and, for this reason, at risk. The risk assessment was performed using the TEF approach for the dioxin-like compounds (PCBs, PCDDs/DFs) (van den Berg et al., 1998, 2006) and TEQ values ranged from 28.23 to 45.15 pg/g lipid wt in sharks from E, S and W Greenland and 1.91–131 pg/g wet wt.

Although it is a top predator, pesticides, PCBs and PBDEs were lower than in the polar bear, the arctic apex predator (e.g. Gebbink, et al., 2008; Verreault et al., 2015).

Interestingly, younger specimens showed higher concentrations than older sharks and lipid content, sex and size were not significantly correlated to the POP concentrations ($p > 0.05$). Longevity may affect bioaccumulation more than diet, in a species where the detoxifying activity may be low: contaminants are transferred from mother to the next generation (Corsolini et al., 2014; Olin et al., 2014). Thus, at birth, offspring tissues already contain contaminants, which can be gradually diluted with body growth (Corsolini & Sarà, 2017), as the elimination rate may be more important than accumulation in a pristine environment, e.g. NE Greenland and deep-sea ecosystems. The presence of POPs in Greenland seawaters may be due to LRT and to release from the dump site located near villages: waste materials have been disposed of in dump sites located near each settlement and close to the sea for decades, and PCBs and other contaminants can be released from old equipment containing them. Contaminants can leach into the soil, beach and the sea, then entering the trophic webs. Concentrations in the Greenland shark from Greenland seawaters were generally lower in specimens collected in NE Greenland with respect to E, S and W (presence of settlements) Greenland seawaters, and all levels were lower than in sharks from other areas (Figure 1.2) (Fisk et al., 2002; Strid et al., 2007, 2010, 2013; Molde et al., 2013); the presence of low levels in top predators from NE Greenland likely confirms this area to be among the most pristine regions of the northern hemisphere.

The EDC concentrations were higher in muscle than in liver samples of the sharks from SE and NE Greenland, and higher in liver than in muscles in those from W and SW Greenland. The 4-NP, NP1-2EO and BPA mean concentrations in liver of SW Greenland specimens were 43.5 ng/g, 288.5 ng/g and 8.2 ng/g wet wt, respectively; in muscle, the mean concentrations were 20.3 ng/g of 4-NP, 171.1 ng/g wet wt of NP1-2EO and 7.9 ng/g of BPA. Further details on concentrations are reported in Ademollo et al. (2018). Our results showed that, despite the highest lipophilicity of 4-NP, its ethoxylates were highly bioconcentrated in these sharks. These findings are in agreement with the evidence that NPEs are the main alkylphenols used in industry, accounting for about 90% of the global production, and that NP1-2EO are inefficiently removed in sewage treatment plants and very slowly degraded in the marine polar environments.

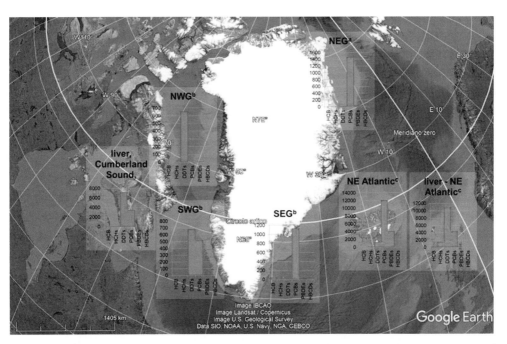

Figure 1.2 Average concentrations of HCB, HCHs, DDTs, PCBs, PBDEs and HBCDs (ng/g lipids) in Greenland shark white muscle (or liver where specified) samples from NE Greenland (NEG), NW Greenland (NWG), SW Greenland (SWG), SE Greenland (SEG), NE Atlantic and Cumberland Sound (data are from: a = Corsolini et al., 2014, 2016; b = Cotronei et al., 2017, 2018a, 2018b; c = Strid et al., 2007; d = Fisk et al., 2002). (A black and white version of this figure will appear in some formats. For the colour version, please refer to the plate section.)

The outcomes of this study suggest an ongoing input of NP1-2EO from human activities, especially from the North American arctic coasts.

The contamination pattern of organic contaminants in different tissues depends on several factors, such as age, gender, metabolism and diet (Corsolini et al., 2014), factors that may be more important than the sampling area of sharks. The liver is vital in detoxifying, storing and redistributing contaminants (Staniszewska et al., 2014), since it is well known that exposure to any contaminant will first result in metabolism by the liver, after which the compounds are stored in muscle (Mita et al., 2011). The usually high concentrations of these EDCs in the shark muscle could be explained by the fact that removal of these compounds from the liver occurs more rapidly with respect to other POPs. These compounds are metabolised and eliminated mostly via bile and faeces, but the high levels detected in muscle can be a result of continuous and chronic exposure. The high standard deviation values are owed to the intraspecific variability of EDC bioaccumulation depending on its opportunistic diet (MacNeil et al., 2012; Nielsen et al., 2014). The growth rate of this long-lived species and its reproductive characteristics are further factors affecting the bioaccumulation of contaminants.

The concentration and organotropism of the non-essential (Hg, Cd, Pb) and essential (Se) elements studied in the Greenland shark from NE Greenland are reported in Corsolini et al. (2014). Cd was found to be mainly accumulated in the liver (10.4±4.87 mg/kg dry wt), although the highest concentration was detected in the only sample of the pancreas (19.6 mg/kg dry wt). The order of Cd-organotropism was: pancreas > liver > spleen > brain > skin > red muscle > gonad > white muscle. The liver shows a particularly high capacity to biotransform pollutants (Klaassen, 1986). Kidney samples, usually rich in specific Cd-binding proteins (metallothioneins) like liver (Roesijadi, 1992), were not available for the analysis. The presence of Cd in marine food webs is related to a diet based on cephalopods, particularly at high latitudes (Bustamante et al., 1998). The Cd levels in the liver of the Greenland shark from NE Greenland were in agreement with data from the North Atlantic (McMeans et al., 2007) and showed values one order of magnitude higher than those reported for other shark species from lower latitudes (e.g. Storelli & Marcotrigiano, 2002). However, the diet of the Greenland shark is mainly based on fishes and mammals, and to a lesser extent on cephalopods and crustaceans (Fisk et al., 2002; Yano et al., 2007; McMeans et al., 2010; Nielsen et al., 2014).

The skin showed the highest Pb concentrations: 0.358±0.172 mg/kg. The order of Pb-organotropism was: skin > red muscle > pancreas > gonad > liver > spleen > white muscle > brain. The high Pb level in the skin may be a consequence of external contamination, as it may be adsorbed onto the skin surface rather than accumulate in the tissue itself: the dermal denticles of the skin provide a rough surface to which particulate matter or sediment can attach, in agreement with studies on other shark species (Vas, 1991).

The highest Hg concentrations were found in red (6.91 ± 3.65 mg/kg dry wt) and white (4.10 ± 1.73 mg/kg dry wt) muscle; concentration in liver was 2.11 ± 0.98 mg/kg dry wt. The Hg-organotropism was: red muscle > white muscle > spleen > pancreas > liver > gonad > brain > skin. High Hg concentrations are usually ascribed to a diet rich in fishes (Domi et al., 2005). Data on Hg levels in sharks are limited, and studies indicate a unique distribution pattern in the liver and muscle (Branco et al., 2007). Hg concentrations were higher in muscle than in the liver of immature tiger shark, *Galeocerdo cuvier,* whereas the opposite pattern was observed in mature specimens (Endo et al., 2008). Direct deposition from the atmosphere is the dominant pathway by which Hg reaches the oceans; exceptions are semi-enclosed basins such as the Mediterranean Sea and the Arctic Ocean, where river runoff and coastal erosion account for about half of Hg inputs (UNEP, 2013).

The highest Se concentration was detected in the pancreas (3.57 mg/kg dry wt) and ranged from 0.105 to 1.95 mg/kg dry wt in the other tissues. The order of Se-organotropism was: pancreas > spleen > gonad > brain > red muscle > liver > skin > white muscle. Selenium is known to detoxify Hg in marine organisms (Koeman et al., 1975). The mechanisms involved in this process are species- and tissue-specific (Cuvin-Aralar & Furness, 1991) and include the formation of inert Hg–Se complexes, the binding of Hg with selenoproteins, the indirect action of Se in preventing oxidative damage by Hg by increasing glutathione peroxidase activity, and the induction of metallothioneins. A positive correlation between Hg and Se is usually found in marine organisms and a molar ratio Hg:Se close to 1 is common in marine mammals and seabirds with high Hg concentrations in the liver (Koeman et al., 1975; Cuvin-Aralar & Furness, 1991; Dietz et al., 2000; Storelli & Marcotrigiano, 2002). It has been suggested that a Hg:Se ratio close to 1 occurs when Hg concentrations increase to >1000 mg/kg (Koeman et al., 1975). A ratio close to 1 has also been reported at Hg > 2 mg/kg wet wt (Dietz et al., 2000). Although the Hg concentration in the Greenland shark liver was 2.11 ± 0.98 mg/kg dry wt, the mean Hg:Se ratio was 1.57 ± 0.35, suggesting that the protective mechanism between Se and Hg is valid also in this species.

Hg and POPs showed different biomagnification patterns. McMeans et al. (2007) suggested that these differences may be due to a lower trophic position than expected or, more likely, to the different half-lives of Hg and POPs.

The Greenland shark is confirmed to bioaccumulate persistent inorganic and organic contaminants in all the studied tissues. In contrast to POPs and EDCs, trace elements occur naturally, although they are also released into the environment from anthropogenic sources. The Greenland shark may eliminate Hg more efficiently than POPs by means of continuous exchange between the blood and seawater through the gills (Bacci, 1994): the different physicochemical features of Hg and POPs may allow more efficient Hg elimination with respect to POPs. Ecotoxicological baseline data on marine wildlife in Greenland including the Greenland shark are particularly important due to possible (or already running) prospective oil, gas and mining activities in the fjords and on the shelf (Christiansen et al., 2014), both of which are visited by this shark. Moreover, global warming and the concomitant declines in ice cover will continue to alter emission, transport and bioavailability of contaminants in the Arctic (AMAP, 2011).

1.4. Sensory capability: the sense of olfaction in the Greenland shark

Animals rely on their sensory system to obtain, process and act upon information from the environment where they live, including food availability as well as presence of competitors, predators and mates. Indeed, an animal's behavioural choices are based on its knowledge of resources availability, competition and predation risk, thus largely depending on the organism's capability to perceive and interpret the spatial structure of the environment (Kramer, 2001; Koy & Plotnick, 2007; Ferrari et al., 2010).

Although vision is generally known to play a central role among senses, in the marine realm the sensory system is stimulated by an extremely wide range of environmental cues, including light, sound, odour, water movement, temperature, pressure, and electric and magnetic field strength. In such a multifaceted environment, olfaction is at least as relevant as vision, sometimes being the principal sense through which a marine organism detects environmental cues (Tierney, 2015). The detection and discrimination of chemicals, and the behavioural reactions they evoke, are among the most primordial activities of living organisms, and chemoreception, which evolved 500 million years ago, can be considered the most ancient of sensory systems (Hara, 1992).

Olfaction is especially important as a distant or far-field chemoreception sense, since chemicals can be entrained in currents and transported far away through the water to the sensory cells. The spatial pattern of odours (odour landscape) on the one hand, and the olfactory organ morphology and functionality on the other, determine the fish behaviour in a particular environment and situation, thus indirectly influencing several aspects of its ecology. The extent of encountering olfactory signals depends upon the nature and concentration of the original source, the way the signal is conserved or modified during propagation in the aquatic environment, the swimming mode of the given species, and the morphology and arrangement of its nasal canals. The behavioural response is then determined primarily by the capability of the sensory system to detect a signal, transduce it into a nervous impulse and integrate it at the peripheral and central level. Therefore morphological information on the olfactory system represents the necessary basis for attempting ecological inferences (Collin, 2012).

In cartilaginous fishes, olfaction is relevant for various survival tasks including prey localisation, predator avoidance and communication with conspecifics. Evidence of the importance of chemical sensing to elasmobranchs is abundant (reviewed by Hodgson & Mathewson, 1978; Smeets, 1998), and sharks in particular are often referred to as 'swimming noses' based on their olfactory abilities. In Chondrichthyes, the olfactory cues are received from the environment through a peripheral organ, the olfactory rosette, organised as an array of primary lamellae connected to a central support, named raphe (Meng & Yin, 1981; Meredith & Kajiura, 2010; Cox, 2013). The lamellae are covered by the olfactory epithelium, where the olfactory receptor neurons (ORNs) are located; axons from the ORNs project to the olfactory bulb in the telencephalon, where the signal undergoes a first step of integration (Dryer & Graziadei, 1996).

Despite similar overall morphological organisation, the olfactory sense organ in various species shows a degree of variability in size, shape, position of the nares, number and surface area of lamellae, and extent of the sensory epithelium (Bell, 1993; Schluessel et al., 2008, 2010; Meredith & Kajiura, 2010). The presence and variation in size and shape of secondary folds on the surface of the primary lamellae (Ferrando et al., 2019) is another feature to be taken into account when using the olfaction organ structure to infer information on the sensory capabilities of a species.

Until recently, little was known about the sensory biology of the Greenland shark, besides the inference of a poor relevance of vision among the senses based on this shark's propensity for infection of the cornea by the ectoparasitic copepod *Ommatokoita elongata* (Berland, 1961). The important role of non-visual sensory systems is also consistent with the evidence that the Greenland shark prefers to live in low-light environments such as deep waters or ice-covered areas. This is further confirmed by studies based on baited remote underwater videos (BRUVs), which have reported that the Greenland shark is the primary consumer of the bait (Devine et al., 2018), presumably enabled by a sensitive olfactory system.

Anatomical studies on the peripheral olfactory organ and olfactory bulb of the Greenland shark have recently been performed (Ferrando et al., 2016, 2017a). The olfactory rosette is an oval-shaped structure anatomically organised as an array of primary lamellae, arranged in parallel, attached to a central raphe, and decreasing in size towards both ends (Figure 1.3); such a linear lamellar organisation is widespread among sharks. In Greenland shark the lamellae at the two sides of the raphe do not touch each other, showing a 'type II parallel lamellae array' (*sensu* Cox, 2013) typically found in rather sedentary species. The lamellar

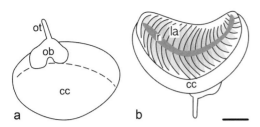

Figure 1.3 Scheme of the olfactory organ and bulb of (Greenland shark) (specimen of 230 cm total length). (a) dorsal view; (b) ventral view. cc=connective capsule; la=primary lamellae; ob=olfactory bulb; ot=olfactory tract; r=raphe. Scale bar 1 cm.

structure is enveloped by a connective capsule, opened ventrally to expose the olfactory lamellae to the water entering the nasal cavity, and in continuity with the olfactory bulb and peduncle in its dorsal part. The olfactory bulbs are subdivided into two sub-bulbs that converge, and are kept together in their central parts, from where a ribbon-like olfactory tract emerges. In a specimen of 230-cm total length, the number of primary lamellae, according to a general definition of primary lamella established in 2017 (Ferrando et al., 2017b), was 44 (Ferrando et al., 2016, 2017b). The total surface area of the primary lamellae for the same specimen was estimated to be 117 cm², a remarkably high value compared to other elasmobranch species (Schluessel et al., 2008). Since most of that area is covered by sensory epithelium, the estimate of surface area is considered a good proxy for olfactory potential. The relatively high value found in the Greenland shark suggested a rather good olfactory capability for the species. Such a sensory potential might sustain the shark scavenging behaviour (Beck & Mansfield, 1969; Leclerc et al., 2011), but also relevant ecological features of the species such as in shore/off shore movements across life stages (Campana et al., 2015), long migrations, as recently documented by using mark-report pop-up archival tags (Hussey et al., 2018), and vertical movements from deep waters to the undersea-ice cover (Skomal & Benz, 2004).

To further support the functional inferences on the Greenland shark olfactory potential made on the morphological bases, and in order to evaluate and discern features related to specific ecological traits, comparative analysis with taxonomically related species is needed. The morphology of the olfactory organ of the congeneric little sleeper shark *Somniosus rostratus* has been investigated and compared to that of Greenland shark in a wider comparative context (Ferrando et al., 2019). The gross morphology and relative size of the olfactory organs of the two *Somniosus* species were similar to each other and also to that of several other elasmobranchs. Nevertheless, they share a noteworthy and unique shape of the secondary folds, which are epithelial folds on the surface of the primary lamellae, increasing the sensory surface area in the olfactory organ. In *S. rostratus* and *S. microcephalus* the secondary folds are very branched, with a very complicated and maze-like three-dimensional shape (Figure 1.4a,b). Secondary folds in the olfactory organs of other elasmobranch species investigated to date have little branching (Figure 1.4c) or none at all (Figure 1.4d,e). Such a peculiar organisation found in *S. rostratus* and *S. microcephalus* results in an increased surface area of lamellae in their olfactory peripheral organs, with potential effects on the water dynamics in the olfactory chamber, but this remains to be investigated.

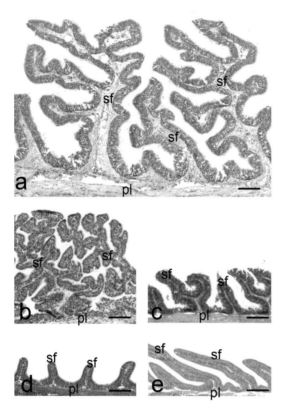

Figure 1.4 Histological sections of olfactory organs of elasmobranchs. Haematoxylin-eosin stain. (a) *S. microcephalus* (230-cm total length). Some secondary folds (sf) on the side of a primary lamella (pl). Almost all of the secondary folds present several branches. (b) *S. rostratus* (96-cm total length). Also in this species the secondary folds are very branched. (c) *Raja miraletus* (41-cm disc width). The secondary folds branch once or not at all. (d) *Pteroplatytrygon violacea* (107-cm disc width). The secondary folds are small, not branched and quite distant from one another. (e) *Scyliorhinus canicula* (27.5-cm total length). The secondary folds are not branched and elongated. Scale bars 200 μm. (A black and white version of this figure will appear in some formats. For the colour version, please refer to the plate section.)

Given the scarcity of information available to date on olfactory system organisation in the family Somniosidae, any conclusion based on the features of the two *Somniosus* species is premature. According to present data, the overall similarity of the peripheral olfactory organ in the Greenland shark and *S. rostratus* might be ascribed to their taxonomic relationship, although a possible ecological explanation for the peculiar organisation of the olfactory secondary folds in both species cannot be excluded. Further comparative studies in the family Somniosidae are needed to fully understand the possible ecological implications of an olfactory organ with the highly complicated structure summarised here in the two congeneric species of *Somniosus*. It would be interesting to explore a possible correlation between olfactory organ structure and relative olfactory bulb sizes. Indeed, according to the present available information on 58 species of elasmobranchs, three bentho-pelagic bathyal species of Somniosidae, the longnose velvet dogfish *Centroselachus crepidater*, the plunked shark *Proscymnodon plunketi* and the roughskin dogfish *Centroscymnus owstonii*, have enlarged olfactory bulbs compared to other squalomorph species of similar ecological niche (Yopak et al., 2015). Other important behavioural aspects should be taken into account in the ecology of the *Somniosus* species when attempting to infer ecological implications from their olfactory organ structure, intraspecific interactions being the most relevant. Unfortunately, as in other elasmobranchs, the social behaviour of these species remains largely unknown.

1.5. Conclusions

Sharks comprise about half of contemporary chondrichthyans, a monophyletic group of predatory fishes that originated about 423 million years ago. Their evolution and life history suggest that sharks have been a relatively stable force in ocean ecosystems over evolutionary time and possess a particular combination of ecological traits. Although sharks are morphologically and phylogenetically related to bony fishes, their life histories resemble those of marine mammals, specifically with respect to their large size, low rate of reproduction and late maturity. These features render the survival of sharks highly sensitive to a number of factors, e.g. predation and fishing. To date, shark populations in the Arctic are still thought to be relatively abundant, although comprehensive survey data to confirm this hypothesis are lacking. The chondrichthyan evolution appears steady, with low origination and extinction rates. Their high resilience might be related to high evolutionary adaptability.

Due to their slow growth rate and late maturity, Greenland sharks are especially vulnerable to human encroachment. There is little knowledge about their physiology and biochemistry. This contribution is an effort to try and fill this gap, by shedding light on parts of the physiology of the Greenland shark and reach better understanding of important biological tools, functions and capacity to withstand increasing loads of environmental contaminants.

Concerning respiration, our results on the mechanisms of oxygen transport indicate the presence of three Hb isoforms in the Greenland shark, which lack significant structural and functional differentiation. However, *in vivo* regulation of oxygen transport can be achieved *via* local blood pH shifts (Bohr effect) and changes in the concentration of physiological allosteric modulators, such as ATP, within erythrocytes. Additionally, differences in gene expression and overall protein synthesis may also contribute to regulate oxygen transport in the short term.

In terms of sensory capability, the overall similarity of the peripheral olfactory organ in the Greenland shark and other sharks, e.g. *S. rostratus*, might be due to a taxonomic relationship, although an ecological meaning of the specific organisation of the olfactory secondary folds cannot be excluded. Further comparative studies in the family Somniosidae are needed, for instance to verify correlations between organ structure and olfactory bulb sizes. Other important behavioural aspects in the ecology of the *Somniosus* species, e.g. largely unknown intraspecific interactions, need to be taken into account.

The Greenland shark accumulates persistent inorganic and organic contaminants in all its tissues. Trace elements occur naturally, although they are also released through anthropogenic sources, similar to POPs and EDCs. Hg can be efficiently eliminated by exchange through the gills (Bacci, 1994). Ecotoxicological baseline data on marine wildlife in Greenland are particularly important due to mining activities in the fjords and on the shelf (Christiansen et al., 2014), both of which are visited by the Greenland shark. Moreover, global warming and declines in ice cover will continue to alter emission, transport and bioavailability of contaminants in the Arctic (AMAP, 2011).

In conclusion, in order to evaluate resilience and vulnerability of this species, we badly need additional information on biochemistry and physiology, including information on contaminant bioaccumulation and organotropism processes in the context of the shark's metabolism. In fact, a detailed understanding of the underlying mechanisms of biochemical and physiological adaptations are key for management and conservation.

1.6 Acknowledgements

Greenland sharks were collected during the TUNU Expeditions as part of the TUNU Programme at the UiT - The Arctic University of Norway. Permissions for samples were given by the Government of Greenland (Licences C-17-129, C-15-17, and C-13-16). This study is in the framework of the project 'TUNU Euro-Arctic Marine Fishes' (TEAM-Fish): Impact of climate change on biodiversity, adaptation, contaminant bioaccumulation. Comparison with the Antarctic, and of the SCAR programme 'Antarctic Thresholds±Ecosystem Resilience and Adaptation' (AnT-ERA). The authors are grateful to colleagues and crew onboard the RV Helmer Hanssen for their kind collaboration. The authors acknowledge grant support from the Italian National Programme for Antarctic Research (PNRA) (Project 2013/AZ1.20).

References

Ademollo, N., Patrolecco, L., Rauseo, J., Nielsen, J., Corsolini, S. (2018). Biaccumulation of nonylphenols and bisphenol A in the Greenland shark *Somniosus microcephalus* from the Greenland seawaters. *Microchemical Journal*, **136**, 106–112; http://dx.doi.org/10.1016/j.microc.2016.11.009.

AMAP (2011). AMAP Assessment 2011: Mercury in the Arctic. Arctic Monitoring and Assessment Programme (AMAP), Oslo.

Aschauer, H., Weber, R.E., Braunitzer, G. (1985). The primary structure of the hemoglobin of the dogfish shark (*Squalus acanthias*). Antagonistic effects of ATP and urea on oxygen affinity of an elasmobranch hemoglobin. *Biological Chemistry Hoppe-Seyler*, **366**, 589–599.

Bacci, E. (1994). *Ecotoxicology of Organic Contaminants*. Lewis Publ., Boca Raton, FL, p. 164.

Beck, B., Mansfield, A.W. (1969). Observations on the Greenland shark, *Somniosus microcephalus*, in northern Baffin Island. *Journal of the Fisheries Research Board of Canada*, **26**, 143–145; https://doi.org/10.1139/f69-013.

Bell, M.A. (1993). Convergent evolution of nasal structure in sedentary elasmobranchs. *Copeia*, **1**, 144–158; doi:10.2307/1446305.

Berland, B. (1961). Copepod *Ommatokoita elongata* (Grant) in the eyes of the Greenland shark, a possible cause of mutual dependence. *Nature*, **191**, 829–830; https://doi.org/10.1038/191829a0.

Berenbrink, M., Koldkjaer, P., Kepp, O., Cossins, A.R. (2005). Evolution of oxygen secretion in fishes and the emergence of a complex physiological system. *Science*, **307**, 1752–1757; doi.org/10.1126/science.1107793 PMID: 15774753.

Bigelow, H.B., Schroeder, W.C. (1948). Sharks. In: J. Tee-Van (ed.) *Fishes of the Western North Atlantic, Part 1*. Yale University, New Haven, CT, pp. 59–546.

Branco, V., Vale, C., Canàrio, J., Dos Santos, N.M. (2007). Mercury and selenium in blue shark (*Prionace glauca*, L. *Xiphias gladius*, L. 1758) from two areas of the Atlantic Ocean. *Environmental Pollution*, **150**, 373–380.

Bustamante, P., Caurant, F., Flower, S.W., Miramand, P. (1998). Cephalopods are a vector for the transfer of cadmium to top marine predators in the north-east Atlantic Ocean. *Sciences of the Total Environment*, 220, 71–80.

Butler, P.J., Metcalfe, J.D. (1988). Cardiovascular and respiratory systems. In: T. J. Shuttleworth (ed.) *Physiology of Elasmobranch Fishes*. Springer-Verlag, Berlin, pp. 1–47.

Campana, S.E., Natanson, L.J., Myklevoll, S. (2002). Bomb dating and age determination of large pelagic sharks. *Canadian Journal of Fisheries and Aquatic Sciences*, **59**, 450–455.

Campana, S.E., Fisk, A.T., Klimley, A.P. (2015). Movements of Arctic and northwest Atlantic Greenland sharks (*Somniosus microcephalus*) monitored with archival satellite pop-up tags. *Deep-Sea Research Part II*, **115**, 109–115.

Chong, K.T., Miyazaki, G., Morimoto, H., Oda, Y., Park, S.Y. (1999). Structures of the deoxy and CO forms of haemoglobin from *Dasyatis akajei*, a cartilaginous fish. *Acta*

Crystallographica Section D Biological Crystallography, **55**, 1291–1300.

Christiansen, J.S. (2012). The TUNU-Programme : Euro-Arctic Marine Fishes: Diversity and Adaptations. In: G. di Prisco and C. Verde (eds) *Adaptation and Evolution in Marine Environments, Volume 1, From Pole to Pole.* Springer-Verlag Berlin/Heidelberg, pp. 35–50.

Christiansen, J.S., Mecklenburg, C.W., Karamushko, O.V. (2014). Arctic marine fishes and their fisheries in light of global change. *Global Change Biology*, **20**(2), 352–359.

Collin, S. (2012). The neuroecology of cartilaginous fishes: sensory strategies for survival. *Brain Behavior and Evolution*, **80**, 80–96; doi:10.1159/000339870.

Corsolini, S. (2009). Industrial contaminants in Antarctic biota. *Journal of Chromatography A*, **1216**, 598–612.

Corsolini, S., Sarà, G. (2017). The trophic transfer of persistent pollutants (HCB, DDTs, PCBs) within polar marine food webs. *Chemosphere*, **177**, 189–199; https ://doi.org/10.1016/j.chemo sphere.2017.02.116.

Corsolini, S., Ancora, S., Bianchi, N., et al. (2014). Organotropism of persistent organic pollutants and heavy metals in the Greenland shark *Somniosus microcephalus* in NE Greenland. *Marine Pollution Bulletin*, **87** (1), 381–387.

Corsolini, S., Pozo, K., Christiansen, J.S. (2016). Legacy and emergent POPs in a marine trophic web of NE Greenland fjords including the Greenland shark *Somniosus microcephalus. Rendiconti Lincei Scienze Fisiche e Naturali*, **27**(S1), 201–206.

Costantini, D., Smith, S., Killen, S.S., Nielsen, J., Steffensen, J.F. (2016). The Greenland shark: a new challenge for the oxidative stress theory of ageing? *Comparative Biochemistry and Physiology, Part A. Molecular and Integrative Physiology*, **203**, 227–232.

Cotronei, S., Pozo, K., Kohoutek, J., et al. (2017). HBCDs in the top predator Greenland shark (*Somniosus microcephalus*) from Greenland seawaters. 8th International Symposium on Flame Retardants: BFR 2017, May 7–10, 2017, York, UK; www .researchgate.net/publication/318882647 (accessed December 4, 2018).

Cotronei, S., Pozo, K., Audy, O., Přibylová, P., Corsolini, S. (2018a). Contamination profile of DDTs in the shark *Somniosus microcephalus* from Greenland Seawaters. *Bulletin of Environmental Contamination and Toxicology*, **101**(1), 7–13; https://doi.org/10.1007 /s00128-018-2371-z.

Cotronei, S., Pozo, K., Kohoutek, J., et al. (2018b). *Occurrence of PBDEs in the Greenland Shark* Somniosus microcephalus. Proceedings SCAR Open Science Conference 'Where the Poles come together', June 19–23, 2018, Davos, Switzerland, p. 1987.

Cox, J.P.L. (2013). Ciliary function in the olfactory organs of sharks and rays. *Fish and Fisheries*, **14**, 364–390; https://doi.org/10 .1111/j.1467-2979.2012.00476.x.

Cuvin-Aralar, M.L.A., Furness, R.W. (1991). Mercury and selenium interaction: a review. *Ecotoxicology and Environmental Safety*, **21**, 348–364.

Dettaï, A., di Prisco, G., Lecointre, G., Parisi, E., Verde, C. (2008). Inferring evolution of fish proteins: the globin case study. *Methods in Enzymology*, **436**, 539–570; doi.org/10.1016/S0076-6879(08)36030-3 PMID:18237653.

Devine, B.M., Wheeland, L.J., Fisher, J.A. (2018). First estimates of Greenland shark (*Somniosus microcephalus*) local abundances in Arctic waters. *Scientific Reports*, **8**, 974; https://doi.org/10.1038 /s41598-017-19115-x.

Dietz, R., Rigét, F., Born, E.W. (2000). An assessment of selenium to mercury in Greenland marine animals. *Sciences of the Total Environment*, **245**, 15–24.

di Prisco, G., Condò, S.G., Tamburrini, M., Giardina, B. (1991). Oxygen transport in extreme environments. *Trends in Biochemical Science*, **16**, 471–474.

di Prisco, G., Eastman, J.T., Giordano, D., Parisi, E., Verde, C. (2007). Biogeography and adaptation of Notothenioid fish: hemoglobin function and globin-gene evolution. *Gene*, **398**, 143–155; doi.org/10.1016/j.gene.2007.02.047 PMID: 17553637.

Domi, N., Bouquegneau, J.M., Das, K. (2005). Feeding ecology of five commercial shark species of the Celtic sea through stable isotope and trace metal analysis. *Marine Environmental Research*, **60**, 551–569.

Dryer, L., Graziadei, P.P.C. (1996). Synaptology of the olfactory bulb of an elasmobranch fish, Sphyrna tiburo. *Anatomy and Embryology*, **193**, 101–114; doi:10.1007/BF00214701.

Edwards, J.E., Broell, F., Bushnell, P.G., et al. (2018). Advancing our understanding of long-lived species: A case study on the Greenland shark. *Frontiers in Marine Science*, https://www.frontiersin.org/articles/10.3389/fmars.2019.00087/full

Endo, T., Hisamichi, Y., Koichi, H., et al. (2008). Hg, Zn and Cu levels in the muscle and liver of tiger sharks (*Galeocerdo cuvier*) from the coast of Ishigaki island, Japan: relationship between metal concentrations and body length. *Marine Pollution Bulletin*, **56**, 1774–1780.

Fago, A., Wells, R.M.G., Weber, R.E. (1997). Temperature-dependent enthalpy of oxygenation in Antarctic fish hemoglobins. *Comparative Biochemistry and Physiology*, **118B**, 319–326.

Ferrando, S., Amaroli, A., Gallus, L., et al. (2019). Secondary folds determine the surface area in the olfactory organ of Chondrichthyes. *Frontiers in Physiology*. https://doi.org/10.3389/fphys.2019.00245

Ferrando, S., Gallus, L., Ghigliotti, L., et al. (2016). Gross morphology and histology of the olfactory organ of the Greenland shark *Somniosus microcephalus*. *Polar Biology*, **39**, 1399–1409; https://doi.org/10.1007/s00300-015-1862-1.

Ferrando, S., Gallus, L., Ghigliotti, L., et al. (2017a). Anatomy of the olfactory bulb in Greenland shark *Somniosus microcephalus* (Bloch & Schneider, 1801). *Journal of Applied Ichthyology*, **33**, 263–269; https://doi.org/10.1111/jai.13303.

Ferrando, S., Gallus, L., Ghigliotti, L., et al. (2017b). Clarification of the terminology of the olfactory lamellae in Chondrichthyes. *The Anatomical Record*, **300**, 2039–2045; doi:10.1002/ar.23632.

Ferrari, M.C., Wisenden, B.D., Chivers, D.P. (2010). Chemical ecology of predator–prey interactions in aquatic ecosystems: a review and prospectus. *Canadian Journal of Zoology*, **88**, 698–724; https://doi.org/10.1139/Z10-029.

Fisher, W.K., Nash, A.R., Thompson, E.O. (1977). Haemoglobins of the shark, *Heterodontus portusjacksoni*. III. Amino acid sequence of the *β*-chain. *Australian Journal of Biological Science*, **30**, 487–506.

Fisk, A.T., Tittlemier, S., Pranschke, J., Norstrom, R.J. (2002). Using anthropogenic contaminants and stable isotopes to assess the feeding ecology of Greenland shark. *Ecology*, **83**, 2162–2172.

Fisk, A. T., Lydersen, C., and Kovacs, K. M. (2012). Archival pop-off tag tracking of Greenland sharks Somniosus microcephalus in the High Arctic waters of Svalbard, Norway. Mar. Ecol. Prog. Ser. 468, 255–265. doi: 10.3354/meps09962

Fossheim, M., Primicerio, R., Johannesen, E., et al. (2015). Recent warming leads to rapid borealization of fish communities in the Arctic. *Nature Climate Change*, **5**, 673, doi:10.1038/NCLIMATE2647.

Fyhn, U.E.H., Sullivan, B. (1975). Elasmobranch hemoglobins: dimerization and polymerization in various species. *Comparative Biochemistry and Physiology*, **50**, 119–129.

Gebbink, W.A., Sonne, C., Dietz, R., et al. (2008). Tissue-specific congener composition of organohalogen and metabolite contaminants in East Greenland polar bears (*Ursus maritimus*). *Environmental Pollution*, **152**, 621–629; http://dx.doi.org/10.1016/j.envpol.2007.07.001.

Gillen, R., Riggs, A. (1973). The hemoglobins of a fresh-water teleost, *Cichlasoma cyanoguttatum* (Baird and Girard). II. Subunit structure and oxygen equilibria of the isolated components. *Archives of Biochemistry and Biophysics*, **154**, 348–359.

Giordano, D., Russo, R., Coppola, D., di Prisco, G., Verde, C. (2010). Molecular adaptations in haemoglobins of notothenioid fishes. *Journal of Fish Biology*, **76**, 301–318; doi.org/10.1111/j.1095–8649.2009.02528.xPMID: 20738709.

Hara, T.J. (1992). Mechanisms of olfaction. In: T. J. Hara (ed.) *Fish Chemoreception. Fish & Fisheries Series*, vol 6. Springer, Dordrecht, pp. 150–170.

Herbert, N.A., Skov, P.V., Tirsgaard, B., et al. (2017). Blood O_2 affinity of a large polar elasmobranch, the Greenland shark *Somniosus microcephalus*. *Polar Biology*, **40**(11), 2297–2305.

Hodgson, E.S., Mathewson, R.F. (1978). *Sensory Biology of Sharks, Skates, and Rays*. U.S. Office of Naval Research, Arlington.

Hong, E.J., Choi, K.C., Jung, Y.W., Leung, P.C., Jeung, E.B. (2004). Transfer of maternally injected endocrine disruptors through breast milk during lactation induces neonatal Calbindin-D 9 k in the rat model. *Reproductive Toxicology*, **18**(5), 661–668.

Hussey, N.E., Orr, J., Fisk, A.T., et al. (2018). Mark report satellite tags (mrPATs) to detail large-scale horizontal movements of deep water species: First results for the Greenland shark (*Somniosus microcephalus*). *Deep Sea Research Part I: Oceanographic Research Papers*, **134**, 32–40.

IARC (1979). World Health Organization-International Agency for Research on Cancer. WHO-IARC-annual report 1979.

Ikonomou, M.G., Rayne, S., Fischer, M. (2002). Occurrence and congener profiles of polybrominated diphenyl ethers (PBDEs) in environmental samples from coastal British Columbia, Canada. *Chemosphere*, **46**, 649–663.

Ingvaldsen, R., Gjøsæter, H., Ona, E., Michalsen, K. (2017). Atlantic cod (*Gadus morhua*) feeding over deep water in the high Arctic. *Polar Biology*, **40**(10), 2105–2111; DOI 10.1007/s00300-017-2115-2.

Janák, K., Covaci, A., Voorspoels, S., Becher. G. (2005). Hexabromocyclododecane in marine species from the Western Scheldt Estuary: diastereoisomer-and enantiomer-specific accumulation. *Environmental Science Technology*, **39**, 1987–1994.

Kannan, K., Corsolini, S., Focardi, S., Tanabe, S., Tatsukawa, R. (1996). Accumulation pattern of butyltin compounds in dolphin, tuna, and shark collected from Italian coastal waters. *Archives of Environmental Contamination and Toxicology*, **31**, 19–23.

Klaassen, C.D. (1986). Distribution, excretion, and absorption of toxicants. In: C. D. Klaassen, M.O. Amdur, J.M.D. Doull (eds) *Casarett and Doull's Toxicology the Basic Science of Poisons*. 3rd edition. Macmillan Publishing Company, New York, pp. 33–63.

Koeman, J.H., Ven, W.S.M., Goeij, J.J.M., Tijoe, P. S., Haften, J.L. (1975). Mercury and selenium in marine mammals and birds. *Sciences of the Total Environment*, **3**, 279–287.

Komiyama, N.H., Shih, D.T., Looker, D., Tame, J., Nagai, K. (1991). Was the loss of the D helix in α-globin a functionally neutral mutation? *Nature*, **352**, 349–351; doi.org/10.1038/352349a0 PMID: 1852211.

Koy, K., Plotnick, R.E. (2007). Theoretical and experimental ichnology of mobile foraging. In: W. Miller III (ed.) *Trace Fossils: Concepts, Problems and Prospects*. Elsevier, Amsterdam, pp. 428–441.

Kramer, D.L. (2001). Foraging behavior. In: C. W. Fox, D.A. Roff, D.J. Fairbairn (eds) *Evolutionary Ecology*. Oxford University Press, Oxford, pp. 232–246.

Leclerc, L.M., Lydersen, C., Haug, T., et al. (2011). Greenland sharks (*Somniosus microcephalus*) scavenge offal from minke (*Balaenoptera acutorostrata*) whaling operations in Svalbard

(Norway). *Polar Research*, **30**, 7342; doi:10.3402/polar.v30i0.7342.

Leclerc, L. M., Lydersen, C., Haug, T., Bachmann, L., Fisk, A. T., and Kovacs, K. M. (2012). A missing piece in the Arctic food web puzzle? Stomach contents of Greenland sharks sampled in Svalbard, Norway. Polar Biol. 35, 1197–1208. doi: 10.1007/s00300-012-1166-7

Lynghammar, A., Christiansen, J.S., Mecklenburg, C.W., et al. (2013). Species richness and distribution of chondrichthyan fishes in the Arctic Ocean and adjacent seas. *Biodiversity*, **14**, 57–66.

MacNeil, M.A., McMeans, B.C., Hussey, N.E., et al. (2012). Biology of the Greenland shark *Somniosus microcephalus*. *Journal of Fish Biology*, **80**, 991–1018.

Malins, D.C., Barone, L. (1970). The ether bond in marine lipids. In: F. Snyder (ed.) *Ether Lipids, Chemistry and Biology*. Academic Press, Boca Raton, FL, pp. 297–312.

Manwell, C., Baker, C.M.A. (1970). *Molecular Biology and the Origin of Species: Heterosis, Protein Polymorphism and Animal Breeding*. Sidwick and Jacson, London.

McMeans, B.C., Börga, K., Bechtol, W.R., Higginbotham, D., Fisk, A.T. (2007). Essential and non-essential element concentrations in two sleeper shark species collected in Arctic waters. *Environmental Pollution*, **148**, 281–290.

McMeans, B.C., Svavarsson, J., Dennard, S., Fisk, A.T. (2010). Diet and resource use among Greenland sharks (*Somniosus microcephalus*) and teleosts sampled in Icelandic waters, using δ13C, δ15 N and mercury. *Canadian Journal of Fisheries and Aquatic Sciences*, **67**, 1428–1438.

Mecklenburg, C.W., Møller, P.R., Steinke, D. (2011). Biodiversity of arctic marine fishes: taxonomy and zoogeography. *Marine Biodiversity*, **41**, 109–140.

Mecklenburg, C.W., Lynghammar, A., Johannesen, E., et al. (2018). *Marine Fishes of the Arctic Region, Vol I*. CAFF Monitoring Series Report 28, February 14, 2018.

Meng, Q., Yin, M. (1981). A study of the olfactory organ of skates, rays and chimaeras. *Journal of Fisheries of China*, **5**, 209–228.

Meredith, T.L., Kajiura, S.M. (2010). Olfactory morphology and physiology of elasmobranch. *Journal of Experimental Biology*, **213**, 3449–3456; doi:10.1242/jeb.045849.

Mita, L., Bianco, M., Viggiano, E., et al. (2011). Bisphenol A content in fish caught in two different sites of the Tyrrhenian Sea (Italy). *Chemosphere*, **82**, 405–410.

Molde, K., Ciesielski, T.M., Fisk, A.T., et al. (2013). Associations between vitamins A and E and legacy POP levels in highly contaminated Greenland sharks (*Somniosus microcephalus*). *Science of the Total Environment*, **442**, 445–454.

Muir, D.G., de Wit, C.A. (2010). Trends of legacy and new persistent organic pollutants in the circumpolar arctic: overview, conclusions, and recommendations. *Science of Total Environment*, **408**, 3044–3051.

Naoi, Y., Chong, K.T., Yoshimatsu, K., et al. (2001). The functional similarity and structural diversity of human and cartilaginous fish hemoglobins. *Journal of Molecular Biology*, **307**, 259–270; doi.org/10.1006/jmbi.2000.4446 PMID: 11243818.

Nash, A.R., Fisher, W.K., Thompson, E.O. (1976). Haemoglobins of the shark, *Heterodontus portusjacksoni*. II. Amino acid sequence of the α-chain. *Australian Journal of Biological Science*, **29**, 73–97.

Nielsen, J. (2018). The Greenland shark (*Somniosus microcephalus*): Diet, tracking and radiocarbon age estimates reveal the world's oldest vertebrate. PhD thesis, University of Copenhagen; DOI:10.13140/RG.2.2.35883.49448.

Nielsen, J., Christiansen, J.S., Grønkjær, P., et al. (2019) Greenland shark (*Somniosus microcephalus*) stomach content and stable isotope values reveal ontogenetic dietary shift. *Frontiers in Marine Science*. https://doi.org/10.3389/fmars.2019.00125

Nielsen, J., Hedeholm, R.B., Simon, M., Steffensen, J.F. (2014). Distribution and feeding ecology of the Greenland shark (*Somniosus microcephalus*) in Greenland waters. *Polar Biology*, **37**, 37–46.

Nielsen, J., Hedeholm, R.B., Heinemeier, J., et al. (2016). Eye lens radiocarbon reveals centuries of longevity in the Greenland shark (*Somniosus microcephalus*). *Science*, **353** (6300), 702–704.

Nielsen, J., Hedeholm, R.B., Lynghammar, A., McClusky, L.M., Berland, B., Steffensen, J.F., Christiansen, J.S. Assessing the reproductive biology of the Greenland shark (*Somniosus microcephalus*). Plos One, in press.

Olin, J.A., Beaudry, M., Fisk, A.T., Paterson, G. (2014). Age-related polychlorinated biphenyl dynamics in immature bull sharks (*Carcharhinus leucas*). *Environmental Toxicology and Chemistry*, **33**, 35–43; https://doi.org/10.1002/etc.2402.

Pedersen, S.A., Madsen, J., Dyhr-Nielsen, M. (2004). *Global International waters assessment: Arctic Greenland, East Greenland Shelf, West Greenland Shelf. United Nations Environment Programme, GIWA Regional Assessment 2b, 15, 16.* University of Kalmar, Kalmar, Sweden.

Perutz, M.F. (1983). Species adaptation in a protein molecule. *Molecular Biology and Evolution*, **1**, 1–28.

Perutz, M.F. (1998). The stereochemical mechanism of the cooperative effects in haemoglobin revisited. *Annual Reviews of Biophysical and Biomolecular Structure*, **27**, 1–34.

Perutz, M.F., Brunori, M. (1982). Stereochemistry of cooperative effects in fish and amphibian haemoglobins. *Nature*, **299**, 421–426.

Porteiro, F.M., Sutton, T.T., Byrkjedal, I., et al. (2017). Fishes of the northern Mid-Atlantic Ridge collected during the MAR-ECO cruise in 522 June–July 2004: an annotated checklist. *Arquipelago Life and Marine Sciences* Supplement 10.

Rigét, F., Bignert, A., Braune, B., Stow, J., Wilson, S. (2010). Temporal trends of legacy POPs in Arctic biota, an update. *Science of Total Environment*, **408**, 2874–2884.

Riggs, A. (1970). Properties of fish hemoglobins. In: W.S. Hoar, D.J. Randall (eds) *Fish Physiology*, Vol. 4. Academic Press, New York, pp. 209–252.

Righton, D.A., Andersen, K.H., Neat, F., et al. (2010). Thermal niche of Atlantic cod *Gadus morhua*, limits, tolerance and optima. *Marine Ecology Progress Series*, **420**, 1–13.

Roesijadi, G. (1992). Metallothioneins in metal regulation and toxicity in aquatic animals. *Aquatic Toxicology*, **22**, 81–113.

Russo, R., Giordano, D., Paredi, G., et al. (2017). The Greenland shark *Somniosus microcephalus*: Hemoglobins and ligand-binding properties. *PLoS ONE*, **12**(10), e0186181; doi.org/10.1371/journal.pone.0186181.

Schluessel, V., Bennett, M.B., Bleckmann, H., Blomberg, S., Collin, S.P. (2008). Morphometric and ultrastructural comparison of the olfactory system in elasmobranchs: the significance of structure–function relationships based on phylogeny and ecology. *Journal of Morphology*, **269**, 1365–1386; https://doi.org/10.1002/jmor.10661.

Schluessel, V., Bennett, M.B., Bleckmann, H., Collin, S.P. (2010). The role of olfaction throughout juvenile development: functional adaptations in elasmobranchs. *Journal of Morphology*, **271**, 451–461; https://doi.org/10.1002/jmor.10809.

SC-POPs (2013). Stockholm convention on persistent organic pollutants. http://chm.pops.int/Implementation/Exemptions/AcceptablePurposesDDT/tabid/456/Default.aspx.

Shadwick, R.E., Bernal, D., Bushnell, P.G., Steffensen, J.F. (2018). Blood pressure in the Greenland shark as estimated from ventral aortic elasticity. *Journal of Experimental Biology*, 221, 1–6. doi:10.1242/jeb.186957.

Skomal, G.B., Benz, G.W. (2004). Ultrasonic tracking of Greenland sharks (*Somniosus* microcephalus) under Arctic ice. *Marine Biology*, **145**, 489–498.

Smeets, W.J.A.J. (1998). Cartilaginous fishes. In: R. Nieuwenhuys, H.J. Donkelaar ten, C. Nicholson (eds) *The Central Nervous System of Vertebrates*. Springer, Berlin, pp. 551–654.

Soares, A., Guieysse, B., Jefferson, B., Cartmell, E., Lester, J.N. (2008). Nonylphenol in the environment: a critical review on occurrence, fate, toxicity and treatment in wastewaters. *Environmental International*, **34**(7), 1033–1049.

Speers-Roesch, B., Richards, J.G., Brauner, C.J., et al. (2012a). Hypoxia tolerance in elasmobranchs. I. Critical oxygen tension as a measure of blood oxygen transport during hypoxia exposure. *Journal of Experimental Biology*, **215**, 93–102.

Speers-Roesch, B., Brauner, C.J., Farrell, A.P., et al. (2012b). Hypoxia tolerance in elasmobranchs. II. Cardiovascular function and tissue metabolic responses during progressive and relative hypoxia exposures. *Journal of Experimental Biology*, **215**, 103–114.

Staniszewska, M., Falkowska, L., Grabowski, P., et al. (2014). Bisphenol A, 4-tert-octylphenol, and 4-nonylphenol in the Gulf of Gdańsk (E. Southern Baltic). *Archives of Environmental Contamination and Toxicology*, **67**, 335.

Storelli, M.M., Marcotrigiano, G.O. (2002). Mercury speciation and relationship between mercury and selenium in liver of *Galeus melastomus* from the Mediterranean Sea. *Bulletin of Environmental Contamination and Toxicology*, **69**, 516–522.

Strid, A., Athanassiadis, I., Athanasiadou, M., et al. (2010). Neutral and phenolic brominated organic compounds of natural and anthropogenic origin in northeast Atlantic Greenland Shark (*Somniosus microcephalus*). *Environmental Toxicology and Chemistry*, **29**, 2653–2659.

Strid, A., Jörundsdóttir, H., Päpke, O., Svavarsson, J., Bergman, Å. (2007). Dioxins and PCBs in Greenland shark

(*Somniosus microcephalus*) from the North-East Atlantic. *Marine Pollution Bulletin*, **54**, 1514–1522.

Strid, A., Bruhn, C., Sverko, E., et al. (2013). Brominated and chlorinated flame retardants in liver of Greenland shark (*Somniosus microcephalus*). *Chemosphere*, **91**(2), 222–228.

Tetens, V., Wells, R.M. (1984). Oxygen binding properties of blood and hemoglobin solutions in the carpet shark (*Cephaloscyllium isabella*): roles of ATP and urea. *Comparative Biochemistry and Physiology*, **79A**, 165–168.

Tierney, K.B. (2015). Olfaction in aquatic vertebrates. In: R.L Doty (ed.) *Handbook of Olfaction and Gustation*. Wiley-Blackwell, USA, pp. 547–564.

UNEP (2013). *Global Mercury Assessment 2013, Sources, Emissions, Releases and Environmental Transport*. UNEP Chemicals Branch, Geneva, Switzerland.

van den Berg, M., Birnbaum, L., Bosveld, A.T.C., et al. (1998). Toxic equivalency factors (TEFs) for PCBs, PCDDs, PCDFs for humans and wildlife. *Environmental Health Perspectives*, **106**, 775–792.

van den Berg, M., Birnbaum, L.S., Denison, M., et al. (2006). The 2005 World Health Organization reevaluation of human and mammalian toxic equivalency factors for dioxins and dioxin-like compounds. *Toxicological Sciences*, **93**(2), 223–241; http://dx.doi.org/10.1093/toxsci/kfl055.

Vas, P. (1991). Trace-metal levels in sharks from British and Atlantic Waters. *Marine Pollution Bulletin*, **22**, 67–72.

Verde, C., De Rosa, M.C., Giordano, D., et al. (2005). Structure, function and molecular adaptations of haemoglobins of the polar cartilaginous fish *Bathyraja eatonii* and *Raja hyperborea*. *Biochemical Journal*, **389**, 297–306; doi.org/10.1042/BJ20050305 PMID: 15807670.

Verde, C., Balestrieri, M., de Pascale, D., et al. (2006). The oxygen-transport system in three species of the boreal fish family

Gadidae. Molecular phylogeny of hemoglobin. *Journal of Biological Chemistry*, 281, 22073–22084; doi.org/10.1074/jbc. M513080200 PMID: 16717098

Verreault, J., Gabrielsen, G.W., Chu, S., Muir, D.C.G., Andersen, M. (2005). Flame retardants and methoxylated and hydroxylated polybrominated diphenyl ethers in two Norwegian Arctic top predators: glaucous gulls and polar bears. *Environmental Science and Technology*, **39**, 6021–6028.

Weber, R.E., Campbell, K.L. (2011). Temperature dependence of haemoglobin-oxygen affinity in heterothermic vertebrates: mechanisms and biological significance. *Acta Physiologica, Oxford*, **202**, 549–562.

Weber, R.E., Wells, R.M., Rossetti, J.E. (1983a). Allosteric interactions governing oxygen equilibrium in the haemoglobin system of the spiny dogfish *Squalus acanthias*. *Journal of Experimental Biology*, **103**, 109–120.

Weber, R.E., Wells, R.M., Tougaard, S. (1983b). Antagonistic effect of urea on oxygenation-linked binding of ATP in an elasmobranch hemoglobin. *Life Sciences*, **32**, 2157–2161.

Yano, K., Stevens, J.D., Compagno, L.J.V. (2007). Distribution, reproduction and feeding of the Greenland shark *Somniosus (Somniosus) microcephalus*, with notes on two other sleeper sharks, *Somniosus (Somniosus) pacificus* and *Somniosus (Somniosus) antarcticus*.*Journal of Fish Biology*, **70**, 374–390.

Yopak, K., Lisney, T.J., Collin, S.P. (2015). Not all sharks are 'swimming noses': Variation in olfactory bulb size in cartilaginous fishes. *Brain Structure and Function*, **220**, 1127–1143; https://doi.org/10.1007/s00429-014-0705-0.

Zhou, J., Cai, Z.H., Zhu, X.S. (2009). Endocrine disruptors: an overview and discussion on issues surrounding their impact on marine mammals. *Journal of Marine Animal Ecology*, **2**, 7–12.

CHAPTER TWO

Metazoan adaptation to deep-sea hydrothermal vents

STÉPHANE HOURDEZ

Centre National pour la Recherche Scientifique (CNRS)

and

DIDIER JOLLIVET

Centre National pour la Recherche Scientifique (CNRS)

2.1 Introduction

Fauna inhabiting the deep-sea usually obtains its nutrition from sinking organic matter formed by photosynthesis in the photic zone. This photosynthetic organic matter is degraded during its fall and, as a result, these great depths are typically host to a high biodiversity but low biomass. The discovery of deep-sea hydrothermal vents in the late 1970s (Corliss & Ballard, 1977) brought an end to this paradigm. In the abyssal desert, these ecosystems look like oases of life, with a very high biomass compared to the surrounding deep-sea at similar depths (Figure 2.1). At hydrothermal vents, the primary production is local, based on chemosynthesis performed by bacteria that can be found either free-living or symbiotic (Fisher, 1995). Symbiosis with autotrophic bacteria itself is an adaptation to the very food-poor deep-sea (see Dubilier et al., 2008 for a review). Although local production and biomass are very high, the diversity of animals is low compared to the general deep-sea, with a limited number of specialised species at a given site, and about 95% of these species are usually endemic (Tunnicliffe, 1991). The very high proportion of endemic species and low specific diversity are likely the result of the challenging conditions that also characterise hydrothermal vents. The fluids that come out of the ocean crust contain reduced chemicals (sulphide, methane, ferrous iron and sometimes hydrogen at some hydrothermal vents) that can be used by the great variety of prokaryotes at the base of this chemosynthesis-based ecosystem (Zierenberg et al., 2000). The reduced chemicals are oxidised by the prokaryotes, and the resulting energy is used to produce organic matter (Fisher, 1995). Although deep-sea hydrothermal vents look like oases, the very same fluids that are essential for local primary

Figure 2.1 Hydrothermal vent fauna represents a high biomass in the deep-sea. The fauna is distributed according to its tolerance and requirements in the mixing zone of hydrothermal fluids and deep-sea water. In this picture, the gastropods *Ifremeria nautilei* and *Alviniconcha* sp. form dense patches. Copyright Ifremer/Chubacarc 2019. (A black and white version of this figure will appear in some formats. For the colour version, please refer to the plate section.)

production render the environment challenging for animals to thrive. At hydrothermal vents, the hydrothermal fluid is hot (up to 400°C) and anoxic, with high concentrations of CO_2 and sulphide. Other toxic compounds, including heavy metals, are also present. Although metazoan life is not encountered in pure vent fluids, mixing of these fluids with the surrounding deep-sea water is very chaotic, both in space and in time (Johnson et al., 1988a, 1988b; Matabos et al., 2015) and the resulting concentrations of the mentioned compounds are very variable. Even if temperature does not reach extreme values, it can vary by up to 10°C in a matter of minutes.

This chapter aims to review adaptations found in hydrothermal vent endemic taxa to some of the challenges they encounter: highly variable and sometimes high temperature, chronic hypoxia, and exposure to sulphide and high concentration of some metals.

2.2 Temperature

Since the discovery of active hydrothermal vent sites in the Eastern Pacific, most eukaryotic animals living close to the fluid emissions have been considered as potentially thermophilic mostly because prokaryotes sampled there, sometimes in close association with invertebrates, are usually thermophilic (optimal growth rate up to 80°C) or hyperthermophilic (optimal growth rate above 80°C) (Lowe et al., 1993; Prieur et al., 1995; Horikoshi, 1998). Among these emblematic and fascinating eukaryotes, the Pompeii worm *Alvinella pompejana* builds tubes on the walls of active still hot smokers (>80°C). Its thermal tolerance to vent fluids was highly debated since the first *in situ* observations with a temperature probe indicating that the worm was able to withstand over brief periods of time temperatures as high as 105°C (Chevaldonné et al., 1992). Further probing of inhabited tubes for longer periods of time (i.e. > 1 hour) supported this first observation with temperatures ranging between 60 and 80°C inside the tube (Cary et al., 1998). At least, two non-filamentous epsilon proteobacteria isolated from *A. pompejana* are moderately thermophilic, sulfur-reducing heterotrophs and use formate as a carbon source (Campbell et al., 2009), suggesting that the worm was able to withstand such temperatures. However, these measurements were heavily disputed as most of the physiological parameters dealing with the thermal tolerance of the worm were below this range (Chevaldonné et al., 2000). Recently, physiological experiments within thermoregulated pressurized aquaria following their hyperbaric recovery have shown that the Pompeii worm was not able to survive temperatures greater than 50°C for periods of time longer than 2 hours (Ravaux et al., 2013). This fits well with additional *in situ* probing of the worm's tube over several *in situ* assays suggesting that the thermal environment of the microbial component of *Alvinella's* epibiosis mostly ranged from 30 to 55°C with short thermal bursts up to 80°C no longer than a few minutes (Di Meo-Savoie et al., 2004), a situation also depicted by other measurements done by Le Bris and Gaill (2007) within and above *Alvinella* tubes. Similar thermal preference was also observed for *Paralvinella sulfincola*, another tube-builder on hydrothermal chimney walls of the Northeastern Pacific (Tunnicliffe et al., 1993). A similar situation was also depicted for the sulphide worm *Paralvinella sulfincola* for which temperatures up to 45°C have been recorded around the tubes of the worms in the northeastern Pacific (Bates et al., 2013). The worm preferred temperatures ranging between 40 and 50°C when offered a thermal gradient in pressurized aquaria (Girguis & Lee, 2006). As a consequence, although not experiencing temperatures as high as thermophilic prokaryotes, the tube-building alvinellid worms still represent one of the first thermophilic eukaryotes in marine ecosystems.

In contrast, other vent taxa such as gastropods, mussels, siboglinid tube-worms or scaleworms seem to prefer colder habitats. They are highly responsive to heat and prefer much cooler temperatures than most of the upper temperatures recorded in the field (Bates et al., 2010). Escaping high temperatures by moving away from the source seems to represent an adaptive way to maintain a safety margin against rapid temperature fluctuations and the fluid toxicity. As stated in Bates et al. (2010), behavioural avoidance of hot conditions is a primary defence used by ectotherms from the tropics and deserts but not specifically displayed by temperate marine fauna, which is more likely to experience tidal or seasonal thermal variations but not rapid and chaotic spikes of temperature. As a consequence, most vent molluscs such as provan-nid gastropods, limpets or bivalves are mainly located at a distance of several to tens of centimetres from vent flows with temperature variability oscillating between 5 and 15°C, while others represent true stenothermal abyssal species with temperatures not greater than 4°C (Bates et al., 2005, 2010). Temperature has therefore pronounced implications in the microspatial distribution of vent species with sharp boundaries in the delimitation of species (Sarrazin et al., 2002; Lee, 2003; Bates et al., 2005). Even animals living at the upper end of the thermal range of the vent environment are not specifically thermophi-lic and usually avoid hot conditions when they are mobile. For example, peltospirid gastropods, which are found on alvinellid tubes, exhibit thermal preferences below 25°C on chimneys of the Eastern Pacific (Matabos et al., 2008). The polychaete *Hesiolyra bergi*, also found in close association with *Alvinella* colonies, probably experiences temperatures above 40°C *in situ*, but for no more than a few minutes, and certainly not long enough to challenge *Alvinella*'s tolerance. This caterpillar worm is however likely to avoid bursts of high temperatures by crawling away rapidly from the hot source (Shillito et al., 2001). Unlike the mobile fauna associated with hot chimneys, *Alvinella* worms are able to control the temperature of their microenvironment by a thermoregulatory behaviour, which allows them to create a water flow within the tube (Chevaldonné & Jollivet, 1993; Le Bris & Gaill, 2007). This may explain why the analysis of the epibiotic metagenome of the worm reveals a relatively high metabolic flexibility of the epibionts with a clear eurythermal signal of adaptation (Grzymski et al., 2008).

   With the noteworthy exception of alvinellid worms, most *in vivo* physiolo-gical experiments at *in situ* pressures reveal that vent animals have thermal preferences quite similar to those of closely related temperate intertidal spe-cies. As an example, Girguis and Childress (2006) showed that the strongest limitation to metabolic fluxes associated with symbiosis in the siboglinid tubeworm *Riftia pachyptila* is temperature with a sharp increase in CO_2, H_2S and O_2 uptakes occurring at 25°C. Optimal temperature for symbiont primary productivity was thus assumed to fall between 25 and 27°C, whereas prolonged exposure to temperatures greater than 32°C is lethal for the tubeworm, at least under such controlled conditions. At atmo-spheric pressure, the oxygen consumption rate for the hydrothermal vent shrimp *Mirocaris fortunata* (commonly exposed to steep thermal gradients on the wall of active chimneys) seems less affected by temperature than that of the intertidal species *Palaemonetes varians* (Q_{10}=1.96 and Q_{10}= 2.57, respec-tively; Smith et al., 2013). Experimentation at *in situ* pressure on the vent-endemic crab *Bythograea thermydron* showed that although mobile, crabs are only able to regulate their oxygen uptake and heart rate over temperatures as high as 32°C for brief periods of time (Mickel & Childress, 1982a, 1982b).

Based on cardiac frequency data in pressure aquaria, Hourdez (2018) showed however that the Atlantic bythograeid crab *Segonzacia mesatlantica* had an upper physiological thermal limit of 22°C.

Based on *Km* values for malate dehydrogenases, Dahlhoff et al. (1991; Dahlhoff & Somero, 1991) suggested that hydrothermal vent species were on average less affected by temperature variations than close relatives that inhabit the typical deep-sea. Other molecular studies demonstrated reduced temperature variation sensitivity in proteins produced by the vent fauna. In respiratory pigments in general, there usually is a decrease of oxygen affinity when temperature increases as a consequence of the exothermic nature of this reaction (Truchot, 1992). Interestingly, all studied hemocyanins from hydrothermal vent species exhibit either no effect or a slightly reverse effect of temperature on oxygen affinity (Sanders et al., 1988; Lallier & Truchot, 1997; Lallier et al., 1998; Chausson et al., 2001, 2004). This temperature insensitivity could represent an adaptation to the highly variable conditions encountered at deep-sea hydrothermal vents. The lack of temperature sensitivity of oxygen binding has also been observed for shallow water crustacea that are exposed to highly variable temperatures such as the hermit crab *Pagurus bernhardus* (Jokumsen & Weber, 1982) and the shrimp *Palaemon elegans* (Morris et al., 1985). Based on malate dehydrogenase studies on various species, Dahlhoff et al. (1991) showed that species inhabiting chimney walls exhibit predicted maximal sustainable body temperatures 20–25°C higher than their relatives that live in colder (diffuse fluid) areas. This difference, leading to a greater sensitivity to temperature of vent animals living at the periphery of chimneys (when compared to chimney species), is probably the result of a recent colonisation of the vent habitat from the fauna living in the abyssal surroundings or in cold-seep environments. This is especially the case of bathymodiolin mussels, which seem maladapted to chronic exposure to temperatures greater than 15°C for long periods of time. Heat shock experiments on the shallow (800 m) vent species *Bathymodiolus azoricus* indeed showed that this mussel responds with a global metabolic depression, including a large number of genes of both the transcriptional and translational machineries down-regulated after exposure to a temperature greater than 15°C (Boutet et al., 2009b).

The first studies about the molecular mechanisms by which vent species may cope with relatively high temperatures have been mainly conducted on *Bathymodiolus* mussels, alvinocaridid shrimp and alvinellid worms. The latter two are both believed to experience the hottest conditions of the vent environment. The shrimp species *Mirocaris fortunata* and *Rimicaris exoculata* both exhibit constitutive up-regulated heat shock proteins 70 that may be used to repair constantly damaged proteins when bathing in the vent emissions to fuel their epibionts or graze on bacterial mats (Ravaux et al., 2007; Cottin et al., 2010). Most of the work dealing with thermal adaptation of vent organisms was however done on the two thermophilic polychaetes *A. pompejana* and *P. sulfincola*, as these species stand at the upper limit of the accepted range for metazoans. The first studies were performed in the early 90s on the tube and the cuticle of the Pompeii worm (Gaill & Hunt, 1991; Gaill et al., 1995; Sicot et al., 2000). The tube has been suggested to represent the first barrier of the worm to cope with the great temporal and spatial variability of temperature (Le Bris & Gaill, 2007). This parchment-like, often mineralised, tube represents a remarkable adaptation not only for *A. pompejana* and *A. caudata* but also for other species of *Paralvinella* that live on the wall of active chimneys. The tube secreted by *Alvinella* has a concentric multilayered structure with a very high thermal and chemical stability, when compared to other annelid tubes. It is made of biopolymers containing about 50% of proteins that form a liquid crystalline-like organisation, which cannot be denatured when exposed to temperatures up to 100°C and acidic conditions (Gaill & Hunt, 1991). As *A. pompejana* acts as a pioneer species and is the first to colonise still hot anhydrite edifices, tube secretion represents a very efficient way to coat the edifice while reducing exposure to both temperature and the toxic fluid.

Another striking adaptation of the worm is the composition of the cuticle that covers its epidermis. Collagen, which represents one of the main components of the epidermis of the worm, is indeed one the most thermostable proteins ever encountered in the animal kingdom. The temperature at which the cuticle collagen is denatured (*Tm*) is 46°C, a temperature very close to the *in vivo* thermal limit of the worm (Ravaux et al., 2013). The thermal stability of this molecule has been characterised by Sicot et al. (2000) who proposed that the presence of a hydroxylated proline at a high frequency on the third position of tandemly repeated amino-acid triplets plays a crucial role to this extent. The vent worm also possesses membranes with a specific lipid composition containing a greater amount of cholesterol and polyunsaturated fatty acids when compared to other worms living in colder conditions, and a very high ratio of eicosapentaenoic acid (EPA)/arachidonic acid (AA) in the anterior versus the posterior half of its body (Phleger et al., 2005). This latter variation of lipid composition of *A. pompejana* along its body wall may correlate with the strong temperature gradient reported in its tube (Cary et al., 1998). Among other metabolic and physiological adaptations of the worm, most enzymes have a half-life of about 50°C after 1 hour of incubation at challenge temperatures (Figure 2.2; Jollivet et al., 1995; Piccino et al., 2004), and mitochondrial respiration is inhibited at temperatures above 45°C (Dahlhoff et al., 1991). Finally, its extracellular hexagonal bilayer haemoglobin loses its ability to carry oxygen at nearly the same thermal limit (Toulmond et al., 1990).

With the rise of the '-omics' era, analyses of molecular processes associated with evolution of adaptation in thermophilic eukaryotes were initiated at the genome level in hydrothermal vent species. Several transcriptomes and

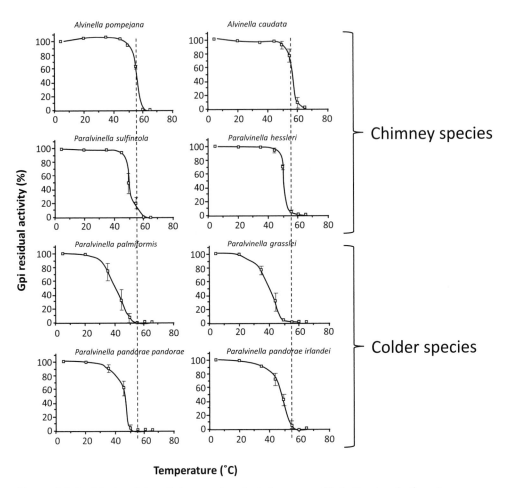

Figure 2.2 Residual activity of glucose phosphate isomerase (*Gpi*) after incubation at different challenge temperatures in eight different species of alvinellid worms, four from chimney walls (exposed to high temperature) and four from colder habitat. The data shown are for the most common allozyme of each species. The dashed line is placed at 55°C to allow easy comparison between species. Modified after Jollivet et al. (1995).

genomes have already been sequenced and assembled, starting with the emblematic *Alvinella pompejana* and *Paralvinella sulfincola* (Gagnière et al., 2010; Dilly et al., 2012). The annotation of the transcriptome of *A. pompejana* reveals a good coverage of most animal metabolic pathways and networks with a prevalence of transcripts involved in oxidative stress resistance, detox-ification, antibacterial defence and heat shock protection (Gagnière et al., 2010). In parallel, quantitative proteomics, expressed sequence tag (EST) libraries and glutathione assays revealed that *P. sulfincola* specimens subjected to increasing temperatures exhibited up-regulated genes associated with the synthesis and recycling of glutathione, the maintenance of high levels of heat shock proteins and the unexpected inhibition of the nicotinamide adenine dinucleotide (NADH) and succinate dehydrogenases pathways involved in oxidative phosphorylation of proteins (Dilly et al., 2012). These findings allowed them to propose that *P. sulfincola* was able to withstand high temperatures by mitigating oxidative stress via an increased synthesis of antioxidants and the decrease of the oxidation flow through the mitochondrial electron transport chain. Comparative studies on the chimney species *P. sulfincola* and its relative *P. palmiformis*, from colder conditions, exhibited strikingly different responses to heat shock (Dilly et al., 2012). While chaperone expression levels changed in the *P. palmiformis* (a normal response), expression levels remained constant in the chimney species *P. sulfincola*, suggesting this latter constantly expresses heat shock response proteins and lives on the edge of its temperature tolerance.

In parallel, the first analyses of the transcriptome and translated proteome of *A. pompejana* pointed to adaptive trends in the amino acid composition of proteins analogous to that found in thermophilic prokaryotes and the mould *Chaetomium thermophilum* (Jollivet et al., 2012; Holder et al., 2013). In particular, the ribosomal proteins of *A. pompejana* displayed thermophilic features similar to those of ultrathermophilic bacteria such as a high proportion of positively charged and of large side-branched hydrophobic residues (Ile, Leu, Tyr: Jollivet et al., 2012), a marked charged versus polar bias (CvP_bias) similar to *C. thermophilum* (Holder et al., 2013) and, more surprisingly, a strong increase in the number of alanine residues, a characteristic that is only shared with the thermophilic fungi. All these characteristics were shared by other thermotolerant alvinellid worms such as *A. caudata*, *Paralvinella sulfincola* or *P. fijiensis* but lost by species that live in colder conditions (Fontanillas et al., 2017). A careful examination of sequence evolutionary rates indicated that the genes encoding proteins of thermophilic (chimney) species were under strong purifying selection with the maintenance of some charged and aromatic residues along with alanine and proline residues (PAYRE) at high frequencies during the course of evolution, suggesting that the ancestor to this unusual worm family was thermophilic (Fontanillas et al., 2017). Additionally, most of the preferred codons of alvinellid worms were GC-ended, despite an overall genome GC content around 45%.

Most of the biochemical indices suggesting that *A. pompejana* represents a true thermophilic eukaryote led to a series of analyses on the thermostability of its proteins using the overexpression of recombinant proteins obtained from cDNA libraries. Most of them clearly indicated that these proteins have a greater stability to temperature than those obtained in homeotherms (Henscheid et al., 2005; Kashiwagi et al., 2010; Shin et al., 2009). This seems to be due to an increase in both ionic and hydrogen bounds in proteins together with an accumulation of cap-prolines in loops to rigidify the molecule (Shin et al., 2009). Optimal activities of *Alvinella* enzymes are often found several degrees above those observed for humans and fall in the range of 42–45°C (Piccino et al., 2004; Kashiwagi et al., 2010; Shin et al., 2009). These values greatly exceeded those measured in other annelids.

Deciphering the mechanisms underlying eukaryotic adaptation to high temperatures is still in its infancy, but more information will be provided during the coming decade with the recent improvements in genome sequencing.

2.3 Hypoxia

Hypoxia is one of the most basic challenges encountered by deep-sea hydrothermal vent species. Nearly all metazoans known to date require oxygen for their metabolism at least during part of their life cycle (Hourdez, 2012). Although oxygen levels are noticeably and chronically lower in deep-sea hydrothermal vent communities, animals from these environments require as much oxygen as their close relatives from well-oxygenated environments (Hourdez & Weber, 2005). Species with symbiotic bacteria, such as tubeworms, require additional oxygen to meet their symbionts' needs (e.g. Girguis and Childress, 2006). This has stimulated interest from different research groups studying adaptations to low levels of oxygen. Different organisational levels have been studied, from the morphology, to the physiology, and to the molecular level. In the past, Hourdez and Lallier (2007) reviewed the adaptations of metazoans from hydrothermal vents and cold seeps to chronic hypoxia, and Hourdez and Weber (2005) reviewed the adaptation in haemoglobins from hydrothermal vent and cold-seep animals.

2.3.1 Fish

Most of the hydrothermal vent taxa are invertebrates but a few fish are encountered near vent communities. Most of them are occasional visitors that tolerate the conditions for the limited time of their visit. The zoarcid fish, *Thermarces cerberus*, is a noteworthy exception that specifically inhabits aggregations of the giant tubeworm, *Riftia pachyptila*, on the East Pacific Rise. *T. cerberus* exhibits much higher oxygen affinities compared to other members of the same family that live in shallow water (Weber et al., 2003). Fish typically possess several haemoglobins in their erythrocytes and one of the major haemoglobin components in *T. cerberus* exhibits a decreased chloride affinity (compared to shallow water relatives), maintaining it in a constantly high affinity state (Weber et al., 2003). On the other hand, *Symenchelis parasitica* is not endemic of hydrothermal vents and only casually visits this environment. Contrary to *T. cerberus* which displayed specific adaptations, in *S. parasitica*, the haemoglobin system is not very different from that of other similar fish from non-vent environments, indicating that it has less need for a high affinity haemoglobin system (Weber et al., 2003; Hourdez & Weber, 2005).

2.3.2 Invertebrates

Invertebrates comprise the vast majority (both in biomass and diversity) of vent-endemic species. With their more variable body plan, one could expect them to be more plastic and adaptations to hypoxia could also affect other organisational levels than in vertebrates.

Morphological adaptations to chronic hypoxia have been studied in a variety of invertebrate groups from hydrothermal vents. All polychaetes studied to date exhibit greatly enlarged gill surface areas compared to relatives that live in well-oxygenated water (Hourdez & Lallier, 2007). In crustaceans, however, there is usually not an increased gill surface area compared to non-vent species (Decelle et al., 2010). Instead, the scaphognathite, a mouth part appendage responsible for renewing the water in the gill chamber, is enlarged compared to relatives from shallow water. As a result, one beat of the scaphognathite propels more water through the gill chamber. Recent work on the hydrothermal vent crab *Segonzacia mesatlantica* has shown that oxygen consumption remained unchanged over a wide range of partial pressures (Hourdez, 2018). This regulation did not involve changes in heart-beat, leaving only ventilation as a compensatory mechanism. Annelid worms do not have gill chambers but possess cilia that can renew the diffusion layer at the surface of the body. Scanning electron microscopy observations of gills surfaces has revealed the presence of numerous cilia in all the species studied to date (Hourdez & Lallier, 2007). It is however difficult to evaluate the efficiency of these cilia. For example, the shallow water species *Terebellides stroemi* also possesses numerous cilia on its gills that resemble those from the hydrothermal vent Alvinellidae (Jouin-Toulmond & Hourdez, 2006). Once again departing from the norm for Polynoidae, some hydrothermal vent species exhibit segmental gills while species from other environments are almost all devoid of these structures (Hourdez & Jouin-Toulmond, 1998). In *Branchipolynoe*, these gills are coelomic gills, communicating with the general coelomic cavity, without the blood vessels that are usually found in other annelids' gills (Hourdez & Jouin-Toulmond, 1998). In hydrothermal vent species with vascular gills, the diffusion distances through the gills are also reduced compared to their shallow water relatives, thereby facilitating gas diffusion (Hourdez & Lallier, 2007). This shortening of diffusion distances is obtained through intraepidermic loops that have never been reported in shallow-water species (Jouin & Gaill, 1990; Jouin-Toulmond & Hourdez, 2006). In hydrothermal vent shrimp, however, there is no significant shortening of diffusion distances, possibly because crustacean gills have already reached a physical limit in all species (Decelle et al., 2010).

In the hydrothermal vent-endemic polychaete family Alvinellidae, further morphological adaptation has been found. An internal gas exchange organ is greatly developed compared to the closely related family Terebellidae (Jouin-

Toulmond et al., 1996). A network of fine capillaries in an anterior pouch that contains numerous erythrocytes allows gas exchange between the haemoglobins contained in the vascular and coelomic compartments (Jouin-Toulmond et al., 1996; Hourdez & Lallier, 2007). The very similar affinities of both vascular and coelomic respiratory pigments allow bidirectional exchange, likely allowing storage of oxygen in the coelomic system and buffering outside oxygen variations. The cerebral ganglion, located downstream of this structure may be the main beneficiary of this strategy (Hourdez & Lallier, 2007).

Respiratory pigments are found in most (but, interestingly, not all) species from deep-sea hydrothermal vents (Hourdez & Lallier, 2007). This includes some taxonomic groups that, in other environments, do not possess respiratory pigments, suggesting a role in adaptation to hypoxia. Hydrothermal vent species of scaleworms are unusual in that they possess large amounts of circulating haemoglobins in their coelomic fluid while all non-hydrothermal vent scaleworms are devoid of this circulating respiratory pigment and only have neuroglobin (a globin expressed in the nervous system) and small amounts of myoglobin (Hourdez et al., 1999b; Projecto-Garcia et al., 2010). This situation has generated interest in the evolution of haemoglobins in this group of annelids. Although extracellular (a common occurrence in annelids), the origin of this haemoglobin is distinct from that of other annelids. In the ancestor to hydrothermal vent species, a gene coding for a myoglobin-like (hence intracellular) protein was likely duplicated and, after molecular tinkering, was expressed as an extracellular haemoglobin (Projecto-Garcia et al., 2010, 2017).

Similar to oxygen-minimum zones (OMZs) invertebrates, respiratory pigments (haemoglobins and hemocyanins) from hydrothermal vent species exhibit very high intrinsic oxygen affinities (Figure 2.3; see Hourdez & Lallier, 2007). This high affinity for oxygen facilitates the diffusion of oxygen from the environment to the body, even from surrounding water with very low partial pressure of oxygen. The release of oxygen near metabolically active tissues is enhanced through a strong Bohr effect that decreases oxygen affinity with pH (local decrease mostly due to CO_2 production through respiration). These high affinities are key in maintaining a relatively constant oxygen consumption down to very low partial pressures of oxygen in the environment. In vent crabs, the resulting critical partial pressures are very low compared to relatives that live in well-oxygenated waters (Gorodezky & Childress, 1994; Hourdez, 2018).

In the scaleworm *Branchipolynoe symmytilida*, two main haemoglobins differ in their functional properties. This offers an interesting parallel to fish haemoglobins that cooperate for the task of oxygen transport and, through specific functional differences, allow the proper function of oxygen delivery over a wider range of conditions (e.g. Weber, 1990; 2000; Fago et al., 2002). In

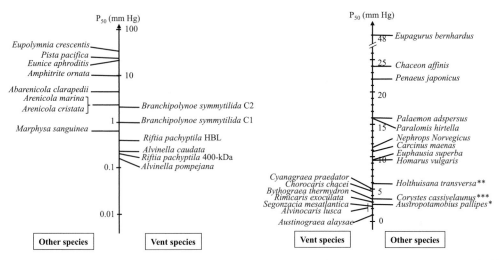

Figure 2.3 Representation of P_{50} (partial pressure of oxygen necessary to saturate half the binding sites of a respiratory pigment) for annelid haemoglobins (log scale, left) and decapod hemocyanins (right) from hydrothermal vent species compared to non-vent species. Note the high affinities (low P_{50}) for vent species compared to most non-vent species. Modified after Hourdez & Lallier (2007), and including unpublished data.

one of the two haemoglobins from *B. symmytilida*, there is a specific effect of CO_2 (i.e. independent of pH) that further decreases its affinity for oxygen and therefore increases release of oxygen near CO_2-producing tissues (Hourdez et al., 1999a).

In addition to facilitating oxygen diffusion from the hypoxic environment to the tissues of the animals, the sheer presence of respiratory pigments could serve as an oxygen storage system that buffers extreme hypoxia or anoxia that could last for long periods of time in such environments (Hourdez & Lallier, 2007). In the scaleworms *Branchipolynoe*, assuming an initially fully saturated pigment, this storage could represent up to 1.5 hours of aerobic metabolism, one order of magnitude more than in shallow water species (Hourdez & Lallier, 2007). In copepods, another group which usually does not have respiratory pigments, studies on two species from different deep-sea hydrothermal vent areas of the world (Sell, 2000; Hourdez et al., 2000) have shown the presence of haemoglobin in their haemolymph. In both cases, although significant (60% of soluble proteins, Hourdez et al., 2000), the amount of haemoglobin only represents a limited storage of oxygen and its function is more likely to facilitate oxygen diffusion from the hypoxic environment to the tissues of the copepod. This view is supported by the very high oxygen affinity that was measured for one of these species (Hourdez et al., 2000). Red copepods have also been observed at other hydrothermal vent sites in the world (SH, pers.

obs.). This common occurrence of large amounts of haemoglobin in very abundant vent-endemic species strongly suggests that it has an adaptive function that has evolved in parallel in these kinds of hypoxic environments.

2.4 Sulphide

Hydrogen sulphide (H_2S) is a very potent poison of aerobic metabolism. Many species from hydrothermal vents are exposed to sulphide concentrations of up to at least 300 μM (Johnson et al., 1988b), although only a small portion of it may be free instead of bound to iron (Luther et al., 2001; Le Bris et al., 2003; Le Bris & Gaill, 2007). Concentration in the low micromolar range can however be toxic to most animals (Smith et al., 1979). This is an especially problematic situation for the organisms that have increased gill surface areas to improve oxygen extraction from the environment as there is no way to exclude sulphide (uncharged in its fully protonated form). The cytochrome oxidase systems of vent animals, like in their non-vent relatives, are sensitive to sulphide poisoning (see Somero et al., 1989). The first line of defence against sulphide poisoning is to oxidise it to less toxic forms such as thiosulfate (Vetter et al., 1987; Somero et al., 1989). When provided with enough oxygen, the detoxification system is so efficient that no free sulphide can be found in the body fluids of vent animals exposed to high levels. The crab *Bythograea thermydron* can maintain its aerobic metabolism when exposed to sulphide concentrations of up to 800 μM (Vetter et al., 1987). This is then accompanied by a 10-fold increase of oxygen consumption rate (Childress & Fisher, 1992). Mitochondria are usually the main site of sulphide oxidation in metazoans and, in the shallow-water lugworm *Arenicola marina*, this oxidation is accompanied by the production of ATP (Grieshaber & Völkel, 1998). This process could also take place in hydrothermal vent organisms. In *Riftia pachyptila*, however, Wilmot & Vetter (1990) demonstrated that although the body wall and branchial plume possess a sulphide-oxidising activity, isolated mitochondria do not oxidise sulphide *in vitro*. The bioproduct of this oxidation, thiosulfate, could be reused by the symbionts in the Benson–Calvin cycle (in particular in mussels see Felbeck et al., 1981).

Organisms that are symbiotic with sulphide-oxidising bacteria (siboglinid tubeworms, vesicomyid clams, bathymodiolin mussels) need to provide this reduced and toxic chemical for their symbionts without experiencing its toxic effects. In mussels, the bacteria are hosted in specialised cells of the gill filaments (bacteriocytes) that are directly bathed by water containing both sulphide and oxygen, used as the final electron acceptor by bacteria. The consumption of sulphide by the symbiotic bacteria hence offers some protection to the rest of the host. In clams and tubeworms however, the acquisition of sulphide is physically distant from the location of the symbionts. In clams, like in mussels, the symbionts are contained in the gills that are bathed in seawater that contains oxygen but sulphide is acquired from the substrate in which the host extends its foot. Sulphide is transported by the blood that contains erythrocytes and an extracellular compound binding sulphide. The carrier is not haemoglobin but a large (several million Dalton) zinc-rich molecule (reviewed in Childress & Fisher, 1992) that does not contain heme groups, but a glycosylated protein that could be a lipoprotein (Zal et al., 2000). In siboglinid tubeworms, both sulphide and oxygen are acquired from the surrounding seawater at the level of the gills and need to be transported to the organ that hosts the symbionts (trophosome), deep in the body of the animal (Childress & Fisher, 1992). In symbiotic animals, sulphide is therefore effectively sequestered, preventing it from reacting with oxygen and avoiding poisoning of aerobic metabolism (Zierenberg et al., 2000). The sulphide-binding mechanism in *Riftia pachyptila* haemoglobin has been investigated relatively early after its role was demonstrated (Arp & Childress, 1983; Arp et al., 1987). Zal et al. (1998) proposed disulphide exchange as the main mechanism of sulphide binding by *Riftia pachyptila* haemoglobins. However, the resolution of the crystal structure of the 400-kDa haemoglobin pointed to the possible role of zinc atoms at specific locations in the edifice in sulphide binding (Flores et al., 2005). This hypothesis was successfully tested using a zinc-specific blocking agent and pointed to possible experimental issues in the disulphide exchange paper (Flores & Hourdez, 2006). For both the tubeworms and the clams, zinc seems to play a central role in sequestering and transporting sulphide to provide this molecule to their symbionts.

In the tissues of various hydrothermal vent annelid species, including the alvinellid species *Alvinella caudata* and *Paralvinella grasslei* (Jouin & Gaill, 1990; Jouin-Toulmond et al., 1996), the scaleworms *Branchypolynoe symmytilida* and *B. seepensis* (Hourdez & Jouin-Toulmond, 1998), electron-dense organelles (EDOs) have been reported. These have been interpreted as secondary lysosomes resulting from the intracellular digestion of mitochondria. EDOs have also been reported in other annelids from environments with high sulphide concentrations (see Wohlgemuth et al., 2007). In the shallow water Maldanidae (Annelida, Polychaeta), *Branchioasychis americana*, Wohlgemuth et al. (2007) showed that exposure to free sulphide in the environment led to the rapid induction of EDOs. These structures are then eliminated after 24 hours, showing their transient nature. Chronic exposure to sulphide in vent organisms may have added further constraint on the evolution of the mitochondrial genome that will have consequences on our use of these markers in calculations of divergence times.

2.5 Metals

Hydrothermal vent fluids are commonly enriched in metals and metal sulphides. The composition of the fluid varies from site to site, mostly based on the composition of the basement rock through which the hot hydrothermal fluid percolates (Von Damm, 1995; Demina & Galkin, 2008). If some of these metals are essential to life (e.g. iron, copper, zinc, etc.), in large amounts they can have adverse effects on living organisms. Other metals common in hydrothermal fluids, such as cadmium, mercury, arsenic and lead, are always toxic. There are two main ways for metals to enter the body of animals: as dissolved ions or complexes, and as ingested particulates. Studies have shown that hydrothermal vent invertebrates accumulate metals (e.g. Roesijadi et al., 1985; Cosson-Mannevy, Cosson & Gaill 1986; Geret et al., 1998; Cosson et al., 2008). The accumulated amounts of metals do not always reflect the environmental levels, suggesting that some of them are specifically eliminated. In addition, environmental exposure can be quite variable. Measurements made on shells of the Mid-Atlantic Ridge mussel *Bathymodiolus azoricus* show variations according to site, time and from individual to individual (Cravo et al., 2007, 2008). Overall, the levels observed in some vent molluscs would be interpreted as a severe contamination in shallow-water relatives (Roesijadi et al., 1985).

The presence of metals in excess can generate an imbalance in the reactive oxygen species (ROS), and these ROS can in turn cause lipid peroxidation, protein modifications and DNA damage (Lushchak, 2011). Metal ions with changeable valence, including iron (very abundant in hydrothermal fluids), can react with H_2O_2 and generate hydroxyl radicals in the Fenton reaction.

Metals can also directly or indirectly interfere with various metabolic processes, including metabolism, membrane transport and protein synthesis, and may act on DNA by interference with genetic control and repair mechanisms (Yamada et al., 1993; Hartwig, 1994; Hassoun & Stohs, 1996; Company et al., 2006). Two strategies are usually followed in response to metal exposure: sequestration of metal ions by specific molecules and oxidative stress response. Sequestration can also rescue metalloproteins that have been inactivated by metals such as cadmium (Isani & Carpenè, 2014).

2.5.1 Sequestration

Metal ions are usually not free in the cells but found in complexes with biomolecules. Iron is by far the most abundant metal found in hydrothermal fluid (Von Damm, 1995) and is essential to life. Ferritin is a specialised protein family usually expressed in the cytosol of cells and whose function is to reversibly bind iron to regulate its concentration (Aisen et al., 2001). It has been shown that its expression level increases after exposure to some heavy metals (Zapata et al., 2009; Chen et al., 2015). In the hydrothermal vent amphipod *Steleuthera ecoprophycea*, Moore and Rainbow (1997) have shown that ferritin forms crystals in the gut caeca and that, in addition to iron, it can bind other metals in lesser amounts. Its function in hydrothermal vent animals has not been investigated in detail, although its function in the stress response has been suggested in the mussel *Bathymodiolus azoricus* (Boutet et al., 2009b). Seven different ferritin isoforms have been shown to coexist in this species, and their response to metal exposure varies, revealing a complex system of iron homeostasis (Bougerol et al., 2015; Fuenzalida, 2017).

Metallothioneins are relatively small proteins, rich in cysteines with a high affinity for d10 configuration metals, including essential ones such as Cu and Zn, and non-essential trace metals (e.g. Cd, Hg) (Isani & Carpenè, 2014). They are usually found in high concentrations in hydrothermal vent taxa. In the mussel *Bathymodiolus platifrons* (a close relative of vent species but that lives at cold seeps), they are found at higher concentration in the gills and digestive gland but are also expressed in the foot (Wong et al., 2015). In the polychaete *Alvinella pompejana*, high concentrations of metallothionein were found in the epidermis and in the digestive tract tissues (Desbruyères et al., 1998). In the symbiotic Siboglinidae *Riftia pachyptila*, which is devoid of a digestive tract, the highest concentration of metallothionein was found in the trophosome, the organ that contains the symbiotic bacteria (Cosson-Mannevy et al., 1988). Its function in the trophosome remains unclear (metal storage or antioxidant). The recent sequencing of amplicons, complete transcriptome and genome sequencing of some *Bathymodiolus* have allowed the study of the genetic diversity of metallothioneins in these cold-seep and hydrothermal vent mussels. As for shallow water bivalves, two main MT types, MT-10 (10-kDa proteins) and MT-20 (20-kDa proteins) have been reported in *Bathymodiolus* spp. (Leignel et al., 2005; Wong et al., 2015). Despite a chronic exposure to metals, metallothionein synthesis is not constitutive and has been shown to be dependent on metal exposure in the vent species *B. thermophilus* (Hardivillier et al., 2006).

Phytochelatins were first identified in plants as the major components of cadmium-binding compounds (Grill et al., 1987) and in nematodes. These molecules have more recently been found in bivalves, including the cold-seep mussel *Bathymodiolus platifrons* (Wong et al., 2015). Its biosynthesis involves the enzymes γ-glutamylcystein synthetase, glutathione synthetase and phytochelatin synthetase, which are all expressed throughout the body of *B. platifrons* (Wong et al., 2015). The product is an oligomer of glutathione subunits that binds metals on its cysteine side chains. The transcript encoding for the phytochelatin synthetase has also been found in the transcriptome of the vent mussel *B. azoricus* (Wong et al., 2015; Fuenzalida et al., unpublished data). This chelation system would be a good avenue to explore for annelids as well, as it has been shown to be present in the earthworm *Eisenia fetida* within which its expression depends on Cd exposure (Brulle et al., 2008).

2.5.2 Oxidative stress response

Assuming *in situ* exposure to oxidative stress, some studies have shed light on the antioxidant response in several species (Figure 2.4). In the alvinellids *Alvinella pompejana* and *Paralvinella grasslei*, studies have shown that catalase activity is very low compared to shallow water annelids. SOD activity is however very high in most tissues. Glutathione peroxidase (GPX) activity correlated with oxidative metabolism enzymes, suggesting this is the main ROS source to which GPX responds. On the contrary, SOD activity is not correlated to oxidative metabolism enzymes and is likely to be modulated by respiration-independent ROS (Marie et al., 2006; Genard et al., 2013).

Experimental work on deep-sea organisms is challenging and studies on metal-induced oxidative stress have only been performed on relatively few hydrothermal vent taxa. These are bivalves, whose shallow-water relatives are typical models for ecotoxicology studies, and a wealth of knowledge is available. Bougerol et al. (2015) investigated the effects of metals on the mussel *B. azoricus* from the Mid-Atlantic Ridge. This species can tolerate maintenance at atmospheric pressure and has therefore been the subject of several studies. After exposure to a cocktail of metals, Bougerol et al. (2015) studied expressed levels of 38 target genes, including some involved in oxidative stress response. Glutathione peroxidase and catalase were differently regulated in response to metal exposure, although no clear pattern of expression could be distinguished.

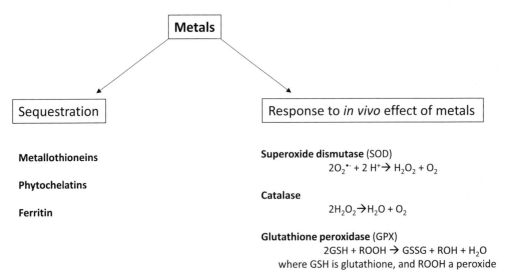

Figure 2.4 Reaction to metals: sequestration and typical enzymes involved in response to oxidative stress in invertebrates. Catalase activity was not detected in the alvinellids *Alvinella pompejana* and *Paralvinella grasslei*. Instead, glutathione peroxidase is a preferred pathway.

When exposed to cadmium and kept under pressure, *B. azoricus* gills exhibited reduced CAT and SOD activity for the first day after exposure, and total oxyradical scavenging capacity (TOSC) and metallothionein levels remained unchanged compared to controls (Company et al., 2006). In the mantle of these mussels, enzymatic activation only developed after 6 days of exposure.

Other proteins can have a function as antioxidant. Metallothioneins bound to zinc have been shown to possess such a potential role (Isani & Carpenè, 2014). Metallothioneins can also help in the antioxidative stress response by rescuing of Cd-inactivated metalloproteins (including antioxidant proteins). Similarly, ferritin heavy chain has a reductase activity that could participate in antioxidant response in addition to its sequestration function (Lau et al., 2004). Finally, in the shallow-water annelid *Hediste diversicolor*, exposure to cadmium induced the production of a myohemerythrin-like protein in the gut (Demuynck et al., 2004). It is unclear whether this protein is involved in sequestration of cadmium or in the antioxidant response. This study however indicates that possible other response mechanisms can be involved in response of marine invertebrates to metal exposure. High-throughput approaches (proteomics, transcriptomics) could shed light on the diversity of response mechanisms in different organisms. Hydrothermal vent invertebrates appear to be good candidates for these approaches.

References

Aisen, P., Enns, C., Wessling-Resnick, M. (2001). Chemistry and biology of eukaryotic iron metabolism. *International Journal of Biochemistry & Cell Biology*, **33**, 940–959.

Arp, A.J., Childress, J.J. (1983). Sulfide binding by the blood of the hydrothermal vent tube worm Riftia pachyptila. *Science*, **219**, 295–297.

Arp, A.J., Childress, J.J., Vetter R.D. (1987). The sulphide-binding protein in the blood of the vestimentiferan tube-worm, Riftia pachyptila, is the extracellular haemoglobin. *Journal of Experimental Biology*, **128**, 139–158.

Bates, A.E., Tunnicliffe, V., Lee, R.W. (2005). Role of thermal conditions in habitat selection by hydrothermal vent gastropods. *Marine Ecology Progress Series*, **305**, 1–15.

Bates, A.E., Lee, R.W., Tunnicliffe, V., Lamare, M.D. (2010). Deep-sea hydrothermal vent animals seek cool fluids in a highly variable thermal environment. *Nature Communications*, **1**, 14.

Bates, A.E., Bird, T.J., Robert, K., et al. (2013). Activity and positioning of eurythermal hydrothermal vent sulphide worms in a variable thermal environment. *Journal of Experimental Marine Biology and Ecology*, **448**, 149–155.

Bougerol, M., Boutet, I., Le Guen, D., Jollivet, D., Tanguy, A. (2015). Transcriptomic response of the hydrothermal mussel Bathymodiolus azoricus in experimental exposure to heavy metals is modulated by the Pgm genotype and symbiont content. *Marine Genomics*, **21**, 63–73.

Boutet, I., Jollivet, D., Shillito, B., Moraga, D., Tanguy, A. (2009a). Molecular identification of differentially regulated genes in the hydrothermal-vent species Bathymodiolus thermophilus and Paralvinella pandorae in response to temperature. *BMC Genomics*, **10** (1), 222.

Boutet, I., Tanguy, A., Le Guen, D., et al. (2009b). Global depression in gene expression as a response to rapid thermal changes in vent mussels. *Proceedings of the Royal Society B: Biological Sciences*, **276**(1670), 3071–3079.

Brulle, F., Cocquerelle, C., Wamalah, A.N., et al. (2008). c-DNA cloning and expression analysis of Eisenia fetida (Annelida: Oligochaeta) phytochelatin synthase under cadmium exposure. *Ecotoxicology & Environmental Safety*, **71**, 47–55.

Campbell, B.J., Smith, J.L., Hanson, T.E., et al. (2009). Adaptations to submarine hydrothermal environments exemplified by the genome of Nautilia profundicola. *PLoS Genetics*, **5**(2), e1000362.

Cary, S.C., Shank, T., Stein, J. (1998). Worms bask in extreme temperatures. *Nature*, **391** (6667), 545.

Chausson, F., Bridges, C.R., Sarradin, P.M., et al. (2001). Structural and functional properties of hemocyanin from Cyanagraea praedator, a deep-sea hydrothermal vent crab. *Proteins*, **45**, 351–359.

Chausson, F., Sanglier, S., Leize, E., et al. (2004). Respiratory adaptations to the deep-sea hydrothermal vent environment: the case of Segonzacia mesatlantica, a crab from the Mid-Atlantic Ridge. *Micron*, **35**, 31–41.

Chen, L., Zhou, J., Zhang, Y., et al. (2015). Preparation and representation of recombinant Mn-Ferritin flower-like spherical aggregates from marine invertebrates. *PLoS One*, **10**, 1–15.

Chevaldonné, P., Jollivet, D. (1993). Videoscopic study of deep-sea hydrothermal vent alvinellid polychaete populations: biomass estimation and behaviour. *Marine Ecology Progress Series*, **95**, 251–262.

Chevaldonné, P., Desbruyères, D., Childress, J.J. (1992). Some like it hot and some even hotter. *Nature*, **359**, 593–594.

Chevaldonné, P., Fisher, C.R., Childress, J.J., et al. (2000). Thermotolerance and the 'Pompeii worms'. *Marine Ecology Progress Series*, **208**, 293–295.

Childress, J.J., Fisher, C.R. (1992). The biology of hydrothermal vent animals: physiology, biochemistry, and autotrophic symbioses. *Oceanography and Marine Biology: An Annual Review*, **30**, 337–441.

Company, R., Serafim, A., Cosson, R.P., et al. (2006). The effect of cadmium on antioxidant responses and the susceptibility to oxidative stress in the hydrothermal vent mussel Bathymodiolus azoricus. *Marine Biology*, **148**, 817–825.

Corliss, J.B., Ballard, R.D. (1977). Oases of life in the cold abyss. *National Geographic*, **152**, 441–454.

Cosson, R.P., Thiébaut, E., Company, R., et al. (2008). Spatial variation of metal bioaccumulation in the hydrothermal vent mussel Bathymodiolus azoricus. *Marine Environmental Research*, **65**(5), 405–415.

Cosson-Mannevy, M.A., Cosson, R.P., Gaill, F. (1986). Mise en évidence de protéines de type metallothionéine chez deux invertébrés des sources hydrothermales, le pogonophore vestimentifère Riftia pachyptila et l'annélide polychète Alvinella pompejana. *Comptes Rendus de l'Académie des Sciences de Paris, Série III*, **302**, 347–352.

Cosson-Mannevy, M.A., Cosson, R.P., Gaill, F., Laubier, L. (1988). Transfert, accumulation et régulation des éléments mineraux chez les organismes des sources hydrothermales. *Oceanologica Acta (Special Issue)*, **8**, 219–226.

Cottin, D., Shillito, B., Chertemps, T., et al. (2010). Comparison of heat-shock responses between the hydrothermal vent shrimp Rimicaris exoculata and the related coastal shrimp Palaemonetes varians. *Journal of Experimental Marine Biology and Ecology*, **393**(1–2), 9–16.

Cravo, A., Foster, P., Almeida, C., et al. (2007). Metals in the shell of Bathymodiolus azoricus from a hydrothermal vent site on the Mid-Atlantic Ridge. *Environment International*, **33**(5), 609–615.

Cravo, A., Foster, P., Almeida, C., et al. (2008). Metal concentrations in the shell of

Bathymodiolus azoricus from contrasting hydrothermal vent fields on the mid-Atlantic ridge. *Marine Environmental Research*, **65**, 338–348.

Dahlhoff, E., Somero, G.N. (1991). Pressure and temperature adaptation of cytosolic malate dehydrogenases of shallow and deep-living marine invertebrates: evidence for high body temperatures in hydrothermal vent animals. *Journal of Experimental Biology*, **159** (1), 473–487.

Dahlhoff, E., O'Brien, J., Somero, G.N., Vetter, R. D. (1991). Temperature effects on mitochondria from hydrothermal vent invertebrates: evidence for adaptation to elevated and variable habitat temperatures. *Physiological Zoology*, **64**(6), 1490–1508.

Decelle, J., Andersen, A.C., Hourdez, S. (2010). Morphological adaptations to chronic hypoxia in deep-sea decapod crustaceans from hydrothermal vents and cold-seeps. *Marine Biology*, **156**(7), 1259–1269.

Demina, L.L., Galkin, S.V. (2008). On the role of abiogenic factors on the bioaccumulation of heavy metals by the hydrothermal fauna of the Mid-Atlantic Ridge. *Oceanology*, **48**(6), 784–797.

Demuynck, S., Bocquet-Muchembled, B., Deloffre, L., Grumiaux, F., Leprêtre, A. (2004). Stimulation by cadmium of myohemerythrin-like cells in the gut of the annelid Nereis diversicolor. *Journal of Experimental Biology*, **207**, 1101–1111.

Desbruyères, D., Chevaldonné, P., Alayse, A.-M., et al. (1998). Biology and ecology of the Pompeii worm (Alvinella pompejana Desbruyères and Laubier), a normal dweller on an extreme deep-sea environment: a synthesis of current knowledge and recent developments. *Deep-Sea Research Part II*, **45**, 383–422.

Dilly, G.F., Young, C.R., Lane, W.S., Pangilinan, J., Girguis, P.R. (2012). Exploring the limit of metazoan thermal tolerance via comparative proteomics: thermally induced changes in protein abundance by two hydrothermal vent polychaetes. *Proceedings of the Royal Society B: Biological Sciences*, **279**(1741), 3347–3356.

Di Meo-Savoie, C.A., Luther III, G.W., Cary, S. C. (2004). Physicochemical characterization of the microhabitat of the epibionts associated with Alvinella pompejana, a hydrothermal vent annelid. *Geochimica et Cosmochimica Acta*, **68**(9), 2055–2066.

Dubilier, N., Bergin, C., Lott, C. (2008). Symbiotic diversity in marine animals: The art of harnessing chemosynthesis. *Nature Reviews Microbiology*, **6**(10), 725–740.

Fago, A., Forest, E., Weber, R. (2002). Hemoglobin and subunit multiplicity in the rainbow trout (Oncorhynchus mykiss) hemoglobin system. *Fish Physiology & Biochemistry*, **24**, 335–342.

Felbeck, H., Somero, G.N., Childress, J.J. (1981). Calvin-Benson cycle and sulphide oxidation enzymes in animals from sulphide rich habitats. *Nature*, **293**, 291–293.

Fisher, C.R. (1995). Toward an appreciation of hydrothermal-vent animals: their environment, physiological ecology, and tissue stable isotope values. In: S. E. Humphris, R.A. Zierenberg, L. S. Mullineaux, R.E. Thomson (eds) *Seafloor Hydrothermal Systems: Physical, Chemical, Biological, and Geological Interactions, Geophysical Monograph 91*. American Geophysical Union, Washington, DC, pp. 297–316.

Flores, J.F., Hourdez, S. (2006). The zinc-mediated sulfide-binding mechanism of hydrothermal vent tubeworm 400-kDa hemoglobin. *Cahiers de Biologie Marine*, **47** (4), 371–377.

Flores, J., Fisher, C.R., Carney, S.L., et al. (2005). Sulfide binding is mediated by zinc ions discovered in the crystal structure of a hydrothermal vent tubeworm hemoglobin. *Proceedings of the National Academy of Sciences USA*, **102**(8), 2713–2718.

Fontanillas, E., Galzitskaya, O.V., Lecompte, O., et al. (2017). Proteome evolution of deep-sea hydrothermal vent alvinellid polychaetes supports the ancestry of thermophily and subsequent adaptation to cold in some lineages. *Genome Biology and Evolution*, **9**(2), 279–296.

Fuenzalida, G. (2017). Transcriptomic approach of the response to metals in the hydrothermal mussel Bathymodiolus azoricus. PhD thesis, Université Pierre and Marie Curie.

Gagnière, N., Jollivet, D., Boutet, I., et al. (2010). Insights into metazoan evolution from Alvinella pompejana cDNAs. *BMC Genomics*, **11**(1), 634.

Gaill, F., Hunt, S. (1991). The biology of annelid worms from high temperature hydrothermal vent regions. *Reviews in Aquatic Sciences*, **4**(2), 107–137.

Gaill, F., Mann, K., Wiedemann, H., Engel, J., Timpl, R. (1995). Structural comparison of cuticle and interstitial collagens from annelids living in shallow sea-water and at deep-sea hydrothermal vents. *Journal of Molecular Biology*, **246**(2), 284–294.

Genard, B., Marie, B., Loumaye, E., et al. (2013). Living in a hot redox soup: antioxidant defenses of the hydrothermal worm Alvinella pompejana. *Aquatic Biology*, **18**, 217–228.

Geret, F., Rousse, N., Riso, R., Sarradin, P.-M., Cosson, R.P. (1998). Metal compartmentalization and metallothionein isoforms in mussels from the Mid-Atlantic Ridge; preliminary approach to the fluid-organism relationship. *Cahiers de Biologie Marine*, **39**(3–4), 291–293.

Girguis, P.R., Childress, J.J. (2006). Metabolite uptake, stoichiometry and chemoautotrophic function of the hydrothermal vent tubeworm Riftia pachyptila: responses to environmental variations in substrate concentrations and temperature. *Journal of Experimental Biology*, **209**(18), 3516–3528.

Girguis, P.R., Lee, R.W. (2006). Thermal preference and tolerance of alvinellids. *Science*, **312**(5771), 231–231.

Gorodezky, L.A., Childress, J.J. (1994). Effects of sulfide exposure history and hemolymph thiosulfate on oxygen-consumption rates and regulation in the hydrothermal vent crab Bythograea thermydron. *Marine Biology*, **120**, 123–131.

Grieshaber, M.K., Völkel, S. (1998). Animal adaptations for tolerance and exploitation of poisonous sulfide. *Annual Review of Physiology*, **60**, 33–53.

Grill, E., Winnacker, E.L., Zenk, M.H. (1987). Phytochelatins, a class of heavy-metal-binding from plants, are functionally analogous to metallothioneins. *Proceedings of the National Academy of Sciences of the USA*, **84**, 439–443.

Grzymski, J.J., Murray, A.E., Campbell, B.J., et al. (2008). Metagenome analysis of an extreme microbial symbiosis reveals eurythermal adaptation and metabolic flexibility. *Proceedings of the National Academy of Sciences of the USA*, **105**(45), 17516–17521.

Hardivillier, Y., Denis, F., Demattei, M.-V., et al. (2006). Metal influence on metallothionein synthesis in the hydrothermal vent mussel Bathymodiolus thermophilus. *Comparative Biochemistry & Physiology, Part C*, **143**, 321–332.

Hartwig, A. (1994). Role of DNA repair inhibition in lead and chromium-induced genotoxicity: a review. *Environmental Health & Perspectives*, **102**, 45–50.

Hassoun E.A., Stohs, S.J. (1996). Cadmium-induced production of superoxide anion and nitric oxide, DNA single strand breaks and lactate dehydrogenase leakage in J774A.1 cell cultures. *Toxicology*, **112**(2–3), 219–226.

Henscheid, K.L., Shin, D.S., Cary, S.C., Berglund, J.A. (2005). The splicing factor U2AF65 is functionally conserved in the thermotolerant deep-sea worm Alvinella pompejana. *Biochimica et Biophysica Acta Gene Structure and Expression*, **1727**(3), 197–207.

Holder, T., Basquin, C., Ebert, J., et al. (2013). Deep transcriptome-sequencing and proteome analysis of the hydrothermal vent annelid Alvinella pompejana identifies the CvP-bias as a robust measure of eukaryotic thermostability. *Biology Direct*, **8**(1), 2.

Horikoshi, K. (1998). Barophiles: deep-sea microorganisms adapted to an extreme environment. *Current Opinion in Microbiology*, **1**(3), 291–295.

Hourdez, S. (2012). Hypoxic environments. In: E. M. Bell (ed.) Life at Extremes: Environments, Organisms and Strategies for Survival, pp. 438–453 CABI Wallingford

Hourdez, S. (2018). Cardiac response of the hydrothermal vent crab Segonzacia mesatlantica to variable temperature and oxygen levels. *Deep Sea Research Part I: Oceanographic Research Papers*, **137**, 57–65.

Hourdez, S., Jouin-Toulmond, C. (1998). Functional anatomy of the respiratory system of Branchipolynoe (Annelida; Polychaeta), commensal with mussels from deep-sea hydrothermal vents. *Zoomorphology*, **118**, 225–233.

Hourdez, S., Lallier, F.H. (2007). Adaptations to hypoxia in hydrothermal vent and cold-seep invertebrates. *Reviews in Environmental Sciences & Biotechnology*, **6**, 143–159.

Hourdez, S., Weber, R.E. (2005). Molecular and functional adaptations in deep-sea hemoglobins. *Journal of Inorganic Biochemistry*, **99**(1), 130–141.

Hourdez, S., Lallier, F.H., Green, B.N., Toulmond, A. (1999a). Hemoglobins from deep-sea scale-worms of the genus Branchipolynoe (Polychaeta, Polynoidae): a new type of quaternary structure. *Proteins*, **34**(4), 427–434.

Hourdez, S., Martin-Jézéquel, V., Lallier, F.H., Weber, R.E., Toulmond, A. (1999b). Characterization and functional properties of the extracellular coelomic hemoglobins from the deep-sea, hydrothermal vent scaleworm Branchipolynoe symmytilida. *Proteins*, **34** (4), 435–442.

Hourdez, S., Lamontagne, J., Peterson, P., Weber, R.E., Fisher, C.R. (2000). Hemoglobin from a deep-sea hydrothermal vent copepod. *The Biological Bulletin*, **199**, 95–99.

Isani, G., Carpenè, E. (2014). Metallothioneins, unconventional proteins from unconventional animals: a long journey from nematodes to mammals. *Biomolecules*, **4**, 435–457.

Johnson, K.S., Childress, J.J., Beehler, C.L. (1988a). Short-term temperature variability in the Rose Garden hydrothermal vent field: an unstable deep-sea environment. *Deep-Sea Research*, **35**, 1711–1721.

Johnson, K.S., Childress, J.J., Hessler, R.R., Sakamoto-Arnold, C.M., Beehler, C.L. (1988b). Chemical and biological interactions in the Rose Garden hydrothermal vent field. *Deep-Sea Research*, **35A**, 1723–1744.

Jokumsen, A., Weber, R.E. (1982). Hemocyanin-oxygen affinity in hermit crab blood is temperature independent. *The Journal of Experimental Zoology*, **221**, 389–394.

Jollivet, D., Desbruyères, D., Ladrat, C., Laubier, L. (1995). Evidence for differences in the allozyme thermostability of deep-sea hydrothermal vent polychaetes (Alvinellidae): a possible selection by habitat. *Marine Ecology Progress Series*, **123**, 125–136.

Jollivet, D., Mary, J., Gagnière, N., et al. (2012). Proteome adaptation to high temperatures in the ectothermic hydrothermal vent Pompeii worm. *PLoS One*, **7**(2), e31150.

Jouin, C., Gaill, F. (1990). Gills of hydrothermal vent annelids: structure, ultrastructure and functional implications in two alvinellid species. *Progress in Oceanography*, **24**, 59–69.

Jouin-Toulmond, C., Hourdez, S. (2006). Morphology, ultrastructure and functional anatomy of the branchial organ of Terebellides stroemii (Polychaeta: Trichobranchidae), with remarks on the systematic position of the genus Terebellides. *Cahiers de Biologie Marine*, **47** (3), 287–299.

Jouin-Toulmond, C., Augustin, D., Desbruyères, D., Toulmond, A. (1996). The gas transfer system in alvinellids (Annelida Polychaeta, Terebellida). Anatomy and ultrastructure of the anterior circulatory system and characterization of a coelomic, intracellular, haemoglobin. *Cahiers de Biologie Marine*, **37**, 135–151.

Kashiwagi, S., Kuraoka, I., Fujiwara, Y., et al. (2010). Characterization of a Y-family DNA polymerase Eta from the eukaryotic thermophile Alvinella pompejana. *Journal of Nucleic Acids*, **2010**, 1–13.

Lallier, F.H., Truchot, J.P. (1997). Hemocyanin oxygen-binding properties of a deep-sea hydrothermal vent shrimp: evidence for a novel cofactor. *Journal of Experimental Zoology*, **277**, 357–364.

Lallier, F.H., Camus, L., Chausson, F., Truchot J.-P. (1998). Structure and function of hydrothermal vent crustacean haemocyanin: an update. *Cahiers de Biologie Marine*, **39**, 313–316.

Lau, A.T., He, Q.Y., Chiu, J.F. (2004). A proteome analysis of the arsenite response in cultured lung cells: evidence for in vitro oxidative stress-induced apoptosis. *Biochemical Journal*, **382**, 641–650.

Le Bris, N., Gaill, F. (2007). How does the annelid Alvinella pompejana deal with an extreme hydrothermal environment? *Reviews in Environmental Science and Biotechnology*, **6** (1–3), 197.

Le Bris, N., Sarradin, P.-M., Caprais, J.C. (2003). Contrasted sulphide chemistries in the environment of 13°N EPR vent fauna. *Deep-Sea Research I*, **50**, 737–747.

Lee, R.W. (2003). Thermal tolerances of deep-sea hydrothermal vent animals from the Northeast Pacific. *The Biological Bulletin*, **205** (2), 98–101.

Leignel, V., Hardivillier, Y., Laulier, M. (2005). Small metallothionein MT-10 genes in coastal and hydrothermal mussels. *Marine & Biotechnology*, **7**, 236–244.

Lowe, S.E., Jain, M.K., Zeikus, J.G. (1993). Biology, ecology, and biotechnological applications of anaerobic bacteria adapted to environmental stresses in temperature, pH, salinity, or substrates. *Microbiology & Molecular Biology Reviews*, **57**(2), 451–509.

Lushchak, V.I. (2011). Environmentally induced oxidative stress in aquatic animals. *Aquatic Toxicology*, **101**, 13–30.

Luther, G.W., Rozan, T.F., Taillefert, M., et al. (2001). Chemical speciation drives hydrothermal vent ecology. *Nature*, **410**, 813–816.

Marie, B., Genard, B., Rees, J.-F., Zal, F. (2006). Effect of ambient oxygen concentration on activities of enzymatic antioxidant defences and aerobic metabolism in the hydrothermal vent worm, Paralvinella grasslei. *Marine Biology*, **150**, 273–284.

Matabos, M., Le Bris, N., Pendlebury, S., Thiébaut, E. (2008). Role of physico-chemical environment on gastropod assemblages at hydrothermal vents on the East Pacific Rise (13 N/EPR). *Journal of the Marine Biological Association of the UK*, **88**(5), 995–1008.

Matabos, M., Cuvelier, D., Brouard, J., et al. (2015). Behavioural study of two hydrothermal crustacean decapods: Mirocaris fortunata and Segonzacia mesatlantica, from the Lucky Strike vent field (Mid-Atlantic Ridge). *Deep-Sea Research Part II*, **121**, 146–158.

Mickel, T.J., Childress, J.J. (1982a). Effects of pressure and temperature on the EKG and heart rate of the hydrothermal vent crab Bythograea thermydron (Brachyura). *The Biological Bulletin*, **162**(1), 70–82.

Mickel, T.J., Childress, J.J. (1982b). Effects of temperature, pressure, and oxygen concentration on the oxygen consumption rate of the hydrothermal vent crab Bythograea thermydron (Brachyura). *Physiological Zoology*, **55**(2), 199–207.

Moore, P.G., Rainbow, P.S. (1997). Ferritin crystals in the gut caeca of a deep-sea hydrothermal vent stegocephalid (Crustacea: Amphipoda). *Journal of the Marine Biological Association of UK*, **77**(1), 269–272.

Morris, S., Taylor, A.C., Bridges, C.R., Grieshaber, M.K. (1985). Respiratory properties of the haemolymph of the intertidal prawn Palaemon elegans (Rathke). *The Journal of Experimental Zoology* **233**, 175–186.

Phleger, C.F., Nelson, M.M., Groce, A.K., et al. (2005). Lipid biomarkers of deep-sea hydrothermal vent polychaetes – Alvinella pompejana, A. caudata, Paralvinella grasslei and Hesiolyra bergi. *Deep Sea Research Part I: Oceanographic Research Papers*, **52**(12), 2333–2352.

Piccino, P., Viard, F., Sarradin, P.-M., et al. (2004). Thermal selection of PGM allozymes in newly founded populations of the thermotolerant vent polychaete Alvinella pompejana. *Proceedings of the Royal Society of London Series B: Biological Sciences*, **271**(1555), 2351–2359.

Prieur, D., Erauso, G., Jeanthon, C. (1995). Hyperthermophilic life at deep-sea hydrothermal vents. *Planetary and Space Science*, **43**(1–2), 115–122.

Projecto-Garcia, J., Zorn, N., Jollivet, D., et al. (2010). Origin and evolution of the unique tetra-domain hemoglobin from the hydrothermal vent scale-worm Branchipolynoe. *Molecular Biology & Evolution*, **27**(1), 143–152.

Projecto-Garcia, J., Le Port, A.-S., Govindji, T., et al. (2017). Evolution of single-domain globins in hydrothermal vent scale-worms. *Journal of Molecular Evolution*, **85**(5–6), 172–187.

Ravaux, J., Toullec, J.Y., Léger, N., et al. (2007). First hsp70 from two hydrothermal vent shrimps, Mirocaris fortunata and Rimicaris exoculata: characterization and sequence analysis. *Gene*, **386**(1–2), 162–172.

Ravaux, J., Hamel, G., Zbinden, M., et al. (2013). Thermal limit for metazoan life in question: in vivo heat tolerance of the Pompeii worm. *PLoS One*, **8**(5), e64074.

Roesijadi, G., Smith, J.S., Crecelius, E.A., Thomas, L.E. (1985). Distribution of trace metals in the hydrothermal vent clam Calyptogena magnifica. *Bulletin of the Biological Society of Washington*, **6**, 311–324.

Sanders, N.K., Arp, A.J., Childress, J.J. (1988). Oxygen binding characteristics of the hemocyanins of two deep-sea hydrothermal vent crustaceans. *Respiration Physiology*, **71**, 57–68.

Sarrazin, J., Levesque, C., Juniper, S.K., Tivey, M. K. (2002). Mosaic community dynamics on Juan de Fuca Ridge sulphide edifices: substratum, temperature and implications for trophic structure. *Cahiers de Biologie Marine*, **43**(3/4), 275–280.

Sell, A. (2000). Life in the extreme environment at a hydrothermal vent: haemoglobin in a deep-sea copepod. *Proceedings of the Royal Society of London B*, **267**.

Shillito, B., Jollivet, D., Sarradin, P.-M., et al. (2001). Temperature resistance of Hesiolyra bergi, a polychaetous annelid living on deep-sea vent smoker walls. *Marine Ecology Progress Series*, **216**, 141–149.

Shin, D.S., DiDonato, M., Barondeau, D.P., et al. (2009). Superoxide dismutase from the eukaryotic thermophile Alvinella pompejana: structures, stability, mechanism, and insights into amyotrophic lateral sclerosis. *Journal of Molecular Biology*, **385**(5), 1534–1555.

Sicot, F.X., Mesnage, M., Masselot, M., et al. (2000). Molecular adaptation to an extreme environment: origin of the thermal stability of the pompeii worm collagen. *Journal of Molecular Biology*, **302**(4), 811–820.

Smith, F., Brown, A., Mestre, N.C., Reed, A.J., Thatje, S. (2013). Thermal adaptations in deep-sea hydrothermal vent and shallow-water shrimp. *Deep-Sea Research Part II: Topical Studies in Oceanography*, **92**, 234–239.

Smith, R.P., Cooper, R.C., Engen, T., et al. (1979). Hydrogen Sulfide. University Park Press, Baltimore.

Somero, G.N., Childress, J.J., Anderson, A.E. (1989). Transport, metabolism and detoxification of hydrogen sulphide in animals from sulphide-rich marine environments. *Critical Reviews in Aquatic Sciences*, **1**, 591–614.

Toulmond, A., Slitine, F.E.I., De Frescheville, J., Jouin, C. (1990). Extracellular hemoglobins of hydrothermal vent annelids: structural and functional characteristics in three alvinellid species. *The Biological Bulletin*, **179** (3), 366–373.

Truchot, J-P. (1992). Respiratory function of arthropod hemocyanins. In: C.P. Mangum (ed.) *Blood and Tissues Oxygen Carriers*. Spinger Verlag, Berlin/Heidelberg, pp. 377–410.

Tunnicliffe, V. (1991). The biology of hydrothermal vents: Ecology and evolution. *Oceanography and Marine Biology: An Annual Review*, **29**, 319–407.

Tunnicliffe, V., Desbruyères, D., Jollivet, D., Laubier, L. (1993). Systematic and ecological characteristics of *Paralvinella sulfincola* Desbruyères and Laubier, a new polychaete (family Alvinellidae) from northeast Pacific hydrothermal vents. *Canadian Journal of Zoology*, **71**(2), 286–297.

Vetter, R.D., Wells, M.E., Kurtsman, A.L., Somero, G.N. (1987). Sulfide detoxification by the hydrothermal vent crab Bythograea thermydron and other decapod crustaceans. *Physiological Zoology*, 60, 121–137.

Von Damm, K.L. (1995). In:S.E. Humphris, R. A. Zierenberg, L.S. Mullineaux, R. E. Thomson (eds) *Controls on the Chemistry and Temporal Variability of Seafloor Hydrothermal Fluids*. Geophysical Monograph Series. American Geophysical Union, Washington, DC, pp. 222–247.

Weber, R.E. (1990). Functional significance and structural basis of multiple hemoglobins with special reference to ectothermic

vertebrates. In: J.-P. Truchot and B. Lahlou (eds) *Animal Nutrition and Transport Processes. 2. Transport, Respiration and Excretion: Comparative and Environmental Aspects*. Karger, Basel, pp. 58–75.

Weber, R.E. (2000). Adaptations for oxygen transport: Lessons from fish hemoglobins. In: G. Di Prisco, B. Giardina, R.E. Weber (eds) *Hemoglobin Function in Vertebrates: Molecular Adaptation in Extreme and Temperate Environments*. Springer-Verlag, Italia, Milan, pp. 23–37.

Weber, R.E., Hourdez, S., Knowles, F., Lallier, F. H. (2003). Hemoglobin function in deep-sea and hydrothermal vent fish: Symenchelis parasitica (Anguillidae) and Thermarces cerberus (Zoarcidae). *Journal of Experimental Biology*, **206**(15), 2693–2702.

Wilmot, D.B.J., Vetter, R.D. (1990). The bacterial symbiont from the hydrothermal vent tubeworm Riftia pachyptila is a sulfide specialist. *Marine Biology*, **106**, 273–283.

Wohlgemuth, S.E., Arp, A.J., Bergquist, D., Julian, D. (2007). Rapid induction and disappearance of electron-dense organelles following sulfide exposure in the marine annelid Branchioasychis americana. *Invertebrate Biology*, **126**(2), 163–172.

Wong, Y.H., Sun, J., He, L.S., et al. (2015). High-throughput transcriptome sequencing of the cold seep mussel Bathymodiolus platifrons. *Scientific Reports*, **5**, 16597.

Yamada, H., Miyahara, T., Sasaki, Y. (1993). Inorganic cadmium increases the frequency of chemically induced chromosome aberrations in cultured mammalian cells. *Mutation Research*, **302**(3), 137–145.

Zal, F., Leize, E., Lallier, F.H., et al. (1998). S-sulfohemoglobin and disulfide exchange: the mechanisms of sulfide binding by Riftia pachyptila hemoglobins. *Proceedings of the National Academy of Sciences of the USA*, **95**(15), 8997–9002.

Zal, F., Leize, E., Oros, D.R., et al. (2000). Haemoglobin structure and biochemical characteristics of the sulphide-binding component from the deep-sea clam Calyptogena magnifica. *Cahiers de Biologie Marine*, **41**(4), 413–423.

Zapata, M., Tanguy, A., David, E., Moraga, D., Riquelme,C. (2009). Transcriptomic response of Argopecten purpuratus post-larvae to copper exposure under experimental conditions. *Gene*, **442**, 37–46.

Zierenberg, R.A., Adams, M.W., Arp, A.J. (2000). Life in extreme environments: hydrothermal vents. *Proceedings of the National Academy of Sciences of the USA*, **97**(24), 12961–12962.

CHAPTER THREE

Extremophiles populating high-level natural radiation areas (HLNRAs) in Iran

Identification of new species and genera with biotechnological interest

FATEMEH HEIDARI
Shahid Beheshti University
HOSSEIN RIAHI
Shahid Beheshti University

and

ZEINAB SHARIATMADARI
Shahid Beheshti University

3.1 Introduction

Recently, much attention has been drawn to the various forms of life existing at the edge of biological limits under extreme physiological conditions. Extremophiles can be defined as organisms thriving in uncommon habitats. All three domains of life (Archaea, Eubacteria and eukaryotes) are represented in extreme environments. Many types of oxygenic phototrophic microorganisms can be found in extreme environments. They occur both at extremely high (hot springs) and extremely low temperatures. Some phototrophs have colonised hypersaline waters or strongly acidic or alkaline habitats. Some extremophilic phototrophs are highly radiation resistant (Seckbach & Oren, 2007).

Ionising radiation (IR) results in severe oxidative stress to all the cell macromolecules (Zupunski et al., 2010; Webb et al., 2013). DNA-associated water molecules that undergo radiolysis become an immediate threat for nucleic acids, generating oxidised DNA bases and sugar moieties, a basic site, DNA single-strand breaks (SSBs) and DNA double-strand breaks (DSBs); a combination of all of these induce the so-called complex (clustered) DNA damage (Hutchinson, 1985; Riley, 1994; Nikitaki et al., 2015). Reactive oxygen species (ROS) generated by ionising radiations cause inactivation of proteins by adding carbonyl residues, cross-linking, mediating amino acid radical chain reactions and ultimately resulting in protein denaturation (Stadtman & Levine, 2003; Imlay, 2006).

Bacteria exhibit different levels of compatibility to damaging radiation and are the simplest model organisms for studying their response and strategies of defence in terms of gene regulation. The study of bacterial response to radiation provides the possibility of using bacteria in various research fields such as fundamental microbiology, microbial ecology, UV-decontamination for water treatment, astrobiology and industrial applications (e.g. cosmetics) (Matallana-Surget & Wattiez, 2013).

Radiation-resistant organisms belong to one of the most interesting categories of extremophiles, i.e. 'Polyextremophiles' (Pavlopoulou et al., 2016). Radiation-resistant microbes are important for astrobiological studies to understand life opportunities in outer space. Kminek et al. (2003) reported that natural halite contains considerable background radiation from the isotope 40 K, which limits the viability of spores in fluid inclusions to 109 million years or less. These considerations have been taken into account for anticipating the chances of microbial life on Mars (Westall et al., 2013).

Various types of radiation-resistant thermophiles have been reported from Archaea including sulfate reducing forms (e.g. *Archaeoglobus fulgidus*), methanogens forms (e.g. *Methanocaldococcus jannaschii*) and the hyperthermophiles *P. furiosus, Thermococcus gammatolerans, Thermococcus radiotolerans*. Also some Actinobacteria have been reported as radiation resistant, e.g. *Rubrobacter radiotolerans* (formerly *Arthrobacter radiotolerans*), *Rubrobacter xylanophilus*, *Rubrobacter taiwanensis*, and members of the genus *Deinococcus*, e.g. *Deinococcus geothermalis* and *Deinococcus murrayi*.

The mechanisms underlying bacterial tolerance to radiation could be classified as: (1) mitigation or amelioration of DNA and protein oxidation by sophisticated antioxidant chemical and enzymatic defence systems, (2) novel and highly adaptive DNA repair mechanisms and (3) nucleoid condensation, which protects DNA from radical damage and enhances DNA repair. A combination of the efficient DNA, protein and membrane lipid repair mechanisms would provide protection to the genome of the organism (Pavlopoulou et al., 2016).

The radiation resistance has been related to desiccation resistance in some non-spore formers such as halophilic Archaea (Stan-Lotter & Fendrihan, 2015). In *Halobacterium salinarum* (a moderate thermophile optimally growing at 50° C), the recovery from irradiation as well as desiccation occur very quickly (Kottemann et al., 2005) as it can regenerate intact chromosome from scattered fragments due to double-strand breaks (Soppa, 2013). The prerequisite for this repair mechanism is overlapping genome fragments, and therefore it can operate only in oligoploid, e.g. *D. radiodurans*, or polyploid species, e.g. Haloarchaea (Soppa, 2014).

The radiation resistance of *Deinococcus geothermalis* can be attributed to channelling of glucose to the pentose phosphate pathway involved in regeneration of NADPH (Liedert et al., 2012). The pentose phosphate pathway and

NADPH are directly involved in repair mechanisms and resistance to oxidative stress. *Deinococcus* avoids NADH formation and thus evades the generation of reactive oxygen species during its recycling by respiratory chain. This property of *Deinococcus* is correlated with its resistance to radiation stress. The role of manganese in radiation tolerance has been very well documented. Mn boosts the protein protection in cells by interacting synergistically with orthophosphates, amino acids, peptides and nucleosides, generating H2O2-scavenging complexes. Compatible solutes like trehalose also play an important role in radiation stress, as well as in heat, osmotic stress, desiccation and oxidation stress. Trehalose along with Mn and phosphate exhibits antioxidant properties which form the basis for high radiation resistance in *R. radiotolerans* and *R. xylanophilus* (Webb & DiRuggiero, 2012).

3.1.1 Radiation resistance in prokaryotic and eukaryotic phototrophs

Cyanobacteria are often found in environments exposed to high light intensities, including high UV levels. They often grow on walls and floors exposed to full sunlight. Some of the most extreme levels of radiation that cyanobacteria encounter are found in Antarctica, where degradation of the ozone layer has brought about an increase in the amount of solar UV radiation that reaches the surface, and cyanobacteria are still very common (Vincent, 2000). Different pigments provide passive protection of cyanobacteria against excessive levels of UV. In the visible light range, communities exposed to high intensities often have high levels of carotenoid pigments (Oren et al., 1995b). In polar areas, cyanobacteria often contain high levels of canthaxanthin, myxoxanthophyll and other carotenoids, the ratio of carotenoids:chlorophyll *a* being maximal under low temperatures, high light regimes and moderate UV radiation (Vincent, 2000). To provide protection against UV many cyanobacteria contain the mycosporine-like amino acids (MAAs), often in large concentrations within their cytoplasm (Castenholz & Garcia-Pichel, 2000). A special case is the glycosylated MAA of *N. commune*, which is excreted and accumulated extracellularly to provide UV protection. An additional sunscreen pigment often encountered is scytonemin, a unique dimeric indole alkaloid located in the sheath surrounding the cells of many species, colouring the cells yellow-brown, and providing light absorption at around 390 nm (Castenholz & Garcia-Pichel, 2000). Moreover, cyanobacteria may possess effective mechanisms to repair UV-induced damage (Rothschild, 1999). An extreme case of radiation resistance in cyanobacteria was provided by a *Synechococcus* isolate from an intertidal evaporitic gypsum crust from Baja California, Mexico. When cells were exposed to high radiation (10^4 kJ/m^2 between 200 and 400 nm) and vacuum in outer space during a 2-week space flight in June 1994, only a slight reduction in viability and activity occurred (Mancinelli et al., 1998).

High levels of carotenoids are also accumulated by many eukaryotic algae exposed to high light intensities. The halophilic green algae *Dunaliella salina* and *D. bardawil* produce ß-carotene up to 8–12% of their dry weight, mainly as a mixture of the all *trans* and the 9-*cis* isomers; ß-carotene is found as globules within the inter-thylakoid space of the chloroplast, and makes the cells resistant to photoinhibition (Ben-Amotz & Avron, 1983). The ß-carotene content of the cells is enhanced when grown at high light, high salinity, low nutrient concentrations and low temperatures. ß-Carotene is valuable as a food-colouring agent, source of provitamin A in food and animal feed, and as a health food antioxidant. *Dunaliella* is presently grown in several countries for ß-carotene production in large nutrient-enriched lagoons or in highly intensive ponds in which all parameters of growth are strictly controlled (Ben-Amotz & Avron, 1983). The snow algae are another well-known case of carotenoid pigments providing protection against damaging high light intensities. Snow is highly reflective to visible radiation, and its high light-scattering properties may create very high photon-fluence rates. At high elevations, the fluence rate of UV-B is also high. The spores of *Chlorogloeopsis nivalis* contain large amounts of astaxanthin esterified with fatty acids which accumulate in extrachloroplastic lipid globules (Bidigare et al., 1993). Aplanospores of snow algae exposed to UV-A (365 nm) and UV-C (254 nm) also produce flavonoids as antioxidant compounds, further reducing the level of free radicals that may damage chlorophyll in the chloroplast thylacoids (Duval et al., 2000). Thanks to these protection mechanisms, the red spores of *C. nivalis* withstand the high UV level to which they are exposed in their natural environment. MAA sunscreen compounds are also widely found in eukaryotic microalgae, as shown by a recent survey of 152 species of marine microalgae from 12 classes (including cyanobacteria). Very high ratios of UV/visible light absorbance (between 2.4 and 6.75) were reported in surface-bloom-forming dinoflagellates, cryptomonads, prymnesiophytes and raphidophytes. Intermediate values (0.9–1.4) were observed in chrysophytes and some prasinophytes and prymnesiophytes. Many diatoms, chlorophytes, eustigmatophytes, rhodophytes, dinoflagellates and prymnesiophytes contain lower concentrations of UV-absorbing compounds, probably MAAs (Jeffrey et al., 1999).

3.1.2 Ionising radiation sources

There are two main components of natural ionising radiation sources, namely cosmic and terrestrial. UNSCEAR (2000a) reported that the annual effective dose due to cosmic and terrestrial sources is 0.39 and 2.03 mSv y^{-1}, respectively. The terrestrial source is likely to contribute the major exposure, consisting of external and internal radiation. Internal radiation due to inhalation of radium (Ra) progenies (e.g. radon and thoron) and ingestion of ^{228}U and ^{232}Th decay products is the major contributor, as compared to external terrestrial gamma radiation (TGR). For inhalation and ingestion exposure, each contributes 1.26 mSv y^{-1} (62%) and 0.29 mSv y^{-1} (14%), respectively. Meanwhile, external (TGR) contributes 0.48 mSv y^{-1}, which is about 24% of the total annual effective dose from terrestrial sources (UNSCEAR, 2000b).

Terrestrial gamma radiation is a major component of the external exposure on the ground. The average received dose due to TGR is 70 µSv y^{-1} (UNSCEAR, 2000b). TGR is mainly caused by various γ radiation emitters from primordial radionuclides of ^{226}Ra, ^{232}Th and 40 K in soil. The contributions from other primordial radionuclides such as ^{87}Rb,^{138}La, ^{147}Sm and ^{176}Lu can be neglected, because of low abundance in the Earth's crust (UNSCEAR, 2000b). These radio-nuclides are extensively dispersed in the Earth's crust (Quindos et al., 1991). In fact, the distribution of radionuclide concentrations is dependent on local geology (Mares, 1984) and geographical factors (Jibiri, 2001). Information about the state of natural radioactivity is necessary for various purposes (El-Gamal et al., 2013). The most crucial part is for radiological assessment of future contamination as a consequence of release of anthropogenic radionuclides from nuclear accidents and global nuclear fallouts (Saito et al., 2012). It is also useful as a baseline for protection against radiation in any related nuclear practices, e.g. for health planning and risk management against radiation sources in medical disciplines.

3.1.3 High background natural radiation areas

Life began in a radiation field that is more intense than that of today (Mortazavi and Mozdarani, 2012); the radiation level today is 10-fold lower than that during the Precambrian era (Møller & Mousseau, 2013). One radio-nuclide that contributes a large proportion of the radiation is radon (Bavarnegin et al., 2013). Nearly 80% of the annual exposure originates from background natural radiation, produced by cosmogenic and primordial radio-nuclides (Schnelzer et al., 2010).

There are several areas of the world known as high-level natural radiation areas (HLNRAs), whose natural radiation sources overcome those considered to have 'normal background'. These and radon-prone areas receive radiation doses relatively higher than in the normal background radiation areas (NBRAs) (Sohrabi, 2013a, 2013b). Based on the inhabitants' annual effective dose (H), Sohrabi (2013a) has classified HLNRAs as follows: low (HE = 5 mSv y^{-1}), medium (HE= 5–10 mSv y^{-1}), high (HE= 20–50 mSv y^{-1}) and very high (HE > 50 mSv y^{-1}). This classification is based on the dose limits of the ICRP and the 2.4 mSv y^{-1} global mean dose value reported by UNSCEAR (ICRP, 1991, 2007; UNSCEAR, 2000a,b). The specific characteristics of HLNRAs depend on the constancy of natural radioactivity. There are areas in which radiation is constant over time and areas in which radiation varies with time (Sohrabi, 2013a). Some of the world's HLNRAs are found in Kerala (India), Guarapari (Brazil), Yangjiang (China), and Ramsar and Mahallat (Iran) (Mortazavi et al., 2002).

3.1.4 HLNRAs in Iran

3.1.4.1 *Ramsar*

The Ramsar region, a northern coastal city of Iran, has been the subject of concern as a highly radioactive region for the past 40 years. The elevated level of natural radioactivity in the Ramsar HLNRAs is caused by ^{226}Ra and its decay products, which have been brought up to the Earth's surface by hot springs. There are nine hot springs with varying concentrations of radium in Ramsar, which are used as spas (Sohrabi, 1993, 1994; Ghiassi-nejad et al., 2002). Surveys at the HLNRs of Ramsar show that radioactivity predominantly originates from the mineral water and secondarily from some travertine deposits with higher thorium than uranium content (Sohrabi, 1993). Sohrabi and Esmaili (2002) provided extensive data from external radiation studies conducted indoors and outdoors, which include the potential annual effective doses to the public in Ramsar, found to range from 0.6 to 131 mSv with a mean value of 6 mSv. In NBRAs, the range was 0.6–1.5 mSv with a mean value of 0.7 mSv.

The source of the high background radioactivity in Ramsar is igneous bedrocks containing high concentrations of uranium. Although uranium is not soluble in anoxic ground water, its daughter nuclide ^{226}Ra is. Dissolved radium is transported to the surface by ground water through pores and fractures in the rock. Water reaches the surface at hot-spring sites, where calcium carbonate precipitates out of solution and ^{226}Ra substitutes for calcium ($RaCo_3$). High concentrations of white radium carbonate are found in the residues of hot springs. In some cases, the residents have used the Ra-enriched rock from the hot springs as building materials (Mortazavi et al., 2001; Dissanayake & Chandrajith, 2009), causing high indoor concentrations of radon. The indoor radon level in Ramsar is as high as 31 kBq m^{-1}, the total external and internal exposure leads to an H value that ranges from 3.0 to 202 mSv y^{-1} and a corresponding 50-year H_E of 10.10 Sv, approximately 10-fold the total 50-year dose limit of 1.00 Sv defined for radiation workers (Sohrabi & Babapouran, 2005; Sohrabi, 2013a). The radon exhalation rate in building materials from Ramsar and the primordial radionuclide contents have been reported (Bavarnegin et al., 2013). Some areas in Ramsar have the highest levels of natural radiation in the world with extraordinarily high radon levels (Sohrabi & Babapouran, 2005; Sohrabi, 2013a). Residents of Talesh Mahalleh in Ramsar receive annual doses as high as 132 mSv from external terrestrial sources, and in general Ramsar residents absorb annual doses of about 260 mSv y^{-1} from background radiation, i.e. 20 and 200 times the permitted limit for radiation and non-radiation workers, respectively. The annual effective dose from internal exposure to ^{222}Rn in Ramsar HBNRAs ranges from 2.4 to 71.74 mSv (Sohrabi & Babapouran, 2005).

3.1.4.2 *Mahallat*

The Mahallat region of Markazi province in Iran is a popular tourist spot due to the occurrence of hot springs and having the greatest geothermal fields in Iran. Five hot springs called 'Abegarm-e-Mahallat', located in the central part of Iran, have a mean water temperature of 45–47°C and are used by visitors as spas. This is an area of high natural radiation background due to the presence of ^{226}Ra and its decay products in the deposited travertine ($CaCO_3$). The mean concentration of ^{226}Ra in these hot springs ranged from 0.48+/−0.05 to 1.35 +/−0.13 Bq l^{-1}. ^{222}Rn concentrations in the hot springs ranged from 145+/−37 to 2731+/−98 Bq l^{-1}. Mean radon concentrations in air were 487+/−160 and 15.4+/−2.7 Bq m^{-3} for indoor and outdoor, respectively. Radiation levels above normal (approximately 100 nGy h^{-1}) were mainly limited to the Quaternary travertine formations in the vicinity of the hot springs (Beitollahi et al., 2007).

3.2 Study of HLNRAs in Iran

3.2.1 Ecological and floristic evaluation of natural microalgae from HLNRAs in Iran

Water and soil samples were collected from 10 thermal springs across the areas of Ramsar (Mazandaran Province, Iran) and Mahallat (Markazi Province, Iran). Chemical analysis revealed different composition of water and soil samples in most elements compared to other ecosystems. The highest water temperature (50°C) was measured in the Shafa hot spring in Mahallat and the lowest water temperature (27°C) was recorded in the Talesh Mahalleh hot spring in Ramsar. The ^{226}Ra content of water samples detected by Radon Emanation technique ranged from 0.1 to 130Bq l^{-1} and in soil samples detected by gamma spectrometry ranged from 54 to 13 000 Bq kg^{-1}. The Ramsar region of Iran is one of the places with the highest natural radioactivity of soil and high-level natural background radiation (HL Hariot NRA) in the world. The results of studies mentioned in this section are given in Table 3.1.

Overall, 23 genera and 79 strains were identified. According to the morphological examination, *Phormidium* spp. and *Leptolyngbya* spp., along with *Thermoleptolyngbya laminosa*, were dominant in hot springs and soil samples. The dominant morphotypes in benthos of the most radioactive spring in Ramsar were *Leptolyngbya* spp. and *Nostoc* spp.; the soil with the highest radioactivity of 13000 Bq kg^{-1} contained *Cylindrospermum licheniforme*. Significant differences were observed in cyanobacterial diversity. The strains identified in the hot springs were 52 and only 27 in the surrounding soil. Morphological and molecular analyses revealed several new taxa to science. Comparison of our results with those of other authors is rather difficult, since we are not aware of any paper on cyanobacterial diversity in similar habitats. Some data are provided by Anitori et al. (2002) on the Paralana hot spring in Australia, but these

Table 3.1 *Temperature and ^{226}Ra activity concentrations in soil and water samples from the hot springs*

Hot spring	Province	Latiude & longitude	Water^{226} Ra (Bq l$^{-1)}$	Soil^{226} Ra (Bq kg^{-1})	Temp (°C)
Abe siah	Mazandaran, Ramsar city	36°89′ 87.53″ N 50°66′ 59.23″ E	130±0.2	4012±316	28
Talesh Mahalleh	Mazandaran, Ramsar city	36° 89′ 29.90″ N 50° 67′ 44.49 ″ E	18.14±0.1	13 000±410	27
Khak sefid	Mazandaran, Ramsar city	36° 89′ 17.30″ N 50°67′ 98.47″ E	16.50±0.1	3620±350	32
Vazir garma	Mazandaran, Ramsar city	36°89′ 87.40″ N 50°66′ 93.37″ E	4.68±0.5	508±47	31
Saddat Mahalleh	Mazandaran, Ramsar city	36°53′ 19.42″ N 50°41′ 51.30″ E	1.77±0.2	3630±320	35
Shafa	Markazy, Mahallat city	34 ° 00′ 22.41″ N 50° 32′ 53.46″ E	1.30±0.1	3000±240	50
Donbe	Markazy, Mahallat city	34 ° 00′ 30.27″ N 50° 32′ 43.47″ E	1.12±0.8	613±48	48
Soleimany	Markazy, Mahallat city	34° 00′ 34.53″ N 50° 33′ 01.75″ E	1.08±0.7	54±4	47
Soda	Markazy, Mahallat city	34 ° 00′ 23.12″ N 50° 32′ 52.55″ E	1.10±0.7	2590±200	46
Hakim	Markazy, Mahallat city	34° 00′ 28.33″ N 50° 33′ 03.06″ E	0.1±0.01	157±14	34

are only superficial. This spring also differs from those studied in this paper in the source of radioactivity (Rn instead of Ra) and has higher temperature.

3.2.2 Phylogenetic evaluation of isolated strains from HLNRAs in Iran: new species/genera

The isolated strains comprise many new taxa to science. In many cases, there are several sequences in public repositories closely related to our sequences, but with uncertain taxonomic assignment, i.e. only attempting genus determination. Phylogenetic analyses have shown that they are often very distant from reference sequences of particular genera and therefore they cannot be classified into these genera. However, there is also a case of sequences of strains assigned to species, e.g. *Oscillatoria acuminata* PCC 6304 (CP003607) or *Coleofasciculus chthonoplastes* KC2 (KY563068). Sequences of clones of one strain may differ. This can be explained by several different operons in the 16S rRNA gene and subsequent ITS region in the genome (Johansen et al., 2017).

New cyanobacterial taxa from thermal springs with high radioactivity are reported in Heidari et al. (2018a). These are *Planktothrix iranica* Heidari &

Hauer sp. nov., *Laspinema* Heidari & Hauer gen. nov., *Laspinema thermale* Heidari & Hauer sp. nov., *Ramsaria* Heidari & Hauer gen. nov., *Ramsaria avicennae* Heidari & Hauer sp. nov., *Klisinema* Heidari & Hauer gen. nov., *Klisinema persicum* Heidari & Hauer sp. nov., *Persinema* Heidari & Hauer gen. nov., *Persinema komarekii* Heidari & Hauer sp. nov., *Nodosilinea radiophila* Heidari & Hauer sp. nov. and *Nodosilinea ramsarensis* Heidari & Hauer sp. nov.

3.2.3 Characterisation of green microalgae (Chlorophyta) from HLNRAs in Iran

We investigated the diversity of green microalgae living in hot springs of high radiation areas in Ramsar. Morphological, ultrastructural and phylogenetic approaches were applied. However, morphological delimitation of green algal species is debatable, particularly in taxa that are morphologically simple and exhibit convergent evolution towards reduced morphology (e.g. coccoid forms). The diversity of green microalgae (Chlorophyta) in the samples was very significant. Only one strain was isolated from the water of Abe-Siah hot spring with the highest ^{226}Ra activity. Eight strains were isolated from soil collected from hot springs in the HBRAs of Ramsar. Most of the isolated strains were non-flagellated unicellular (coccal morphology). From the data analyses, it can be mentioned that diversity of coccid green algae is related to the presence of ^{226}Ra content and radiation in the water and soil. There is a higher number of coccids for radioactive environments (both water and soil of hot springs). As regards the radiological distribution of the recorded species, the highest richness occurs at the highest ^{226}Ra content. Clearly, more radioactive areas (Talesh Mahalleh and Abe-Siah hot springs) exhibit the highest record number. Light microscopic images of green microalgae are shown in Figure 3.1.

3.3 Screening the performance of isolated microalgae for radioisotope/metal adsorption

The capability of the 26 microalgae strains isolated from hot springs in Iranian HLNRAs were investigated for the level of two radionuclides (^{238}U and ^{226}Ra) and the heavy-metal Cd. ^{238}U uptake of a dead biomass of *Graesiella emersonii* isolated from the soil samples at Abe-Siah hot spring (with the highest ^{226}Ra content of 130 Bq l^{-1} in the water, and 4012 Bq kg^{-1} in the soil) showed a 90% removal of ^{238}U (Table 3.2). This strain also showed high capacity for ^{226}Ra sorption, found associated with 95% removal of ^{226}Ra at an initial radioactivity of 30 Bq ml^{-1} in the contact solution after 2 hours (pH 7.0) (Heidari et al., 2017). *Nostoc punctiforme* Hariot (1886) strain A.S/S4 and *Chroococcidiopsis thermalis* Geitler (1933) strain S.M/ S9 had maximum ^{238}U uptake capacities of 73% and 88% dry biomass,

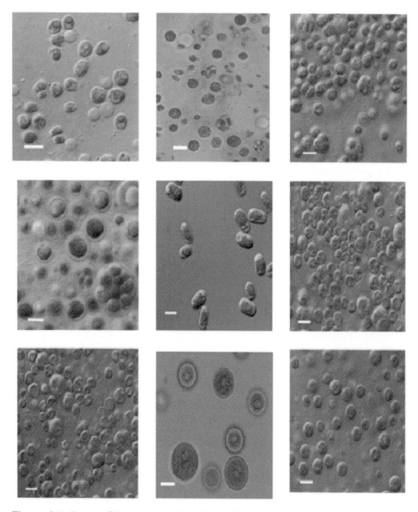

Figure 3.1 Some of the green microalgae (Chlorophyta) isolated from HBRAs in Iran. Scales, 10 μm. (A black and white version of this figure will appear in some formats. For the colour version, please refer to the plate section.)

respectively (Heidari et al., 2018b). The biosorption capacities in *N. punctiforme* (A. S/S4) were 630 mg g^{-1} and 37 kBq g^{-1} for ^{238}U and ^{226}Ra, respectively, and 730 mg g^{-1} and 55 kBq g^{-1} in *C. thermalis* (S.M/S9). Besides ^{238}U and ^{226}Ra, the two cyanobacterial strains are efficient on removing one of the most toxic heavy metals (Cd) from aqueous solution. *N. punctiforme* (A. S/S4) and *C. thermalis* (S.M/S9), with maximum binding capacity of 160 and 225 mg g^{-1}, respectively, showed high capacity of Cd adsorption (225 mg g^{-1}). The main metal-binding groups of the cell walls of cyanobacteria are carboxylate and phosphate groups (Heidari et al., 2018b).

Table 3.2 *The isolated strains, sampling locations, ^{226}Ra activity of water and soil samples, and results of screening for the performance of the isolated strains for ^{238}U sorption (%)*

Strains (No. 26)	Sample location	^{226}Ra conc. (water: Bq l^{-1}) (soil: Bq kg^{-1})	^{238}U Sorption (%) (100 mg ml^{-1})
Nostoc hatei (S.C.Dixit) 1936	W. A.S.[a]	130±0.20	31.18
Nostoc ellipsosporum (Rabenhorst ex Bornet & Flahault)1886,1888	W. A.S.	130±0.20	41.51
Phormidium corium (Gomont ex Gomont)1892	W.S.M.	1.77±0.20	53.69
Phormidium terebriforme (C.Agardh ex Gomont) Anagnostidis & Komárek 1988	W.SO.	1.08±0.70	69.09
Nostoc punctiforme (Vaucher ex Bornet & Flahault, 1886)	W.A.S.	130 ±0.20	72.50
Synechocystis minuscule (Woronichin, 1926)	W.A.S.	130±0.20	58.05
Phormidium nigrum (Vaucher ex Gomont) Anagnostidis & Komárek 1988	W.A.S.	130±0.20	67.00
Chroococcus membraninus (Nägeli, 1849)	W.S.M.	1.77±0.20	87.45
Phormidium terebriforme (C. Agardh ex Gomont) Anagnostidis & Komárek 1988	W.D.	1.12±0.80	31.53
Geitlerinema sulphureum (Strzeszewski) Anagnostidis 2001	W.D.	1.12±0.80	67.20
Synechococcus bigranulatus (Skuja) 1933	W.D.	1.12±0.80	37.96
Nostoc minutissimum (Kützing ex Bornet & Flahault) 1886, 1888	W.SH.	1.30±0.10	58.43
Geitlerinema amphibium (C. Agardh ex Gomont) Anagnostidis 1989	W.HK.	0.1±0.01	76.76
Jaaginema geminatum (Schwabe ex Gomont) Anagnostidis & Komárek	W.HK.	0.1±0.01	33.85
Pseudanabaena limnetica (Lemmermann) Komárek 1974	W.S.	1.10±0.70	10.26
Pseudanabaena catenata (Lauterborn) 1915	W.SO.	1.08±0.70	31.57

Table 3.2 (cont.)

Strains (No. 26)	Sample location	^{226}Ra conc. (water: Bq l^{-1}) (soil: Bq kg^{-1})	^{238}U Sorption (%) (100 mg ml^{-1})
Nostoc paludosum (Kützing ex Bornet & Flahault) 1886, 1888	W.A.S.	130±0.20	61.63
Phormidium paulsenianum (J. B. Petersen) 1930	S.A.S.[b]	4012±31	72.37
Nostoc calcicola (Brébisson ex Bornet & Flahault) 1886	S.SO.	54±4.00	45.21
Nostoc minutum (Desmazières ex Bornet & Flahault) 1886	S.SO.	54±4.00	51.43
Phormidium papyraceum (Gomont ex Gomont) 1892	S.T.M.	13 000±41	61.92
Cylindrospermum licheniforme (Kützing ex Bornet & Flahault) 1886	S. T.M.	13 000±41	55.61
Chlorella sp.	W.A.S.	130±0.10	49.31
Graesiella emersonii (Shihara & R. W. Krauss) H. Nozaki, M. Katagiri, M. Nakagawa, K. Aizawa & M. M. Watanabe 1995	S. A.S.	4012±31	90.90
Chlorella vulgaris (Beyerinck) [Beijerinck] 1890	S. T.M.	13 000±41	37.71
Chlorella sp.	S. T.M.	13 000±41	70.53

[a]W: water.
[b]S: soil.
A.S., Abe Siah; T.M., Talesh Mahhaleh; S.M., Saddat Mahhale; D., Donbe; HK., Hakim; SH., Shafa; S., Soda; SO., Soleimany hot spring.

3.4 Total phenolic content and antioxidant activity

The total phenolic content of HLNRA microalgae revealed that in *Grasiella emersonii* ISB80, *C. thermalis* ISB83, *N. punctiforme* ISB81, *Planktothrix* sp. ISB 69 and *Limnothrix planktonica* ISB71 it is 116.46, 100.30, 39.69, 26.61 and 38.15 µg mg^{-1}, respectively. DPPH scavenging activities (%) of the microalgae extracts and synthetic antioxidant BHT were also calculated at different concentrations ranging between 2 and 100 mg l^{-1}. The 50% inhibitory concentration (IC$_{50}$) values for DPPH scavenging activities of the microalgae extracts was calculated. The highest DPPH inhibitory effects were observed in *G. emersonii* ISB80 and *C. thermalis* ISB83, the unicellular microalgae strains, with values of 10 and 18 μg ml^{-1}, respectively. The lowest DPPH activity was recorded in *N. punctiforme* ISB81 and *L. planktonica* ISB71 (55 and 44 μg ml^{-1}, respectively). Microalgal extracts displayed antioxidant activities, in comparison with BHT synthetic antioxidant (70 μg ml^{-1}).

3.5 Discussion

A number of excellent studies and reviews have been conducted regarding the radiological issues related to HLNRAs. For instance, Sohrabi (1998) reviewed the literature concerning the sources of naturally occurring radioactive material (NORM) and human exposure and has presented criteria for the classification of HLNRAs (Sohrabi, 2013a). Hendry et al. (2009) note that there is an established association between long-term radiation exposure in disease incidence in radon-prone areas. Møller & Mousseau (2013) have assessed the effects of variations in background radioactivity on humans, animals and other organisms. The authors expected that the hormetic effects of radiation should be more pronounced in HLNRAs because of adaptation to such enhanced levels of radiation; they also predicted that exposure to natural radiation should have positive effects on humans and other organisms if hormesis operates at naturally occurring low doses of radiation. A number of epidemiological studies have been conducted to analyse the risk of cancer incidence in the world's HLNRAs. Most of these studies have concluded that there is no link between exposure to high background natural radiation and an increased rate of cancer or mortality. However, the results of these studies should be considered with caution because of the confounding factors associated with their methodology. For instance, Wei et al. (1990) have observed an increase in the incidence of Down's syndrome among Chinese HLNRA residents. However, the findings of this study have also been linked to other factors, such as the difference in the age of maternity between HLNRA and NBRA residents.

In this study, HLNRAs in Iran and the microalgae community were investigated in 10 hot springs in areas with high natural radiation. The taxa were identified based on important morphological traits compared with other studies. In short, the mechanism of resistance to high environmental radiation by cyanobacteria can be attributed to morphological traits and their genetic adaptations. In a similar study, two cyanobacteria strains of the *Oscillatoria* group (5240 m high in India) with a radiation intensity of 5 kGy were investigated. Although their molecular phylogeny was ambiguous, these taxa were closely related to *Leptolyngbya*. The remarkable point in this Cyanophyta is the formation of an extracellular polymeric layer, which facilitates binding of soil particles and actually facilitates the formation of soil biotic shells. This ability to produce extra-cellular polymers is directly related to the tolerance of dry cyanobacteria in the wilderness, and dry tolerance in Cyanophyta is similar to the tolerance of high levels of radiation that results in the formation of the same extracellular polymeric compounds (Roa et al., 2016).

Our findings also showed that the unique features of cyanobacteria in hot-spring irradiation are related to the physicochemical properties of water and the ecological parameters of the region. The presence of radium and radio-activity is high and can have a significant impact on the diversity of cyano-bacteria. Comparing the results of this study with other reports is relatively difficult, even impossible because no articles have ever been published on such habitat. Some information is available in Anitori et al. (2002), who studied the Paralana Springs in Australia, but the source of radiation in Paralana is radon (^{222}Rn) in place of radium (^{226}Ra), and the water temperature in Paralana is much higher.

Recent decades have witnessed an increase of pollutant levels as a result of toxic metal and industrial radioisotope release in nature which have led to serious environmental hazards. To avoid harmful effects, these metals have to be eliminated and removed before discharge to wastewater. Regarding the inefficient conventional physicochemical cleaning techni-ques, development of novel and highly efficient methods for pollutant removal from the environment has gained substantial attention. Advances in biotechnology have generated bioremediation, a widespread, novel, effi-cient and environmentally friendly method for environmental protection and cleaning. In this regard, unique properties of algae have introduced them as an ideal tool for selective removal and concentration of heavy metals, thus attracting considerable attention throughout the globe. It must be mentioned that depending on algal samples and descriptions of each species, biosorption occurs through various mechanisms. In other words, when an alga is selected for biosorption, the origin of the biomass is the most effective factor which needs significant attention. In this tech-nique, properties such as mechanism efficiency, thermal stability and sorp-tion capacity must be taken into consideration. This explains the significance of algal species for heavy-metal sorption from solutions. In Iran, there are some HLNRAs. Based on this study, the highest natural radiation in the country is reported in Ramsar and Mahallat cities, which have more than nine and five hot springs, respectively. The hot springs of these regions transfer the radioisotopes from the deep earth to the open ground and enhance radiation by sedimentation of travertine rock and accumulation of radioisotopes in the soils surrounding the springs. The annual receiving dose reaches 260 mSvy^{-1} in Ramsar, far more than the permitted dosage for people working with these radiations. Owing to their thermal and radiation properties, these hot springs have special ecological features, which allow them to be considered extreme environments.

In short, the aim of this study is to conduct a systematic investigation on the extremophilic algae population of HLNRAs in Iran, including application for radioisotope removal.

Among the radio-resistant cyanobacteria, members of the *Chroococcidiopsis* genus have been identified as highly effective examples. They are resistant to lethal drought effects and can withstand up to 15 KGy of x-rays. *Chroococcidiopsis* cells can withstand drought by means of a rich polysaccharide multilayer sheath. The survival rate of cells of this genus is second only to the most radioresistant organism, *Deinococcus radiodurans*. However, this bacterium, which so far shows the highest resistance to gamma and UV rays, has limitations for use in industry, because it needs some genetic changes to be able to tolerate toxic metals in the environment and to have the ability to clean and absorb them for use in *in situ* cleaning (Tapias et al., 2009).

The isolated and purified ISB83 strain *C. thermalis* described in this study is a successful organism and a leading agent capable of eliminating metals which contaminate water resources, nuclear waste or other radioisotope contamination. Cyanobacteria are capable of photosynthesis and do not require expensive carbon sources in laboratory cultures and large commercial production, have fast and easy growth, and are rich in bioactive compounds. They can be effective in many industries, e.g. pharmaceutical and medical. After extracting the active compounds of this organism, they can be used in cleaning water resources and nuclear and industrial waste, with no concern for environmental secondary contamination.

This study can be considered as the first report on finding radiation-resistant cyanobacteria in geothermal sources and in HLNRAs of the world. It can assist in selecting certain strains with high potentials for biotechnological/industrial applications, e.g. production of bioactive compounds. Based on this idea, this study was also aimed at investigating the effectiveness of microalgae, isolated from HLNRAs, in the purification and absorption of environmental radioactive contaminants and heavy metals. Our data provide the first information on the nature of the microalgae community in the hot springs in the highest background radiation areas in the world. This study will form the basis for a comprehensive study of the cyanobacterial and microalgae community of the hot springs and other unexplored areas of the world, aimed at using science to improve the relationships between extreme microorganisms and biotechnology.

References

Anitori, R.P., Trott, C., Saul, D.J., Berquist, P.L., Walter, M.R. (2002). A culture-independent survey of the bacterial community in a radon hot spring. *Astrobiology*, **2**(3), 255–270.

Bavarnegin, E., Fathabadi, N., Vahabi Moghaddam, M., et al. (2013). Radon exhalation rate and natural radionuclide content in building materials *Enigmatic Microorganisms and Life in Extreme Environments. Journal of Environmental Radioactivity*, **117**, 36–40.

Beitollahi, M., Ghiassi-Nejad, M., Esmaeli, A. (2007). Radiological studies in the hot spring region of Mahallat, Central Iran. *Radiation Protection Dosimetry*, **123**, 505–508.

Ben-Amotz, A., Avron, M. (1983). Accumulation of metabolites by halotolerant algae and its industrial potential. *Annual Review Microbiology*, **37**, 95–119.

Bidigare, R.R., Ondrusek, M.E., Kennicutt, M.C. II, et al. (1993). Evidence for a photoprotective function for secondary carotenoids of snow algae. *Journal of Phycology*, **29**, 427–434.

Castenholz, R.W., Garcia-Pichel, F. (2000). Cyanobacterial responses to UV-radiation. In: B.A. Whitton, M. Potts (eds) *Ecology of Cyanobacteria: Their Diversity in Time and Space*. Springer Science+Business Media B. V.

Dissanayake, C.B., Chandrajith, R. (2009). *Introduction to Medical Geology*. Springer, the Netherlands.

Duval, B., Shetty, K., Thomas, W.H. (2000). Phenolic compounds and antioxidant properties in the snow alga Chlamydomonas nivalis after exposure to UV light. *Journal of Applied Phycology*, **11**, 559–566.

El-Gamal, H., El-Azab Farid, M., Abdel Mageed, A.I., Hasabelnaby, M., Hassanien, H.M. (2013). Assessment of natural radioactivity levels in soil samples from some areas in Assiut, Egypt. *Environment Sciences Pollution Research*, **20**, 8700–8708.

Ghiassi-nejad, M., Mortazavi, S.M., Cameron, J. R., Niroomand-rad, A., Karam, P.A. (2002). Very high background radiation areas of Ramsar, Iran: preliminary biological studies. *Health Physics*, **82**, 87–93.

Heidari, F., Riahi, H., Aghamiri, M.R., Shariatmadari, Z., Zakeri, F. (2017). Isolation of an efficient biosorbent of radionuclides (238U & 226Ra): green algae from high background radiation areas in Iran. *Journal of Applied Phycology*, 29, 2887–2898. DOI 10.1007/s10811-017-1151-1.

Heidari, F., Zima,J, Jr., Riahi, H., Hauer, T. (2018a). New simple trichal cyanobacterial taxa isolated from radioactive thermal springs. *Fottea Olomouc*, **18**(2), 137–149.

Heidari, F., Riahi, H., Aghamiri, M.R., et al. (2018b). 226 Ra, 238U and Cd adsorption kinetics and binding capacity of two cyanobacterial strains isolated from highly radioactive springs and optimal conditions for maximal removal effects in contaminated water. *International Journal of Phytoremediation*, **20**(4), 369–377.

Hendry, J.H., Simon, S.L., Wojcik, A., et al. (2009). Human exposure to high natural background radiation: what can it teach us about radiation risks? *Journal of Radiological Protection*, **29**, A29.

Hutchinson, F. (1985). Chemical changes induced in DNA by ionizing radiation. *Progress in Nucleic Acid Research Molecular Biology*, **32**, 115–154.

ICRP (1991). 1990 Recommendations of the International Commission on Radiological Protection. ICRP Publication 60. *Annals of the ICRP*, **21**(1–3), 1–201.

ICRP (2007). The 2007 Recommendations of the International Commission on Radiological Protection. ICRP Publication 103. *Annals of the ICRP*, **37**.

Imlay, J.A. (2006). Iron-sulphur clusters and the problem with oxygen. *Molecular Microbiology*, **59**(4), 1073–1082.

Jeffrey, S.W., MacTavish, H.S., Dunlap, W.C., Vesk, M., Groenewoud, K. (1999). Occurrence of UV-A- and UV-B-absorbing compounds in 152 species (206 strains) of marine microalgae. *Marine Ecology Progress Services*, **189**, 35–51.

Jibiri, N.N. (2001). Assessments of health risk levels associated with terrestrial gamma radiation dose rates in Nigeria. *Environmental Integrity*, **21**, 21–26.

Johansen, J.R., Mareš, J., Pietrasiak, N., et al. (2017). Highly divergent 16S rRNA sequences in ribosomal operons of Scytonema hyalinum (Cyanobacteria). *PLoS ONE*, **12**(10), e0186393.

Kminek, G., Bada, J.L., Pogliano, K., Ward, J.F. (2003). Radiation-dependent limit for the viability of bacterial spores in halite fluid inclusions and on Mars. *Radiation Research*, **159**, 722–729.

Kottemann, M., Kish, A., Iloanusi, C., Bjork, S., DiRuggiero, J. (2005). Physiological responses of the halophilic archaeon Halobacterium sp. strain NRC-1 to desiccation and gamma irradiation. *Extremophiles*, **9**, 219–227.

Liedert, C., Peltola, M., Bernhardt, J., Neubauer, P., Salkinoja-Salonen, P. (2012). Physiology of resistant Deinococcus geothermalis bacterium aerobically cultivated in low-manganese medium. *Journal of Bacteriology*, **194**, 1552–1156.

Mancinelli, R.L., White, M.R., Rothschild, L.J. (1998). Biopan survival I: Exposure of the osmophiles Synechococcus sp. (Nägeli) and Haloarcula sp. to the space environment. *Advances in Space Research*, **22**, 327–334.

Mares, S. (1984). *Introduction to Applied Geophysics*. Springer-Science+Business Media, B.V., p. 574.

Matallana-Surget, S., Wattiez, R. (2013). Impact of solar radiation on gene expression in bacteria. *Proteomes*, **1**, 70–86.

Møller, A.P., Mousseau, T.A. (2013). The effects of natural variation in background radioactivity on humans, animals and other organisms. *Biological Review*, **88**, 226–254.

Mortazavi, S.M.J., Mozdarani, H. (2012). Is it time to shed some light on the black box of health policies regarding the inhabitants of the high background radiation areas of Ramsar?*Iranian Journal of Radiation Research*, **10**, 111–116.

Mortazavi, S.M.J., Ghiassi-Nejad, M., Beitollahi, M. (2001). Very high background radiation areas (VHBRAs) of Ramsar: Do we need any regulations to protect the inhabitants? 34th Midyear Meeting Radiation Safety and ALARA – Considerations for the 21st Century, California, USA, pp. 177–182.

Mortazavi, S.M.J., Ghiassi-Nejad, M., Ikushima, T. (2002). Do the findings on the health effects of prolonged exposure to very high levels of natural radiation contradict current ultra conservative radiation protection regulations? *International Congress Series*, **1236**, 19–21.

Nikitaki, Z., Hellweg, C.E., Georgakilas, A.G., Ravanat J.-L. (2015). Stress induced DNA damage biomarkers: applications and limitations. *Frontiers in Chemistry*, **3**, 35–55.

Oren, A., Gurevich, P., Anati, D.A., Barkan, E., Luz, B. (1995a). A bloom of Dunaliella parva in the Dead Sea in 1992: biological and biogeochemical aspects. *Hydrobiologia*, **297**, 173–185.

Pavlopoulou, A., Savva, G.D., Louka, M., et al. (2016). *Adaption of Microbial Life to Environmental Extremes*. Springer, Cham

Quindos, L.S., Fernandez, P.L., Soto, J., Rodenas, C. (1991). Terrestrial gamma radiation levels outdoors in Cantabria, Spain. *Journal of Radiological Protection*, **11**, 127–130.

Riley, P.A. (1994). Free radicals in biology: oxidative stress and the effects of ionizing radiation. *International Journal of Radiation Biology*, **65**(1), 27–33.

Roa, S., Chan, O., Lacap-Bugler, D. (2016). Radiation-tolerant bacteria isolated from high altitude soil in Tibet. *Indian Journal of Microbiology*, **56**, 508–512. DOI 10.1007/s12088-016-0604-6.

Rothschild, L.J. (1999). Microbes and radiation. In: J. Seckbach (ed.) *Enigmatic Microorganisms and Life in Extreme Environments*. Kluwer Academic Publishers, Dordrecht, pp. 549–562.

Saito, K., Ishigure, N., Petoussi-Henss, N., Schlattl, H. (2012). Effective dose conversion coefficients for radionuclides exponentially distributed in the ground. *Radiation Environment Biophysics*, **51**, 411–423.

Schnelzer, M., Hammer, G.P., Kreuzer, M., Tschense, A., Grosche, B. (2010). Accounting for smoking in the radon-related lung cancer risk among German uranium miners: results of a nested case-control study. *Health Physics*, **98**, 20–28.

Seckbach, J., Oren, A. (2007). Oxygenic photosynthetic microorganisms in extreme environments: possibilities and limitations. In: J. Seckbach (ed.) *Algae and Cyanobacteria in Extreme Environments*, Vol 11. Springer, pp. 3–25.

Sohrabi, M. (1994). Proceedings of the international conference on high levels of natural radiation. *Radiation Measurement*, **23**, 261–262.

Sohrabi, M. (2013a). Response to the letter of H. Abdollahi. *Radiation Measurement*, **59**, 290–292.

Sohrabi, M. (2013b). World high background natural radiation areas: need to protect public from radiation exposure. *Radiation Measurements*, **50**, 166–171.

Sohrabi, M. (1993). *Recent Radiological Studies of High Level Natural Radiation Areas of Ramsar, International Conference on High Levels of Natural Radiation Areas*. IAEA Publication Series. IAEA, Vienna/Ramsar, Iran.

Sohrabi, M. (1998). The state-of-the-art on worldwide studies in some environments with elevated naturally occurring radioactive materials (NORM). *Applied Radiation Isotopes*, **49**, 169–188.

Sohrabi, M., Babapouran, M. (2005). *New public dose assessment from internal and external exposures in low-and elevated-level natural radiation areas of Ramsar, Iran*. International Congress Services, 169–174. Elsevier.

Sohrabi, M., Esmaili, A.R. (2002). New public dose assessment of elevated natural radiation areas of Ramsar (Iran) for epidemiological studies. *International Congress Services*, **1225**, 15–24.

Soppa, J. (2013). Evolutionary advantages of polyploidy in halophilic archaea. *Biochemistry Society Transactions*, **41**, 339–343.

Soppa, J. (2014). Polyploidy in archaea and bacteria: about desiccation resistance, giant cell size, long-term survival, enforcement by a eukaryotic host and additional aspects. *Journal of Molecular Microbiology Biotechnology*, **24**, 409–419.

Stadtman, E.R., Levine, R.L. (2003). Free radical-mediated oxidation of free amino acids and amino acid residues in proteins. *Amino Acids*, **25**(3–4), 207–218.

Stan-Lotter, H., Fendrihan, S. (2015). Halophilic archaea: life with desiccation, radiation and oligotrophy over geologic times. *Life*, **5**, 1487–1496.

Tapias, A., Leplat, C., Confalonieri, F. (2009). Recovery of ionizing-radiation damage after high doses of gamma ray in the hyperthermophilic archaeon *Thermococcus gammatolerans*. *Extremophiles*, **13**, 333–334.

UNSCEAR (2000a). *Biological Effects at Low Radiation Doses*. United Nations Scientific Committee on the Effects of Atomic Radiation., New York.

UNSCEAR (2000b). *Sources and Effects of Ionizing Radiation, Report to the General Assembly of the United Nations with Scientific Annexes*. United Nations Scientific Committee on the Effects of Atomic Radiation, New York.

Vincent, W.F. (2000). *Cyanobacterial dominance in the polar regions*. Kluwer Academic Publishers, Dordrecht, pp. 591–611.

Webb, K.M., Yu, J, Robinson, C.K., et al. (2013). Effects of intracellular Mn on the radiation resistance of the halophilic archaeon Halobacterium salinarum.*Extremophiles*, **17** (3), 485–497. • Jerry Yu • Courtney K. Robinson

Webb, K.M., DiRuggiero, J. (2012). Role of Mn2+ and compatible solutes in the radiation resistance of thermophilic bacteria and archaea. *Archaea*, **2012**, 1–11. doi:10.1155/ 2012/845756

Wei, L., Zha, Y., Tao, Z., et al. (1990). Epidemiological investigation of radiological effects in high background radiation areas of Yangjiang, China. *Journal of Radiation Research*, **31**, 119–136.

Westall, F., Loizeau, D., Foucher, F., et al. (2013). Habitability on Mars from a microbial point of view. *Astrobiology*, **13**, 887–889.

Zupunski, L., Vesna, S.J., Trobok, M., Vojin, G. (2010). Cancer risk assessment after exposure from natural radionuclides in soil using Monte Carlo techniques. *Environmental Sciences Pollution Research*, **17**, 1574–1580.

PART II Biodiversity, bioenergetic processes, and biotic and abiotic interactions

AD H.L. HUISKES

Introduction

Biodiversity, the amount of diversity in life forms, but also of processes, interactions, genetics of ecosystems, biomes or even the whole planet, has been one of the most popular topics in the last few decades. Scientists study and discuss the biodiversity of a score of ecosystems and biomes, which is readily picked up by journalists and distributed to the public.

On the urge of the United Nations General Assembly, the United Nations Environment Programme (UNEP) drafted a conceptual framework for an Intergovernmental Science-Policy Platform on Biodiversity and Ecosystem Services (IPBES). In 2019 IPBES held a plenary meeting in Paris where 132 IPBES members received a full report (Global Assessment Report on Biodiversity and Ecosystem Services) and adopted a summary for policy makers.

The decrease in biodiversity of the Earth and, in close connection with this decrease, the change in ecosystem services had become an issue of concern and resulted in some political action.

Biodiversity (one of the outcomes of evolution) dictates how ecosystems function and underpins the life-support system of our planet. A thorough understanding of biodiversity, which can be hit by environmental changes, is therefore critical, not least because the survival of mankind depends on it. Will climate change result in relaxation of selection pressure, or will it ultimately lead to extinction of populations and species? With climate change so central to current public, governmental and research priorities, extreme environments provide a focal point for understanding and informing what has happened, is happening and will happen to our planet: namely fundamentally important information for understanding predicted changes and responses in less extreme environments (e.g. ours). Biodiversity is essential for the health of Earth. Threats to biodiversity caused by climate and other environmental changes must be, if not eliminated, at least minimised, in order to benefit society as a whole.

Despite the Global Assessment Report producing a wealth of information on biodiversity and ecosystem services, information on extreme environments or less studied groups of organisms is scarce because of the relatively few studies on these subjects.

Polar and deep-sea biomes are among these. Both biomes are dealt with in this section of the book and offer new information on respectively a little studied biome (anoxic marine sediments), ecophysiological processes in organisms living in extreme environments (polar regions) and less studied organisms in extreme environments (viruses in polar regions).

CHAPTER FOUR

Metazoan life in anoxic marine sediments

ROBERTO DANOVARO
Polytechnic University of Marche
CINZIA CORINALDESI
Polytechnic University of Marche
ANTONIO DELL'ANNO
Polytechnic University of Marche
CRISTINA GAMBI
Polytechnic University of Marche
ANTONIO PUSCEDDU
University of Cagliari

and

MICHAEL TANGHERLINI
Stazione Zoologica Anton Dohrn

4.1 Extreme environments: deep hypersaline anoxic basins

Benthic deep-sea ecosystems (beneath 200 m depth) represent the largest biome on our planet, covering >65% of the Earth's surface and hosting >95% of the global biosphere (Danovaro et al., 2014). Deep-sea ecosystems also contain the largest hypoxic and anoxic regions of the Biosphere such as the oxygen minimum zones (OMZ) and the deep hypersaline anoxic basins (DHAB). The OMZs typically occur at bathyal depths of between 200 and 1000 m, and are major sites of carbon burial along the continental margins (Levin, 2003). The most intense (O_2 <20 µmol kg^{-1}) and largest oxygen minimum zones, known as suboxic layers, are mainly localised in subsurface of the upwelling regions in the Eastern Pacific and Northern Indian open oceans (Paulmier & Ruiz-Pino, 2008). OMZs can be considered as analogues of the primitive anoxic ocean associated with a high CO_2 atmosphere (>1000 ppmv: Royer et al., 2004), because of comparable reduced chemical conditions and similarities between ancient bacteria and those living in the OMZs (e.g. Planctomycetales; Gribaldo & Philippe, 2002). OMZs appear as a refuge of abundant specific microbes capable of chemolithoautrophic carbon assimilation (Walsh et al., 2009) and organisms adapted to low O_2 availability (e.g. zooplankton as specific copepods and euphausiids; Wishner et al., 2008; Antezana, 2009 and benthic fauna as specific taxa of meio- and macrofauna; Levin, 2003).

Permanent anoxic conditions in the oceans are present in the subsurface seafloor (Lipp et al., 2008), and among other areas, in the interior of the Black Sea (at depths >200 m; Kuypers et al., 2003) and in the DHABs of the Mediterranean Sea (van der Wielen et al., 2005; Daffonchio et al., 2006). Six DHABs have been discovered in the Eastern Mediterranean Sea: L'Atalante, Urania, Bannock, Discovery, Tyro and La Medee. These basins lie at depths ranging from 3200 to 3600 m and contain brine, the origin of which has been attributed to the dissolution of 5.9- to 5.3-million-year-old Messinian evaporites (Hsü et al., 1977). Brines enclosed in these basins are characterised by high density, which hamper the mixing with overlying oxic seawater and result in a sharp chemocline and anoxic conditions. L'Atalante, Bannock and Urania brines have similar dominant ion compositions, but the overall salinity of Urania is lower, whereas concentrations of sulfide and methane are considerably higher. The most important difference between the geochemistry of Discovery brine compared with the other three is the extremely high concentration of Mg^{2+} and low concentration of Na^+ (Van der Wielen et al., 2005). The combination of nearly saturated salt concentration and corresponding high density and hydrostatic pressure, absence of light, anoxia and a sharp chemocline makes these basins some of the most extreme habitats on Earth. The bottom of the L'Atalante basin is a relatively flat area bounded to the southwest by the Cleft Basin and it is characterised by a morphological escarpment that is several hundreds of metres high, which is the seabottom expression of the main back thrust of the accretionary ridge. These characteristics originated from the dissolution of buried salt deposits (evaporitic deposits), which remained from the hypersaline waters of the Miocene period (5.5 My before present). The L'Atalante basin is characterised by the presence of a thick brine layer (ca. 400 m) with high density (1.23 g cm^{-3}) and high contents of Na^+ (4.674 mM), Cl^- (5.289 mM) and Mg^+ (410 mM) (van der Wielen et al., 2005). This layer limits the mixing with the overlying oxic deep waters to only the upper 1 to 3 m of the brine, and it additionally acts as a physical barrier for particles settling to the bottom sediments. As a result, the inner part of the L'Atalante basin is completely anoxic since 53,000 years before present (Fusi et al., 1996) and is characterised by elevated methane (0.52 mM) and hydrogen sulfide (2.9 mM) concentrations (Van der Wielen et al., 2005).

Investigations carried out so far in these extreme environments have been primarily focused on the unicellular organisms inhabiting the brine and the seawater–brine interface, revealing the presence of metabolically active and highly diversified prokaryotic assemblages (Van der Wielen et al., 2005; Daffonchio et al., 2006). Van der Wielen & coworkers (2005) investigated prokaryotic communities in four Mediterranean DHABs revealing that Bacteria dominated the Discovery basin and were slightly more abundant in L'Atalante and Bannock basins, whereas Archaea dominated the Urania basin. In all four hypersaline anoxic basins, bacterial diversity was higher than archaeal diversity, and the Urania basin displayed the lowest overall diversity.

Sediments below the brines of all DHABs have been much less investigated so far (Danovaro et al., 2010; Corinaldesi et al., 2014; Bernhard et al., 2015). A long-term study conducted in the anoxic hypersaline sediments of the L'Atalante basin has revealed the presence of specimens belonging to the Phylum of Loricifera that were actually alive at the time of sampling and would be able to spend their entire life cycle under anoxic conditions (Danovaro et al., 2010, 2016). This discovery was particularly surprising because this is the first evidence of metazoans (as meiofauna found in different extreme marine ecosystems; Zeppilli et al., 2018) living permanently in the absence of molecular oxygen. This finding is supported by a large set of independent experimental approaches based on incubations with radioactive tracers and specific fluorogenic probes and analyses on the biochemical composition of the body, quantitative X-ray microanalysis, infrared spectroscopy and transmission electron microscopy observations (Danovaro et al., 2010, 2016).

4.2 Biodiversity of animals living in anoxic conditions

Danovaro et al. (2010) carried out three oceanographical expeditions to search for the presence of living fauna in the anoxic sediments of the L'Atalante basin (Figure 4.1). In all sediments collected from the inner part of the anoxic basin, specimens belonging to three animal Phyla – Nematoda, Arthropoda (only Copepoda) and Loricifera – were identified. All specimens collected from the L'Atalante basin were initially stained with Rose Bengal (a protein-binding stain) and examined under the microscope. All copepods were empty exuviae, and the nematodes were only weakly stained (suggesting that they had been dead for a while), whereas all loriciferans were intensely coloured. Although the Rose Bengal *per se* is not sufficient to prove that loriciferans were alive, further evidence suggests that loriciferans were collected alive. All microscopic methodologies utilised (confocal laser microscopy, phase-contrast microscopy, SEM and TEM) did not show any sign of degraded tissues (Figure 4.2a,b) while all other meiofaunal taxa were largely degraded (Danovaro et al., 2010). The loriciferans were found either fully retracted, or partially retracted or fully extended and these characteristics can be

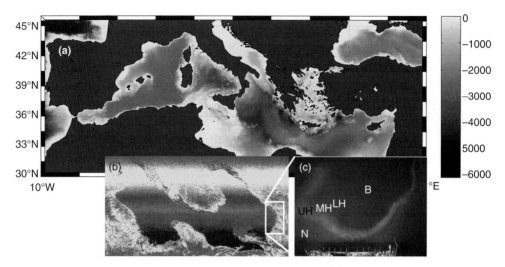

Figure 4.1 Location of the deep hypersaline anoxic basin in (a) the Mediterranean basin, showing details of the top view of the L'Atalante (b) basin and (c) halocline (panel c from Bernhard et al., 2015: UH, Upper Halocline; MH, Mid Halocline; LH, Lower Halocline; N, normoxic normal saline). (A black and white version of this figure will appear in some formats. For the colour version, please refer to the plate section.)

observed only in alive specimens (Neves et al., 2014). Since the discovery of the Loricifera at the beginning of the 1980s (Kristensen, 1983), several hundred species have been collected from coastal areas (off the coast of Fort Pierce, Pacific; off the coast of Roscoff, Atlantic; Tyrrhenian Sea, Mediterranean) to the deep-sea ecosystems (Bay of Biscay, Mediterranean Sea; Izu-Ogasawara Trench, Pacific; Angola basin, Rockall, Faroe Bank and Great Meteor Seamount, Atlantic; New Ireland Basin, North of Papua New Guinea and Antarctica) where Loricifera can display high species diversity (Neves et al., 2016). The loriciferans collected from the L'Atalante basin belong to three species, new to science, members of the genera *Spinoloricus*, *Rugiloricus* and *Pliciloricus* (Figure 4.2c, d,e). More than 80% of all specimens belong to a new species *Spinoloricus cinziae*, recently described in Neves et al. (2014). The three genera found in L'Atalante basin by Danovaro et al. (2010) were also reported by another research team (Bernhard et al., 2015), and this finding makes this DHAB one of the most important biodiversity hotspots of Loricifera known worldwide. Only one coastal species, *Nanaloricus khaitatus*, has been previously described from the entire Mediterranean Sea (Todaro & Kristensen, 1998). Although undescribed deep-sea loriciferans are certainly present in the Mediterranean Sea (Kristensen, 2003), the ecological

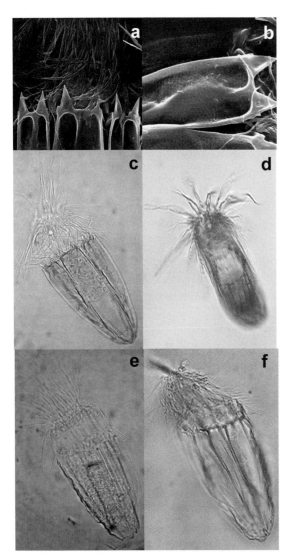

Figure 4.2 Animals retrieved from the deep hypersaline anoxic L'Atalante basin. Scanning electron microscopy images of (a) the lorica and (b) the anterior edge of the lorica showing the spikes of the *Spinoloricus cinziae*; and images of Loricifera belonging to the (c) *Spinoloricus cinziae* (stained with Rose Bengal); (d) *Pliciloricus* (non-stained with Rose Bengal); (e) *Rugiloricus* containing an oocyte (non-stained with Rose Bengal); (f) *Spinoloricus* (a moulting exuvium). (A black and white version of this figure will appear in some formats. For the colour version, please refer to the plate section.)

and evolutionary model known as the 'abundant centre' hypothesis predicts that the higher the level of biodiversity, the higher the probability that species is originated from that system (Sagarin et al., 2006). Loricifera of the L'Atlante basin not only show a high biodiversity but also an exceptional abundance. The abundance of loriciferans inhabiting this anoxic basin, ranging from 75 to 701 individuals per square metre, is by far the highest abundance per unit of surface of the sediment investigated reported worldwide (Danovaro et al., 2016). On the basis of all meiofaunal samples analysed so far over the last 30 years in the entire deep Mediterranean Sea (i.e. more than 500 samples, Gambi et al., 2017) only in a few samples collected off Corsica, some undescribed individuals of loriciferans were observed (in the Western Mediterranean Sea, at ca. 3000 km distance from the L'Atalante basin; Soetaert et al., 1984, 1991). Moreover, the sediment layers of the anoxic basin are perfectly stratified (Jilbert et al., 2010), which allow hypothesising the lack of any relevant physical disturbance caused by major sediment transport by turbidity currents (Polonia et al., 2016) that could allow any transfer of Loricifera from the adjacent oxygenate sediments.

Loricifera are characterised by a complex life cycle, which involves a succession through larval and postlarval stages to the adult forms (Kristensen & Brooke, 2002). Because new species and new life cycle stages are constantly discovered (e.g. Heiner & Kristensen, 2008; Pardos & Kristensen, 2013), the actual biodiversity of Loricifera, as well as the multiplicity of their body plan, seems to be far from being completely understood (Bang-Berthelsen et al., 2012). Additional analyses on Loricifera from the L'Atalante basin also revealed the presence of specimens of two genera (*Spinoloricus* and *Rugiloricus*) containing a large oocyte in their ovary suggesting that some organisms were reproducing (Figure 4.2e). Moreover, the presence of empty postlarval exuviae (exoskeleton) of loriciferans suggests the grow and moulting of these organisms in this anoxic system (Figure 4.2f). These results reported for the first time the presence of nearly all stages involved in the life cycle of this organism in the same place, indicating thus that the species had adapted in the L'Atalante basin and was apparently able to live and complete its entire life cycle without access to free oxygen.

The analysis of the vertical distribution of Loricifera in the anoxic sediments of the L'Atalante basin revealed that loriciferans were found in a sediment layer 10–15 cm beneath the sediment surface that can be dated approximately from 2000 to 3000 years before present (Corinaldesi et al., 2008) (Figure 4.3). The whole body of Loricifera was present (head including mouth cone and introvert, neck, thorax and abdomen) in all investigated sediment layers including the deepest ones (Danovaro et al., 2016). Since the sediment layers of the anoxic basin are perfectly stratified (Jilbert et al., 2010), living loriciferans can move into deeper sediment layers. Active movement of meiofauna (metazoans of size ranging from 20 to 500 μm to which Loricifera belong) across the sediments is a common phenomenon in sediments worldwide (Giere, 2009). The body musculature of adult loriciferans account for muscle fibres related to the scalids and the mouth cone (Neves et al., 2013), which help the animal to burrow deep in the sediments – even in the hadal zone, which is characterised by a high hydrostatic pressure (Kristensen & Shirayama, 1988).

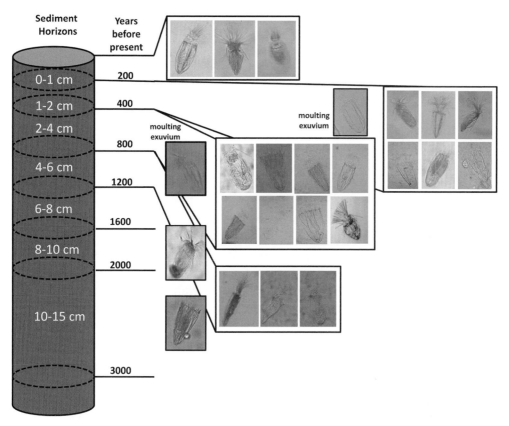

Figure 4.3 Vertical distribution of Loricifera in the anoxic sediments of the L'Atalante basin. The approximate age (years before present) of sediment horizons is provided.

4.3 Adaptation of Metazoans to permanently anoxic conditions

Independent experimental approaches were dedicated to the demonstration of the viability of these loriciferans of the L'Atalante basin (Danovaro et al., 2010). The viability was tested under anoxic conditions (in a N_2 atmosphere), in the dark and at the *in situ* temperature (ca 14°C). Intact and undisturbed sediment cores were injected with (3H)-leucine to investigate the ability of these loriciferans to take up this radiolabelled amino acid. Following multiple and replicated incubations and controls (killed loriciferans), the loriciferans incorporated amounts of this radioactive substrate that were significantly higher than in the controls. The viability of the Loricifera was further investigated by incubating intact and undisturbed sediment cores containing the loriciferans with Cell-TrackerTM Green (CMFDA: Molecular Probes, Inc.,

Eugene, Oregon, US) which has been previously used to identify living uni-cellular eukaryotes in anoxic sediments (Bernhard et al., 2000). Comparative analyses conducted on anoxic sediments by confocal laser microscopy on Loricifera kept alive and others that were killed prior to incubation revealed that fluorescence intensity was, on average, 40% higher in the living Loricifera than in recently killed specimens. The adaptations to the extreme environmental conditions of the L'Atalante basin imply that these organisms have developed specific mechanisms for: (i) tolerating an enormous osmotic pressure (due to the high salinity and hydrostatic pressure); (ii) detoxifying highly toxic compounds (due to the high hydrogen sulfide concentrations); and (iii) living without oxygen. Quantitative X-ray microanalysis and Fourier-transformed infrared spectroscopy on the body composition of the loriciferans collected from the anoxic sediments revealed significant differences from the specimens collected in the oxygenated sediments of the deep Atlantic Ocean. Loriciferans from the L'Atalante basin had a Ca content that was nine-fold lower than in specimens inhabiting oxygenated sediments, on average, and showed Mg, Br and Fe, which were absent in the loriciferans from oxygenated sediments. Moreover, loriciferans from both oxic and anoxic sediments had similar concentrations of Na and S, in spite of the much higher salinity and sulfide concentration present in the deep-anoxic sediments of the L'Atalante basin. Moreover, Fourier-transformed infrared spectroscopy analyses indicated that the *lorica* of the loriciferans inhabiting oxygenated deep-sea sediments was apparently made of chitin, which was replaced by a chitin derivative, similar to chitosan, in the loriciferans inhabiting anoxic sediments. These results suggest the presence of chemical/structural adaptations of these loriciferans that can inhabit the anoxic sediments of the L'Atalante basin. Scanning electron microscopy also revealed the lack of prokaryotes attached to the body surface of the loriciferans while the ultrastructural analyses (carried out by transmission electron microscopy) revealed the lack of mitochondria, which are replaced by hydrogenosome-like organelles (Danovaro et al., 2010). These organelles have been previously encountered in various unrelated unicellular eukaryotes (Hackstein et al., 1999; Boxma et al., 2005), but have never been observed so far in free-living multicellular organisms (including the facultative anaerobes that face extended periods of aerobiosis during their life cycle) (Tielens et al., 2002). Moreover, the Loricifera retrieved from anoxic sediments contained hydrogenosome-like structures similar to those reported in anaerobic ciliates (Akhmanova et al., 1998; Müller, 1993). Since the hydrogenosomes do not coexist with mitochondria and they are present only in obligate anaerobic eukaryotes (type II anaerobes) (Roger, 1999), these data exclude the possibility that the Loricifera encountered in the anoxic basin are carcasses of organisms inhabiting oxygenated sediments and transported/settled into the anoxic basin.

4.4 Interactions between Metazoa and microbes in anoxic sediments

Viruses and prokaryotes were investigated in a deep hypersaline anoxic basin of the Eastern Mediterranean Sea (L'Atalante basin at ca. 3000 m depth; Danovaro et al., 2005). Although anoxic systems characterised by the chronic lack of oxygen are extreme, microbial life flourishes in either the hypersaline brines and their sediments. Prokaryotic life forms, including Archaea and Bacteria, and viruses were indeed found within the anoxic sediments (Corinaldesi et al., 2014). When these systems are compared, deep-sea sites characterised by oxic conditions, viral abundance and virus to prokaryote abundance ratio displayed similar values to those reported in oxic sites. The analysis of vertical profiles of viral abundance in the L'Atalante basin revealed the lack of significant changes with depth in the sediment, suggesting that benthic viruses in these anoxic and hypersaline conditions are preserved or resistant to decay. The anoxic basin also displayed very high concentrations of labile organic components (proteins and lipids) and extracellular DNA. These findings suggest that the DHAB sediments represent a reservoir for long-term preservation of benthic viruses and nucleic acids.

Previous investigations also showed that several morphotypes resembling microbes (cocci-like shape) were present within the tissues of the Loricifera collected in the basin. The metagenomic analysis of Loricifera extracted from various benthic ecosystems (both under anoxic and normoxic conditions) showed the presence of prokaryotes belonging to several taxonomic groups of Bacteria, especially Gammaproteobacteria, were also encountered, with groups representing gut-associated Bacteria belonging to the order of Enterobacteriales or sulfate-oxidising bacteria belonging to the order Thiotrichales. The prokaryotic assemblages identified from the metazoan tissues were different and less diversified than those found in normoxic conditions in the sediments and characterised by completely different species (Corinaldesi & Tangherlini, pers. comm.). These findings open up new perspectives in the exploration of the characteristics and properties of the microbiome associated with metazoans living in anoxic conditions.

References

Akhmanova, A., Voncken, F., van Alen, T., et al. (1998). A hydrogenosome with a genome. *Nature*, **396**, 527–528.

Antezana, T. (2009). Euphausia mucronata: a keystone herbivore and prey of the Humboldt Current System. *Deep-Sea Research II*, **57**(7–8), 652–662.

Bang-Berthelsen, I.H., Schmidt-Rhaesa, A., Kristensen, R.M. (2012). 6. Loricifera. In: A. Schmidt-Rhaesa (ed.) *Hand book of Zoology. Gastrotricha, Cycloneuralia and Gnathifera. Vol. 1: Nematomorpha, Priapulida, Kinorhyncha, Loricifera.* De Gruyter, Berlin, pp. 307–328.

Bernhard, J.M., Buck, K.R., Farmer, M.A., Bowser, S.S. (2000). The Santa Barbara basin is a symbiosis oasis. *Nature*, **403**, 77–80.

Bernhard, J.M., Morrison, C.R., Pape, E., et al. (2015). Metazoans of redoxcline sediments in Mediterranean deep-sea hypersaline anoxic basins. *BMC Biology*, **13**, 105.

Boxma, B., de Graaf, R.M., van der Staay, G.W., et al. (2005). An anaerobic mitochondrion that produces hydrogen. *Nature*, **434**, 74–79.

Corinaldesi, C., Beolchini, F., Dell'Anno, A. (2008). Damage and degradation rates of extracellular DNA in marine sediments: implications for the preservation of gene sequences. *Molecular Ecology*, **17**, 3939–3951.

Corinaldesi, C., Tangherlini, M., Luna, G.M., Dell'Anno, A. (2014). Extracellular DNA can preserve the genetic signatures of present and past viral infection events in deep hypersaline anoxic basins. *Proceedings of the Royal Society Part B*, **281**, 20133299.

Daffonchio, D., Borin, S., Brusa, T., et al. (2006). Stratified prokaryote network in the oxic-anoxic transition of the deep-sea halocline. *Nature*, **440**, 203–207.

Danovaro, R., Dell'Anno, A., Pusceddu, A., et al. (2010). The first metazoa living in permanently anoxic conditions. *BMC Biology*, **8**, 30.

Danovaro, R., Gambi, C., Dell'Anno, A., et al. (2016). The challenge of proving the existence of metazoan life in permanently anoxic deep-sea sediments. *BMC Biology*, **14**, 43.

Danovaro, R., Corinaldesi, C., Dell'Anno, A., Fabiano, M., Corselli, C. (2005). Viruses, prokaryotes and DNA in the sediments of a deep-hypersaline anoxic basin (DHAB) of the Mediterranean Sea. *Environmental Microbiology*, **7**(4), 586–592.

Danovaro, R., Snelgrove, P., Tyler, P. (2014). Challenging the paradigms of deep-sea ecology. *Trends in Ecology and Evolution*, **29** (8), 465–475.

Fusi, N., Aloisi de Larderel, G., Borelu, A., et al. (1996). Marine geology of the Medriff Corridor, Mediterranean ridge. *The Island Arc*, **5**, 420–439.

Gambi, C., Corinaldesi, C., Dell'Anno, A., et al. (2017). Functional response to food limitation can reduce the impact of global change in the deep-sea benthos. *Global Ecology and Biogeography*, **26**(9), 1008–1021.

Giere, O. (2009). *Meiobenthology. The Microscopic Motile Fauna of Aquatic Sediments*. Springer-Verlag, Berlin/Heidelberg.

Gribaldo, S., Philippe, H. (2002). Ancient phylogenetic relationships. *Theoretical Population Biology*, **61**, 391–408.

Hackstein, J.H.P., Akhmanova, A., Boxma, B., Harhangi, H.R., Voncken, G.J. (1999). Hydrogenosomes: eukaryotic adaptations to anaerobic environments. *Trends in Microbiology*, **7**, 441–447.

Heiner, I., Kristensen, R.M. (2008). Urnaloricus gadi nov. gen. et nov. sp. (Loricifera, Urnaloricidae nov. fam.), an aberrant Loricifera with a viviparous pedogenetic life cycle. *Journal of Morphology*, **270**, 129–153.

Hsü, K.J., Montadert, L., Bernoulli, D., et al. (1977). History of the Mediterranean salinity crisis. *Nature*, **267**, 399–403.

Jilbert, T., Reichart, G-J., Mason, P., de Lange, G. J. (2010). Short-time-scale variability in ventilation and export productivity during the formation of Mediterranean sapropel S1. *Paleocenography*, **25**, PA4232.

Kristensen, R.M. (1983). Loricifera, a new phylum with Aschelminthes characters from the meiobenthos. *Zeitschrift für zoologische Systematik und Evolutionforschung*, **21**, 163–180.

Kristensen, R.M. (2003). Loricifera. In: R. Hofrichter (ed). *Das Mittelmeer, Fauna, Flora, Ökologie, II/1 Bestimmungsführer*. Spektrum Akademischer Verlag, Heidelberg/Berlin, pp. 638–645.

Kristensen, R.M., Brooke, S. (2002). Phylum Loricifera. In: C.M. Young, M.A. Sewell, M. E. Rice (eds) *Atlas of Marine Invertebrate*

Larvae. Academic Press,London, pp. 179–187.

Kristensen, R.M., Shirayama, Y. (1988). Pliciloricus hadalis (Pliciloricidae), a new Loriciferan species collected from the Izu-Ogasawara Trench, Western Pacific. *Zoological Sciences*, **5**, 875–881.

Kuypers, M.M., Sliekers, A.O., Lavik, G., et al. (2003). Anaerobic ammonium oxidation by anammox bacteria in the Black Sea. *Nature*, **422**, 608–611.

Levin, L. (2003). Oxygen minimum zone benthos: adaptation and community response to hypoxia. *Oceanography and Marine Biology Annual Review*, **41**, 1–45.

Lipp, J.S., Morono, Y., Inagaki, F., Hinrichs, K-U. (2008). Significant contribution of Archaea to extant biomass in marine subsurface sediments. *Nature*, **454**, 991–994.

Müller, M. (1993). The hydrogenosome. *Journal of General Microbiology*, **139**, 2879–2889.

Neves, R.C., Bailly, X., Leasi, F., et al. (2013). A complete three-dimensional reconstruction of the myoanatomy of Loricifera: comparative morphology of an adult and a Higgins larva stage. *Frontiers in Zoology*, **10**, 19.

Neves, R.C., Gambi, C., Danovaro, R., Kristensen, R.M. (2014). Spinoloricus cinziae (Phylum Loricifera), a new species from a hypersaline anoxic deep basin in the Mediterranean Sea. *Systematics and Biodiversity*, **12**, 489–502.

Neves, R.C., Reichert, H., Søresen, M.V., Kristensen, R.M. (2016). Systematics of phylum Loricifera: identification keys of families, genera and species. *Zoologischer Anzeiger*, **265**, 141–170.

Pardos, F., Kristensen, R.M. (2013). First record of Loricifera from the Iberian Peninsula, with the description of Rugiloricus manuelae sp. nov. (Loricifera, Pliciloricidae). *Helgoland Marine Research*, **67**, 623–638.

Paulmier, A., Ruiz-Pino, D. (2008). Oxygen Minimum Zones (OMZs) in the modern ocean. *Progress in Oceanography*, **80**(3–4), 113–128.

Polonia, A., Vaiani, S.C., de Lange, G.J. (2016). Did the A.D. 365 Crete earthquake/tsunami trigger synchronous giant turbidity currents in the Mediterranean Sea? *Geology*, 44(3): 191–194. doi:10.1130/G37486.1.

Roger, A.J. (1999). Reconstructing early events in eukaryotic evolution. *American Naturalist*, **154**, 146–163.

Royer, D.L., Bemer, R.A., Montanez, I.P., Tabor, N. J., Beerling, D.J. (2004). CO_2 as a primary driver of Phanerozoic climate. *GSA Today*, 14(3), 3–7. doi:10.1130/1052-5173.

Sagarin, R.D., Gaines, S.D., Gaylord, B. (2006). Moving beyond assumptions to understand abundance distributions across the ranges of species. *Trends in Ecology and Evolution*, **21**, 524–530.

Soetaert, K., Heip, C., Vincx, M. (1984). Meiofauna of a deep-sea transect off Corsica. *Annales de la Société royale zoologique de Belgique*, **114**, 323–324.

Soetaert, K., Heip, C., Vincx, M. (1991). The meiobenthos along a Mediterranean deep-sea transect off Calvi (Corsica) and in an adjacent canyon. *P.S.Z.N.I: Marine Ecology*, **12**(3), 227–242.

Tielens, A.G., Rotte, C., van Hellemond, J.J., Martin, W. (2002). Mitochondria as we don't know them. *Trends in Biochemical Sciences*, **27**, 564–572.

Todaro, M.A., Kristensen, R.M. (1998). A new species and first report of the genus Nanaloricus (Loricifera, Nanaloricida, Nanaloricidae) from the Mediterranean Sea. *Italian Journal of Zoology*, **65**, 219–226.

van der Wielen, P.W., Bolhuis, H., Borin, S., et al. (2005). BioDeep Scientific Party: the enigma of prokaryotic life in deep hypersaline anoxic basins. *Science*, **307**, 121–123.

Walsh, D.A., Zaikova, E., Howes, C.G., et al. (2009). Metagenome of a versatile chemolithoautotroph from expanding oceanic dead zones. *Science*, **326**, 578–582.

Wishner, K.F., Gelfman, C., Gowing, M.M., et al. (2008). Vertical zonation and distributions of calanoid copepods through the lower oxycline of the Arabian Sea oxygen minimum zone. *Progress in Oceanography*, **78**, 163–191.

Zeppilli, D., Leduc, D., Fontanier, C., et al. (2018). Characteristics of meiofauna in extreme marine ecosystems: a review. *Marine Biodiversity*, **48**, 35–71.

CHAPTER FIVE

How to survive winter?

*Adaptation and acclimation strategies of eukaryotic algae
from polar terrestrial ecosystems*

MARTINA PICHRTOVÁ

Charles University

EVA HEJDUKOVÁ

Charles University

LINDA NEDBALOVÁ

Charles University

and

JOSEF ELSTER

University of South Bohemia

5.1 Introduction

The polar regions are of outstanding international scientific and environmental significance as they support important components of the global biogeochemical cycles. They comprise a whole range of habitats with extreme environmental conditions, which challenge living organisms with multiple environmental stresses. At the same time, they are vulnerable to disturbances and have long recovery times (Robinson et al., 2003; Elster & Benson, 2004; Thomas et al., 2008). Moreover, the Arctic is especially undergoing a particularly rapid climate change compared to the rest of the planet, including changes in temperature and precipitation (Thomas et al., 2008). However, predicting the impacts of climate change on arctic ecosystems is difficult (Bokhorst et al., 2015), because (i) climate change is not uniform across the Arctic (AMAP, 2011), and (ii) at local and regional scales, ecosystem responses to warming are not necessarily the same due to variations driven by other biotic and climatic factors (Post et al., 2009; Callaghan et al., 2013). Warming of the Arctic is also expected to result in an increasing frequency of stochastic climatic events (Saha et al., 2006; Bokhorst et al., 2009, 2011, 2015; Callaghan et al., 2013; Bjerke et al., 2014), such as extreme winter warming.

Photosynthetic microorganisms are major primary producers in the most hostile polar habitats and have successfully adapted to a wide range of extreme conditions, including winter freezing and desiccation (Thomas et al., 2008). A thorough understanding of their adaptation and acclimation strategies is crucial for evaluating and predicting their response to environmental changes. However, polar terrestrial eukaryotic microalgae, in contrast to cyanobacteria, usually form extensive, yet only annual, mats (Pichrtová et al., 2016b). It seems that, despite various stress tolerance mechanisms, winter conditions have a strong effect on the survival of whole populations, which appear to depend on a small fraction of surviving cells (Hawes, 1990; Pichrtová et al., 2016b).

In this chapter, we review the present knowledge on winter survival strategies of photosynthetic micro-eukaryotes in polar terrestrial ecosystems. Throughout this review, the term terrestrial also includes hydroterrestrial environments as defined by Elster (2002) unless further specified. Hydroterrestrial habitats have liquid water available almost throughout the entire growing season, but freeze-dry during winter (Elster, 2002). They include not only wetlands, shallow pools, snow-fed streams and rivers, seepages and springs, but also supraglacial and snow field habitats (Elster, 2002).

5.2 Physical settings

The most important feature limiting life in the polar regions is the frigid climate. The severity and complexity of polar environments are a consequence of the Earth's geometry. The North and South Poles, instead of having a daily alternation of diurnal cycles, have nightless summer months, followed by sunless winters and receive on average less solar radiation than the equatorial regions. However, this is not the only reason for the frigidity of the polar regions. Reflection losses in areas covered by ice or snow are higher in comparison with a snow-free terrestrial landscape or open waters (e.g. Youssouf et al., 2016; Lembo et al., 2017). Nevertheless, these properties affecting the heat balance in the polar regions have changed over geological time (Thomas et al., 2008).

Frost-free periods are usually quite short and diurnal fluctuations in temperature causing repeated freeze–thaw cycles can occur at any time (Davey & Rothery, 1992; Elster & Komárek, 2003). This overnight freezing is restricted to the vegetation surface, with the temperature dropping only to a few degrees below zero. In winter, air temperatures fall far below 0°C. The ground surface temperature of a study site on Svalbard reached down to temperatures between −30 and −35°C several times during winter (Láska et al., 2012). Hence, the terrestrial algal communities are completely desiccated and frozen until liquid water is available again when spring returns. However, terrestrial algae can be protected by

Figure 5.1 Stream with well acclimated (hardened) dry and frozen biomass of viable *Klebsormidium* sp. Central Svalbard, Endalen, late autumn. Photo J. Elster. (A black and white version of this figure will appear in some formats. For the colour version, please refer to the plate section.)

insulating snow cover and do not experience temperatures far below zero, even during winter (Hawes, 1989; Davey, 1991).

Seasonal and less pronounced diurnal variations of the polar terrestrial environments result in a series of water availability and freezing/melting gradients ranging from aquatic and semiaquatic to dry/ice habitats. These gradients result in different patterns, which change over geological time and are also influenced by current climate change (Elster, 2002). The widespread warming and increasing or decreasing winter precipitation in the Arctic and in some areas in the Antarctic have resulted in different regional snow-cover responses (Callaghan et al., 2011). Snow cover and its properties are principal ecological parameters for the terrestrial polar environment and influence their surface energy balance, water balance and thermal regimes.

Three different habitat categories were described in polar non-marine environments based on the variability in their ecological characteristics, mainly the availability of liquid water during the growing season (Broady, 1996; Vincent, 1988; Vincent & James, 1996; Elster, 2002) (Figures 5.1, 5.2 and 5.3). Terrestrial and hydroterrestrial environments are regarded as unstable; the third category is represented by stable lacustrine habitats.

5.3 Stress factors in the polar terrestrial environment

The polar environment is characterised by several interrelated stress factors. Low temperatures have several effects on living organisms. Membrane fluidity decreases with decreasing temperatures, and the rate of metabolic reactions slows down as described by the Arrhenius

Figure 5.2 Moss stream with dead *Tribonema* sp. biomass, central Svalbard, Endalen, beginning of July. Because of water erosion during spring melt, the stream started to flow in a different direction resulting in *Tribonema* biomass drying and not surviving the summer period. Photo J. Elster. (A black and white version of this figure will appear in some formats. For the colour version, please refer to the plate section.)

Figure 5.3 Detail of moss stream with dead *Tribonema* sp. biomass. Photo J. Elster. (A black and white version of this figure will appear in some formats. For the colour version, please refer to the plate section.)

equation. This slowdown also affects repair processes, which makes various stress factors more harmful at low temperatures (Roos & Vincent, 1998). Freezing itself is stressful in two different ways. First, formation of ice crystals mechanically disrupts the cells, particularly if the crystals are formed intracellularly (Hawes, 1990; Fuller, 2004). Second, extracellular freezing increases intracellular solute concentrations, which leads to osmotic stress. This phenomenon makes the physiological effect of freezing similar to that of desiccation or salt stress; they all lead to osmotic dehydration and a lower intracellular water potential (Bisson & Kirst, 1995). Therefore, these stressors have to be discussed together since the adaptation mechanisms of algae are similar.

Algae are poikilohydric organisms that are able to tolerate desiccation to different capacities. The external environment directly manages their metabolic activity by affecting the presence or absence of water. In addition to poikilohydry, on a physiological level, changes in temperature and water status provoke a series of adaptive responses such as developing resistance and tolerance to cold, freezing, drought, desiccation and salinity stress (Elster & Benson, 2004).

Naturally, other stresses play important roles in polar habitats that are not addressed here in detail – namely, high solar irradiance (including deleterious effects of UV radiation) during the polar days, long periods of darkness during the polar nights, the short vegetation season or low nutrient availability.

5.4 Life strategies and adaptation mechanisms

Polar organisms have developed a wide range of adaptive strategies that allow them to avoid, or at least minimise, the injurious effects of extreme and fluctuating environmental conditions. However, such strategies are not exclusive to polar algae, as most of them can also be found in selected habitats at other latitudes (e.g. various aeroterrestrial and mountainous habitats). Nevertheless, Elster & Benson (2004) suggested that the physical and chemical conditions that occur uniquely in polar environments have selected for specific and resilient sets of adaptive biological characteristics.

Therefore, we present an overview of the most important adaptive strategies enabling winter survival of polar terrestrial algae together with specific examples from various studies performed on polar microalgae. First, we focus on the formation of specialised cells. The three main phylogenetic groups of algae using this strategy in polar terrestrial environments are introduced. Then, we present examples of algae which, in contrast to the previous subchapter, survive stress conditions in a vegetative state. In addition, we describe the phenomena of acclimation and mixotrophy. Finally, various biochemical adaptations are presented.

5.4.1 Production of specialised stages

Complex life cycles are a major avoidance strategy and usually involve the development of specialised stages that enable algae in general to survive the most stressful periods. Agrawal (2009) distinguished between dormant cells that require a period of dormancy before germination and other types of specialised cells that are also resistant but not dormant and thus can germinate immediately after formation. Both types of specialised cells are, however, important not only for stress tolerance, but also for dispersal in time and space (Rengefors et al., 1998).

5.4.1.1 Snow algae (Chlamydomonadaceae)

Snow algae are a diverse group of extremophilic microorganisms that form massive red, pink, orange or green blooms on snow in the polar and mountain regions worldwide. They are mostly represented by species from the genera *Chloromonas*, *Chlamydomonas* and *Chlainomonas*, from the green algal order Chlamydomonadales, that have optimised their physiology to the snow environment (Komárek & Nedbalová, 2007). The key feature of their life cycle that has been described in detail in many species is the presence of both flagellated and immotile stages (e.g. Hoham et al., 1979, 1983).

The formation of resistant coccoid stages (cysts or aplanozygotes) can be viewed as one of the major adaptations to the snow habitat, which allows them to survive long periods that are not suitable for active growth. As the nature of these cells is frequently not clear, we use the term cysts through the following text. In contrast to flagellates, cyst-like stages are believed to be resistant to freezing, high irradiances and any environmental changes in general (Remias, 2012). They are characterised by a resistant cell wall. The detailed structure of the ornamented secondary cell wall of *Chloromonas* cysts is used for determination on a fine scale taxonomic level (Procházková et al., 2018; Matsuzaki et al., 2019). For

Figure 5.4 Scanning electron microscopy picture of a snow alga identified as a cyst of *Chloromonas* cf. *nivalis*. The field sample was collected in Nathorst Land, Svalbard. Scale bar, 5 μm. Courtesy of L. Procházková.

example, the cysts of the Antarctic species *Chloromonas polyptera* are characterised by many flanges-like structures, which enables to differentiate them from other taxa (Remias et al., 2013). A detailed study of arctic species of *Chloromonas* is still at its beginning (Figure 5.4). In an extensive field study performed in Svalbard, Müller et al. (1998) showed that cysts of snow algae contained high amounts of lipids and starch that serve as food reserves.

Mature cysts of different species were repeatedly shown to be metabolically active, which contradicts the view that they are solely dormant stages (Remias et al., 2005, 2010). In a study performed in Svalbard, Stibal et al. (2007) observed detailed dynamics of photosynthetic activity of different types of snow algal cysts which were associated with red snow. However, these measurements were performed during the summer period when light for photosynthesis is available. To our knowledge, the physiological status of snow algal cysts during winter has not been studied yet.

5.4.1.2 *Zygnematophyceae*
Zygnematophyceae is a class of streptophyte algae closely related to land plants (Embryophyta) (de Vries et al., 2016, 2018). Therefore, understanding their stress tolerance mechanisms is also important for research related to the transition of plants to life on land. The Zygnematophyceae represent the most species-rich lineage of charophyte algae and are characterised by a unique mode of sexual reproduction, namely conjugation, which leads to the formation of highly specialised, stress-resistant zygospores containing algaenan (a sporopollenin-like substance) in their cell walls (Poulíčková et al., 2007). In addition to zygospores, other specialised cell types have been described in Zygnematophyceae, namely parthenospores, aplanospores and akinetes.

Members of this class are very common in polar terrestrial habitats. Several filamentous genera form extensive mucilaginous mats in slow-flowing meltwater streamlets or shallow puddles both in the Arctic (Sheath et al., 1996; Kim et al., 2008; Holzinger et al., 2009) and Antarctica (Hawes, 1989, 1990; Davey, 1991; Skácelová et al., 2013). Other Zygnematophyceae, such as *Mesotaenium berggrenii* and *Ancylonema nordenskioeldii*, are typical inhabitants of bare glacier surfaces.

Despite their abundance in the polar regions, the conjugation process and formation of highly resistant zygospores in these regions has only very rarely been observed nor are there reports of other types of specialised cells. Elster et al. (1997) reported zygospore formation of *Zygnema* cf. *leiospermum* occurring in multiple sites of a glacial stream in Central Ellesmere Island. Pichrtová et al. (2016b) observed conjugation only twice (species *Zygnemopsis lamellata* and *Zygnema* cf. *calosporum*) in Svalbard, despite intensive research (Pichrtová et al., 2018). Remias et al. (2011) reported the conjugation process and young zygotes in *Ancylonema nordens-kiöldii*, which were collected from the Midtre Lovenbreen glacier in Svalbard. Therefore, it has been suggested that a trade-off between sexual reproduction and growth plays an important role in these extreme environments (Holzinger et al., 2009).

5.4.1.3 Diatoms

Diatoms (Bacillariophyceae) are a group of algae which are characterised by a golden-brownish pigmentation and a siliceous shell wall (Round et al., 1990). They represent one of the most abundant algal groups in many polar freshwater and terrestrial habitats (Jones, 1996; Van de Vijver & Beyens, 1999; Van de Vijver et al., 2002). Diatoms are known to be able to form two types of resting stages, namely morphologically distinct resting spores and vegetative-looking resting cells. Both stages are characterised by reduced metabolic activity and differences in their cellular components compared with vegetative cells. During this resting state only small amounts of cellular carbon are required to survive (Kuwata et al., 1993; Jewson et al., 2008). The resting stages are characterised by dense and dark cytoplasmic matter, rounder plastids (Round et al., 1990), condensed organelles, larger vesicles of storage products, granular cytoplasm, enlarged vacuoles or oil droplets and contracted chloroplasts (McQuoid & Hobson, 1996).

Resting spores are morphologically different from actively growing or vegetative cells. They are characterised by a rounder shape, thicker cell wall and different ornamentation (McQuoid & Hobson, 1995, 1996). Spore formation is usually associated with limiting nutrient concentrations and winter survival (McQuoid & Hobson, 1995, 1996; Kuwata & Takahashi, 1999). The most common occurrence of diatom resting spores is in centric marine diatoms from temperate neritic habitats (McQuoid & Hobson, 1996; Kuwata and Takahashi, 1999) and are generally relatively rare in non-marine (observed mostly in centric) species (Edlund et al., 1996; McQuoid & Hobson, 1996; Jewson et al., 2008). To our knowledge, resting spores have not yet been observed in terrestrial diatoms.

5.4.2 Survival in the vegetative state

As mentioned above, many algae survive stress conditions in a vegetative state without the production of any morphologically distinct cells (Agrawal, 2009). In this case, the vegetative cells, similarly to dormant cells, have reduced physiological activity combined with other adaptations, e.g. biochemical. Sheath et al. (1996), who studied tundra stream algae in the Arctic, including members of the Chlorophyta,

Streptophyta, Rhodophyta and stramenopiles (e.g. diatoms), observed that most species do not form specialised cells, but instead, their vegetative cells are adapted to withstand prolonged freezing. They possess however thick cell walls and accumulate storage materials (Sheath et al., 1996).

Recently, stress tolerance of polar hydroterrestrial algae of the genera *Zygnema* and *Zygnemopsis* (class Zygnematophyceae) has been intensively studied. Since these algae rarely form truly specialised stages, they are a typical example of algae with the ability to survive multiple stress conditions in a vegetative state: the modified, resistant vegetative cells are called pre-akinetes (McLean & Pessoney, 1971; Pichrtová et al., 2014a). Pre-akinetes are formed directly from young vegetative cells that stop dividing and gradually accumulate storage compounds and thicken their cell walls; their chloroplasts are structurally reduced, and their physiological activity is diminished (Pichrtová et al., 2014b, 2016a; Herburger et al., 2015, Figure 5.5). These cells are not dormant and gradually recover their vegetative appearance and physiological activity immediately after transfer to favourable conditions, which enables rapid growth of *Zygnema* populations during early spring (Fuller, 2013; Pichrtová et al., 2016b).

Such pre-akinetes in *Zygnema* were observed at the end of summer in Svalbard (Pichrtová et al., 2014a, 2016b) and viable pre-akinetes were even found frozen in solid ice during winter (Pichrtová et al., 2016b). After

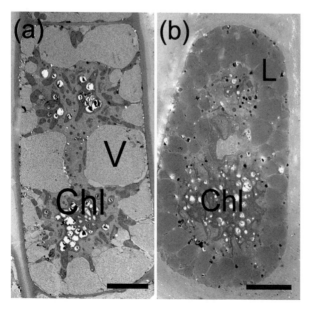

Figure 5.5 Transmission electron microscopy picture of Antarctic *Zygnema* sp. cells. (a) An overview of a young vegetative cell with a high degree of vacuolisation and stellate chloroplasts, and (b) an overview of a pre-akinete with reduced chloroplasts and massive accumulation of lipid bodies. Chl, chloroplast; L, lipid body; V, vacuole. Scale bar, 10 µm. Reprinted from Pichrtová et al. (2016a).

melting and transfer into a culture medium, they immediately restored their photosynthetic activity and gradually developed back into young and dividing vegetative cells (Pichrtová et al., 2016b). Pre-akinetes can also be found as senescent or stationary-phase-like cells in old cultures with depleted nutrients. Thus, it has been suggested that nutrient limitation is involved in pre-akinete formation (Fuller, 2013; Pichrtová et al., 2014a). This hypothesis was later supported experimentally in which pre-akinetes were formed much faster in medium lacking any form of nitrogen (Pichrtová et al., 2014b). Such experimentally induced pre-akinetes were tolerant to desiccation, whereas young vegetative cells or intermediate stages did not survive any of the desiccation treatments (Pichrtová et al., 2014b). The phenomenon of stress tolerance in response to nutrient depletion is also well known for bacteria (Siegele & Kolter, 1992) and eukaryotic algae from non-polar regions (McLean & Pessoney, 1971; Nagao et al., 1999).

Hawes (1990) conducted a series of freezing experiments with an Antarctic *Zygnema* species. He used both newly collected material from the field described as 'resembling pre-akinetes' as well as cultures in the exponential growth phase and hence young vegetative cells. The newly collected samples showed no decrease in photosynthetic rate during repeated freeze–thaw cycles (temperature ranging between +5 and −4°C), and several cells were still alive even after 120 days at −20°C. In addition, the cultures were able to survive at −15°C for 60s when very slow cooling rates were applied, which indicates a certain level of frost tolerance even in young cells (Hawes, 1990). This was confirmed by our observations with an Arctic *Zygnema* species. Young cells survived slightly subzero temperatures and were not able to recover from frost damage only when exposed to −8°C. By contrast, pre-akinetes survived even at −70°C during 8 hours of exposure (Trumhová et al., 2019).

Klebsormidium is a globally distributed algal genus which is ecologically important in the polar terrestrial environment (Ryšánek et al., 2016). No specialised dormant stages are known in members of this genus, yet they were repeatedly shown to be outstandingly tolerant of stress in the vegetative state, being able to survive both desiccation and freezing (Elster et al., 2008; Nagao et al., 2008; Karsten et al., 2010). Freezing to −4°C had no effect on the survival of the investigated strains, and they still retained 80% viability at −40°C (Elster et al., 2008).

Šabacká & Elster (2006) studied the effect of freezing and desiccation on several strains of antarctic terrestrial microalgae from the genera *Chlorella*, *Chlorosarcina*, *Pseudococcomyxa* and *Klebsormidium*. Whereas minor subzero temperatures (−4°C) during experimentally induced summer diurnal freeze–thaws did not cause significant damage, low subzero temperatures representing an annual winter freeze (−40°C) were fatal for more than 50% of the populations. However, individual members survived also in −100°C and −196°C treatments, but with markedly reduced viability (Šabacká & Elster, 2006).

   Most diatom species do not form morphologically different resting spores, but to ensure survival under stress conditions, they create physiologically adapted resting cells (Figure 5.6). They are morphologically identical to vegetative cells (Lund, 1954; Anderson, 1975; Sicko-Goad et al., 1986), yet are characterised by dense and dark cytoplasmic matter and rounder plastids (Round et al., 1990) as well as lower pigmentation and contracted chloroplasts (Kuwata et al., 1993). Resting cells have been observed in many pennate diatoms (McQuoid & Hobson, 1996), including terrestrial and freshwater species (Souffreau et al., 2013). It has been suggested that resting cell formation in terrestrial diatoms might be important for their survival during freezing and desiccation, which might represent a specific adaptation to the varying environmental conditions in terrestrial habitats (Souffreau et al., 2013). By contrast, both vegetative and laboratory-induced resting cells of freshwater benthic diatoms were shown to be highly sensitive to desiccation, abrupt heating or freezing (Souffreau et al., 2010; Hejduková et al., 2019). A recent study of diatom survival in freshwater and hydroterrestrial habitats in the High Arctic observed remarkable viability of diatoms during the winter season with more than 20% of the cells on

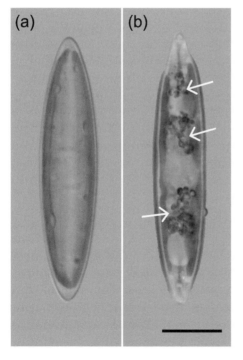

Figure 5.6 Vegetative (a) and resting (b) cells of the diatom *Navicula radiosa*. The laboratory strain was isolated from a sample collected on Svalbard. Note the shrinkage of the chloroplast and numerous lipid droplets (marked by arrows) in the resting cell that was induced by a 2-week incubation in dark at 5°C. Scale bar, 10 μm. Photo: E. Hejduková.

average found in a resting stage and almost 5% becoming active after thawing. The populations were, however, dominated by dead or damaged cells (Hejduková, in press).

5.4.3 Acclimation to cold conditions and desiccation

Acclimation also plays an important role in the development of tolerance for winter survival. The period of declining temperatures during autumn may be important in promoting cold-hardiness. Various antarctic hydroterrestrial algae were found to be psychrotrophs rather than psychrophiles, which might suggest that their ability to grow at cold temperatures is not genetically fixed but a result of acclimation (Seaburg et al., 1981).

The positive effect of acclimation has already been described in a non-polar strain of *Klebsormidium flaccidum*, in which survival rates increased by 15% after exposure to 2°C for 7 days (Nagao et al., 2008). Unfortunately, this phenomenon has not been experimentally studied on polar isolates of *Klebsormidium*. Nevertheless, a recent phylogenetic analysis demonstrated unlimited dispersal and intensive gene flow, together with remarkable ecological plasticity, of this genus (Ryšánek et al., 2016). Moreover, laboratory strains of *Klebsormidium* were shown to be highly resistant to both freezing and desiccation injuries and no statistical differences were observed between strains originating from different regions and habitats (Elster et al., 2008). All these findings support the ecological role of acclimation.

The effect of acclimation, in this case to osmotic stress and desiccation, was also studied in Arctic *Zygnema* pre-akinetes (Pichrtová et al., 2014a, 2014b). Populations sampled from partly desiccated environments were much more resistant to osmotic stress than populations from pools, even though all samples consisted of well-developed pre-akinetes (Pichrtová et al., 2014a). This finding, later confirmed by a laboratory study, shows that acclimation plays an important role in stress tolerance of *Zygnema* pre-akinetes (Pichrtová et al., 2014b). As mentioned above, the pre-akinetes were induced by nitrogen depletion and subsequently experimentally desiccated (Pichrtová et al., 2014b). In general, the pre-akinetes tolerated desiccation at 86% relative humidity, but survival at 10% relative humidty only occurred after additional acclimation induced either by very slow experimental desiccation or precultivation on agar plates (Pichrtová et al., 2014b).

Acclimation is also known to increase the tolerance of diatoms to unfavourable conditions (Mock & Valentin, 2004). Resting cells of terrestrial diatoms, induced in laboratory conditions by acclimation at a lower temperature and in dark conditions as well as under nitrogen limitation, showed a higher degree of stress tolerance, especially to desiccation (Souffreau et al., 2013). The same methodological approach for the induction of resting cells was applied by Hejduková et al. (2019), who emphasised their importance for the survival of mild freezing (–4°C treatment) in a set of polar and temperate diatom strains originating from terrestrial and freshwater environments.

Acclimation might also be involved in speciation. Physiological differences among various populations of stress-acclimatised organisms might be related to life cycle timing, germination, dormancy, frost resistance and many other characteristics. As there are many environmental gradients, so there are gradients of evolutionary progression from induced acclimation, to temporary ecoforms, to more stable ecotypes and finally to a new species (Svoboda, 2009).

5.4.4 Mixotrophy

The shift from photoautotrophy towards mixotrophy or heterotrophy can be an important process for the successful survival of algal populations in low nutrient and low-light polar environments. During long periods of darkness, mixotrophic algae can remain metabolically active and rapidly resume photosynthetic activity when light becomes available. The utilisation of organic compounds has been documented in many algal taxa belonging to most of the ecologically important taxonomic groups (e.g. diatoms, chrysophytes, green algae, cryptophytes, dinoflagellates). Actually, photosynthetic eukaryotes have a wide range of strategies for mixotrophy, ranging from phagotrophy to the uptake of dissolved organic carbon (Jones, 2000).

Among polar lacustrine communities, the phenomenon of mixotrophy was shown to be a widely used survival strategy that plays a major role in carbon cycling (Laybourn-Parry et al., 2000). Decades ago it was found that the high proportion of mixotrophic algae in boreal humic lakes might compensate for the poor light environment (Jansson et al., 1996). The current climate warming in both polar regions is associated with increased snowmelt that can occur even during winter months. Such events could support protists capable of mixotrophic growth whose adaptive capacity is supposed to shape the structure and function of microbial communities (Vrionis et al., 2013). However, in contrast to freshwater and marine environments, the possible role of mixotrophic nutrition in polar terrestrial habitats is still poorly understood.

5.4.5 Biochemical acclimation and adaptation mechanisms

The above discussed acclimation and adaptation mechanisms, production of resting stages and acclimation, are closely connected with a shift in the biochemical composition of the cells. In the following, we introduce the most widespread strategies against desiccation and cryoinjury stresses.

5.4.5.1 Lipids

Lipid accumulation is considered as a general response of algae to unfavourable conditions, such as nutrient depletion and other stresses, e.g. high light, alkaline pH or heavy-metal exposure (Guschina & Harwood, 2006). The accumulation of nitrogen free lipids under stress conditions is thought to be the result of a shift in photosynthate allocation from growth to energy storage (Vítová et al., 2014). For example, an extensive occurrence of lipid bodies was observed in the terrestrial green alga *Prasiola crispa* from Antarctica when exposed to desiccation (Jacob et al., 1992). High lipid content is a major feature of specialised and vegetative cells that are resistant to seasonal changes of environmental factors, and play a crucial role in over-wintering of populations. This strategy is shared by various unrelated groups of algae as lipid accumulation is typical for resting cells of diatoms, *Zygnema* pre-akinetes (Pichrtová et al., 2016a), snow algal cysts (Remias et al., 2010) and various other algae that survive unfavourable conditions in the vegetative state (Sheath et al., 1996). This lipid accumulation serves as an energy supply and carbon source, and enables rapid growth early in the spring, which gives these species a competitive advantage (Davey, 1988; Pichrtová et al., 2016a).

In snow algae, the transformation into cysts and lipid production is accompanied by progressive incorporation of esterified secondary carotenoids, namely astaxanthin (Remias et al., 2010). The decrease in water content because of the high content of lipid bodies containing astaxanthin is considered as an important mechanism which protects the cysts from the deleterious effects of subzero temperatures. The accumulation of astaxanthin is particularly high in the red snow forming species *Chlamydomonas nivalis*, where the astaxanthin/chlorophyll a ratio frequently reaches values of around 20 (Remias et al., 2005). Large amounts of astaxanthin diglukoside diesters were detected in cysts of this species suggesting that astaxanthin accumulation also has an additional role in carbon and hence energy storage (Řezanka et al., 2008).

Apart from the total lipid content, the changes in lipid and fatty acid composition also represent an important process in enhancing cell survival under stress conditions in polar habitats. Perhaps the best studied metabolic adaptation of microorganisms to low temperatures is the desaturation of fatty acids in membrane lipids that ensures the maintenance of membrane fluidity (Morgan-Kiss et al., 2006). A high proportion of polyunsaturated fatty acids, reaching up to 70% of total fatty acids, was repeatedly reported in field samples of psychrophilic algae collected from various cold habitats (e.g. Procházková et al., 2018), as well as in their laboratory cultures (Teoh et al., 2004). Several long-chain polyunsaturated fatty acids were detected in antarctic *Zygnema* sp., although in low concentrations, likely because their content was not investigated at low cultivation temperatures (Pichrtová et al., 2016a). A high proportion of unsaturated fatty acids enhancing physiological performance of snow algal cysts at low temperatures is also often detected. For example, 80% of the total fatty acid pool consisted of monounsaturated fatty acids in red cells of *Chlamydomonas nivalis* from Antarctica (Bidigare et al., 1993). Spijkerman et al. (2012) found that between 45 and 55% of total fatty acids were polyunsaturated fatty acids in field samples from Svalbard that contained red and orange cysts, which were also attributed to *Chlamydomonas nivalis*.

The complex adaptation of polar algae to their extreme habitats makes them suitable candidates for low temperature biotechnologies. Low requirements for cultivation temperature and light, combined with relatively high growth rates and a high lipid/polyunsaturated acid content were demonstrated for example in a cold-adapted green alga belonging to the genus *Monoraphidium*, which was isolated from Antarctica (Řezanka et al., 2017).

5.4.5.2 *Accumulation of inorganic and organic osmolytes*

One of the most widespread protective adaptations is the accumulation of osmolytes. These compounds prevent water loss and thus help to maintain homeostasis (Bisson & Kirst, 1995). As such, they contribute to tolerance to desiccation (Crowe et al., 2002), freezing-drying (Tanghe et al., 2003) and osmotic stress (Reed et al., 1984). Osmolytes are either simple inorganic ions or various organic substances, such as sugars, polyols or amino acids. Some organic osmolytes, termed compatible solutes, play different additional roles in cellular protection as they protect and stabilise membranes, proteins and DNA (Crowe et al., 2002, 2004; Yancey, 2005). They can prevent intracellular freezing by decreasing the freezing point as well as through water replacement. This process is termed vitrification and is involved in both desiccation and freezing tolerance (Welsh, 2000; Yancey, 2005; Clarke et al., 2013). It is generally accepted that especially sucrose and trehalose bind to dried membranes, proteins and nucleic acids as water substitutes, and as such preserve the native structures of the molecules (Crowe et al., 2002).

Several compounds (glucose, sucrose, an additional unidentified sugar and several amino acids) were shown to accumulate during cold acclimation in *Klebsormidium flaccidum* (Nagao et al., 2008). Moreover, a recent transcriptomics study on *Klebsormidium crenulatum* showed the up-regulation of enzymes connected with the biosynthesis of the sucrose and raffinose family of oligosaccharides in response to desiccation stress (Holzinger et al., 2014).

In an antarctic *Zygnema* sp., sucrose was identified as the main carbohydrate with traces of glucose, fructose and mannitol (Hawes, 1990). Moreover, Kaplan et al. (2013) revealed quite low (negative) values of cellular osmotic potentials in arctic and antarctic *Zygnema* spp., but the composition of soluble carbohydrates was not investigated. Comparison of metabolic composition of vegetative cells and pre-akinetes showed accumulation of several sugars and sugar alcohols, mainly raffinose, in pre-akinetes. Sucrose was one of the most abundant sugars, but only showed a slight increase during pre-akinete forma- tion (Arc et al., 2020). Furthermore, a recent transcriptomics study confirmed a strong up-regulation of sucrose synthesis in response to desiccation stress in *Zygnema* (Rippin et al., 2017).

Prasiola (Trebouxiophyceae) synthesises free proline as a cryoprotectant, and accumulates it at the onset of winter (Jackson & Seppelt, 1995). Therefore, it can continue to photosynthesise down to -7°C (Davey, 1989) or -15°C (Becker, 1982), even though the thallus appears to be frozen. In addition, *Prasiola* is also tolerant to salt (Jacob et al., 1992) and desiccation stress (Kosugi et al., 2010).

5.4.5.3 *Secretion of extracellular polysaccharides*

The secretion of extracellular polysaccharides (EPS) is typical for cyanobacteria where it helps in increasing their freezing and desiccation tolerance (Tamaru et al., 2005). However, the production of hydrophilic polysaccharides, i.e. mucilage, is also widespread in eukaryotic algae (Domozych, 2011). Mucilage helps to retain water around the cells and protects them from freezing and desiccation stress. In several habitats across the Arctic, there are species which produce mucilage in quite high quantity e.g. *Hydrurus foetidus* belonging to the Chrysophyceae (Klaveness et al., 2011). Also *Zygnema* sp. pre-akinetes are often embedded in thick mucilaginous sheaths. Fuller (2013) showed that these secondary pectic layers of the cell walls act as a sponge and provide extracel- lular protection against desiccation. Recently, homogalactouronan was revealed as the major component of these pectic layers and its crucial role in desiccation resistance was confirmed (Herburger et al., 2019).

Apart from desiccation avoidance, mucilage layers also enable the algae to aggregate into mats, biofilms and crusts. In these structures, numerous other microorganisms are embedded. As such, these algae are true ecosystem engi- neers, because their mucilage layers protect whole communities against multiple environmental stresses (Knowles & Castenholz, 2008). In general, the formation of these aggregates is itself considered as a stress avoidance strategy (Holzinger & Pichrtová, 2016). In addition, the secretion of EPS may serve as a source of organic matter for bacteria (Bashenkhaeva et al., 2015).

5.4.5.4 *Ice-binding proteins (IBPs) and ice nucleation agents*

Ice-binding proteins (IBPs) are a diverse group of proteins that bind to ice surfaces and control ice growth to avoid freezing injury. They have been detected in many cold-adapted organisms including bacteria, fungi, algae, vas- cular plants and animals (Dolev et al., 2016). In polar algae, IBPs that are able to distort the shape of growing ice crystals and thus prevent mechanical damage to the cells were first discovered in sea-ice diatoms (Raymond, 2000; Janech et al., 2006). Also another type of IBP that is secreted by sea-ice-associated chlamydomonad algae helps to maintain a liquid environment in brine pockets by structuring the external ice (Raymond et al., 2009). Reports on the occurrence of IBPs from terrestrial cold-adapted algae are limited to species associated with snow. Genes for more than 20 isoforms of IBP were identified in *Chloromonas brevispina* indicating the importance of this protein for this species. Their close relationship to bacterial and fungal proteins provided evidence that they were possibly acquired by horizontal gene transfer (Raymond, 2014). A screening of snow algal strains from the CCCryo Culture Collection of Cryophilic Algae (Fraunhofer, Postdam-Golm, Germany), which contains mainly isolates from the Arctic, revealed that only psychrophilic strains from the Chlamydomonadales were able to produce IBPs. Tests of recrystallisation activity revealed a rather high variability of active concentrations of IBPs, even within this small taxonomic group (Leya, 2013). It can be expected that the secretion of IBPs is common not only in snow algae, but also in other terrestrial groups of polar eukaryotic algae that have to cope with freezing.

Another way for preventing ice crystal formation is controlled freezing at relatively high temperatures due to ice nucleation agents. Ice nucleation proteins can be considered as a subset of IBPs (Dolev et al., 2016). However, current knowledge on ice nucleation activity in algae is limited. It was suggested that ice nucleation agents are produced by fungi and bacteria that are associated with the algae in field samples. For example, field samples of the green alga *Prasiola crispa* had a mean freezing point that was 10°C higher compared to axenic culture (Worland & Lukešová, 2001).

5.5 Conclusions

Even though eukaryotic algae are the most important primary producers in many polar terrestrial habitats, their stress tolerance with respect to winter survival has hardly been described. In this chapter, we give a complex overview of the present knowledge about stress tolerance strategies investigated in polar terrestrial algae. Different mechanisms have been described that enable the cells to survive extreme conditions and several of these mechanisms are present in phylogenetically unrelated taxa. These include survival in only a slightly modified vegetative state. This is a very advantageous strategy because it is energetically less demanding than sexual reproduction connected with the formation of true specialised stages.

Due to the lack of studies, it is often difficult to state which strategies have developed as unique traits of polar organisms and which are shared also by their non-polar relatives. Nevertheless, the evaluation of winter survival strategies is important to predict the effect of global climate change on polar terrestrial microbial communities, their biodiversity and ability to cope with potential invasive species.

Acknowledgement

This work was supported by the following projects of the Ministry of Education, Youth and Sports – ECOPOLARIS (CZ.02.1.01/0.0/0.0/16_013/0001708) and Polyphasic assessment of diversity of phototrophic microorganisms from cold environments and their bioprospection potential (INTER-EXCELLENCE, LTAIN 19139).

This work has also been supported by Charles University Research Centre program No. 204069

References

Agrawal, S.C. (2009). Factors affecting spore germination in algae – review. *Folia Microbiologica*, **54**, 273–302.

AMAP (2011). *Snow, Water, Ice and Permafrost in the Arctic (SWIPA): Climate Change and the Cryosphere*. Arctic Monitoring and Assessment Programme (AMAP), Oslo, 1–538.

Anderson, O.R. (1975). Ultrastructure and cytochemistry of resting cell formation in Amphora coffeaeformis (Bacillariophyceae). *Journal of Phycology*, **11**(3), 272–281.

Arc, E., Pichrtová, M., Kranner, I., Holzinger, A. (2020). Pre-akinete formation in Zygnema sp. from polar habitats is associated with metabolite re-arrangement. *Journal of Experimental Botany*, doi:10.1093/jxb/eraa12

Bashenkhaeva, M.V., Zakharova, Y.R., Petrova, D.P., et al. (2015). Sub-ice microalgal and bacterial communities in freshwater Lake Baikal, Russia. *Microbial Ecology*, **70**(3), 751–765.

Becker, E.W. (1982). Physiological studies on Antarctic Prasiola crispa and Nostoc commune at low temperatures. *Polar Biology*, **1**(2), 99–104.

Bidigare, R.R., Ondrusek, M.E., Kennicutt, M.C., et al. (1993). Evidence for a photoprotective function for secondary carotenoids of snow algae. *Journal of Phycology*, **29**(4), 427–434.

Bisson, M.A., Kirst, G.O. (1995). Osmotic acclimation and turgor pressure regulation in algae. *Naturwissenschaften*, **82**(10), 461–471.

Bjerke, J.W., Karlsen, S.R., Høgda, K.A., et al. (2014). Record-low primary productivity and high plant damage in the Nordic Arctic Region in 2012 caused by multiple weather events and pest outbreaks. *Environmental Research Letters*, **9**, 084006.

Bokhorst, S., Bjerke, J.W., Tømmervik, H., Callaghan, T.V., Phoenix, G.K. (2009). Winter warming events damage sub-Arctic vegetation: consistent evidence from an experimental manipulation and a natural event. *Journal of Ecology*, **97**(6), 1408–1415.

Bokhorst, S., Bjerke, J.W., Street, L., Callaghan, T.V., Phoenix, G.K. (2011). Impacts of multiple extreme winter warming events on sub-Arctic heathland: phenology, reproduction, growth, and CO_2 flux responses. *Global Change Biology*, **17**(9), 2817–2830.

Bokhorst, S., Phoenix, G.K., Berg, M.P., et al. (2015). Climatic and biotic extreme events moderate long-term responses of above- and belowground sub-Arctic heathland communities to climate change. *Global Change Biology*, **21**(11), 4063–4075.

Broady, P.A. (1996). Diversity, distribution and dispersal of Antarctic algae. *Biodiversity & Conservation*, **5**(11), 1307–1335.

Callaghan, T.V., Johansson, M., Brown, R.D., et al. (2011). The changing face of Arctic snow cover: a synthesis of observed and projected changes. *Ambio*, **40**(1), 17–31.

Callaghan, T.V., Jonasson, C., Thierfelder, T., et al. (2013). Ecosystem change and stability over multiple decades in the Swedish subarctic: complex processes and multiple drivers. *Philosophical Transactions of the Royal Society B: Biological Sciences*, **368**(1624), 20120488.

Clarke, A., Morris, G.J., Fonseca, F., et al. (2013). A low temperature limit for life on Earth. *PLoS One*, **8**, e66207.

Crowe, J.H., Oliver, A.E., Tablin, F. (2002). Is there a single biochemical adaptation to anhydrobiosis? *Integrative and Comparative Biology*, **42**(3), 497–503.

Crowe, J.H., Crowe, L.M., Tablin, F., et al. (2004). Stabilization of cells during freeze-drying: the trehalose myth. In: B. J. Fuller, N. Lane, and E. E. Benson (eds) *Life in the Frozen State*. CRC Press, London, pp. 581–602.

Davey, M.C. (1988). Ecology of terrestrial algae of the fellfield ecosystems of Signy Island, South Orkney Islands. *British Antarctic Survey B*, **81**, 69–74.

Davey, M.C. (1989). The effects of freezing and desiccation on photosynthesis and survival of terrestrial antarctic algae and Cyanobacteria. *Polar Biology*, **10**(1), 29–36.

Davey, M.C. (1991). The seasonal periodicity of algae on Antarctic fellfield soils. *Holarctic Ecology*, **14**(2), 112–120.

Davey, M.C., Rothery, P. (1992). Factors causing the limitation of growth of terrestrial algae in maritime Antarctica during late summer. *Polar Biology*, **12**(6–7), 595–601.

de Vries, J., Stanton, A., Archibald, J.M., Gould, S.B. (2016). Streptophyte terrestrialization in light of plastid evolution. *Trends in Plant Science*, **21**(6), 467–476.

de Vries, J., Curtis, B.A., Gould, S.B., Archibald, J. M. (2018). Embryophyte stress signaling evolved in the algal progenitors of land plants. *Proceedings of the National Academy of Sciences of the USA*, **115**(15), E3471–E3480.

Dolev, M.B., Braslavsky, I., Davies, P.L. (2016). Ice-binding proteins and their function. *Annual Review of Biochemistry*, **85**, 515–542.

Domozych, D.S. (2011). Algal cell walls. eLS; https://doi.org/10.1002/9780470015902.a0000315.pub4

Edlund, M.B., Stoermer, E.F., Taylor, C.M. (1996). Aulacoseira skvortzowii sp. nov. (Bacillariophyta), a poorly understood diatom from Lake Baikal, Russia. *Journal of Phycology*, **32**(1), 165–175.

Elster, J. (2002). Ecological classification of terrestrial algae communities of polar environment. In: L. Beyer and M. Bölter (eds) *GeoEcology of Terrestrial Oases, Ecological Studies*, **154**. Springer, Berlin, pp. 303–326.

Elster, J., Benson, E.E. (2004). Life in the polar environment with a focus on algae and Cyanobacteria. In: B. Fuller, N. Lane and E. Benson (eds) *Life in the Frozen State*. Taylor and Francis, London, pp. 109–150.

Elster, J., Komárek, O. (2003). Periphyton ecology of two snow-fed streams in the vicinity of H. Arctowski station, King George Island, South Shetlands, Antarctica. *Antarctic Science*, **15**(2), 189–201.

Elster, J., Svoboda, J., Komárek, J., Marvan, P. (1997). Algal and cyanoprocaryote communities in a glacial stream, Sverdrup Pass, 79° N, Central Ellesmere Island, Canada. *Algological Studies*, **85**, 57–93.

Elster, J., Degma, P., Kováčik, Ľ., et al. (2008). Freezing and desiccation injury resistance in the filamentous green alga Klebsormidium from the Antarctic, Arctic and Slovakia. *Biologia*, **63**(6), 839–847.

Fuller, B.J. (2004). Cryoprotectants: the essential antifreezes to protect life in the frozen state. *CryoLetters*, **25**(6), 375–388.

Fuller, C. (2013). Examining morphological and physiological changes in Zygnema irregulare during a desiccation and recovery period. PhD thesis, California State University San Marcos.

Guschina, I.A., Harwood, J.L. (2006). Lipids and lipid metabolism in eukaryotic algae. *Progress in Lipid Research*, **45**(2), 160–186.

Hawes, I. (1989). Filamentous green algae in freshwater streams on Signy Island, Antarctica. In: W.F. Vincent, J.C. Ellis-Evans (eds) *High Latitude Limnology. Developments in Hydrobiology*, **49**. Kluwer, Dordrecht, pp. 1–18.

Hawes, I. (1990). Effects of freezing and thawing on a species of Zygnema (Chlorophyta) from the Antarctic. *Phycologia*, **29**(3), 326–331.

Hejduková, E., Pinseel, E., Vanormelingen, P., et al. (2019). Tolerance of pennate diatoms (Bacillariophyceae) to experimental freezing: comparison of polar and temperate strains. *Phycologia*, **58**(4), 382–392; doi:10.1080/00318884.2019.1591835.

Hejduková, E., Elster J., Nedbalová, L. (in press). Annual cycle of freshwater diatoms in the High Arctic revealed by multiparameter fluorescent staining. Microbial Ecology.

Herburger, K., Lewis, L.A., Holzinger, A. (2015). Photosynthetic efficiency, desiccation tolerance and ultrastructure in two phylogenetically distinct strains of alpine Zygnema sp. (Zygnematophyceae, Streptophyta): role of pre-akinete formation. *Protoplasma*, **252**(2), 571–589.

Herburger, K., Xin, A., Holzinger, A. (2019). Homogalacturonan accumulation in cell walls of the green alga Zygnema sp. (Charophyta) increases desiccation resistance. *Frontiers in Plant Science*, **10**, 540. doi:10.3389/fpls.2019.00540

Hoham, R.W., Roemer, S.C., Mullet, J.E. (1979). The life history and ecology of the snow alga Chloromonas brevispina comb. nov. (Chlorophyta, Volvocales). *Phycologia*, **18**(1), 55–70.

Hoham, R.W., Mullet, J.E., Roemer, S.C. (1983). The life history and ecology of the snow alga Chloromonas polyptera comb. nov. (Chlorophyta, Volvocales). *Canadian Journal of Botany*, **61**(9), 2416–2428.

Holzinger, A., Pichrtová, M. (2016). Abiotic stress tolerance in charophyte green algae: new challenges for omics techniques. *Frontiers in Plant Science*, **7**(678), 1–17.

Holzinger, A., Roleda, M.Y., Lütz, C. (2009). The vegetative arctic freshwater green alga Zygnema is insensitive to experimental UV exposure. *Micron*, **40**(8), 831–838.

Holzinger, A., Kaplan, F., Blaas, K., et al. (2014). Transcriptomics of desiccation tolerance in the streptophyte green alga Klebsormidium reveal a land plant-like defence reaction. *PLoS ONE*, **9**(10), e110630.

Jackson, A.E., Seppelt, R.D. (1995). The accumulation of proline in Prasiola crispa during winter in Antarctica. *Physiologia Plantarum*, **94**(1), 25–30.

Jacob, A., Wiencke, C., Lehmann, H., Kirst, G.O. (1992). Physiology and ultrastructure of desiccation in the green alga Prasiola crispa from Antarctica. *Botanica Marina*, **35**, 297–303.

Janech, M.G., Mock, T., Kang, J.S., Raymond, J.A. (2006). Ice-binding proteins from sea ice diatoms (Bacillariophyceae). *Journal of Phycology*, **42**(2), 410–416.

Jansson, M., Blomqvist, P., Jonsson, A., Bergström, A.-K. (1996). Nutrient limitation of bacterioplankton, autotrophic and mixotrophic phytoplankton and heterotrophic nanoflagellates in Lake Örträsket. *Limnology and Oceanography*, **41**(7), 1552–1559.

Jewson, D.H., Granin, N.G., Zhdanov, A.A., et al. (2008). Resting stages and ecology of the planktonic diatom Aulacoseira skvortzowii in Lake Baikal. *Limnology and Oceanography*, **53**(3), 1125–1136.

Jones, J. (1996). The diversity, distribution and ecology of diatoms from Antarctic inland waters. *Biodiversity & Conservation*, **5**(11), 1433–1449.

Jones, R.I. (2000). Mixotrophy in planktonic protists: an overview. *Freshwater Biology*, **45**(2), 219–226.

Kaplan, F., Lewis, L.A., Herburger, K., Holzinger, A. (2013). Osmotic stress in Arctic and Antarctic strains of the green alga Zygnema (Zygnematales, Streptophyta): effects on photosynthesis and ultrastructure. *Micron*, **44**, 317–330.

Karsten, U., Lütz, C., Holzinger, A. (2010). Ecophysiological performance of the aeroterrestrial green alga Klebsormidium crenulatum (Klebsormidiophyceae, Streptophyta) isolated from an alpine soil crust with an emphasis on desiccation stress. *Journal of Phycology*, **46**(6), 1187–1197.

Kim, G.H., Klochkova, T.A., Kang, S.H. (2008). Notes on freshwater and terrestrial algae from Ny-Ålesund, Svalbard (High Arctic sea area). *Journal of Environmental Biology*, **29**(4), 485–491.

Klaveness, D., Brte, J., Patil, V., et al. (2011). The 18S and 28S rDNA identity and phylogeny of the common lotic chrysophyte Hydrurus foetidus. *European Journal of Phycology*, **46**(3), 282–291.

Knowles, E.J., Castenholz, R.W. (2008). Effect of exogenous extracellular polysaccharides on the desiccation and freezing tolerance of rock-inhabiting phototrophic microorganisms. *FEMS Microbiology Ecology*, **66**(2), 261–270.

Komárek, J., Nedbalová, L. (2007). Green cryosestic algae. In: J. Seckbach (ed.) *Algae and Cyanobacteria in Extreme Environments*. Springer, Dordrecht, pp. 321–342.

Kosugi, M., Katashima, Y., Aikawa, S., et al. (2010). Comparative study on the photosynthetic properties of Prasiola (Chlorophyceae) and Nostoc (Cyanophyceae) from Antarctic and non-Antarctic sites. *Journal of Phycology*, **46**(3), 466–476.

Kuwata, A., Takahashi, M. (1999). Survival and recovery of resting spores and resting cells of the marine planktonic diatom Chaetoceros pseudocurvisetus under fluctuating nitrate conditions. *Marine Biology*, **134**(3), 471–478.

Kuwata, A., Hama, T., Takahashi, M. (1993). Ecophysiological characterization of two life forms, resting spores and resting cells, of a marine planktonic diatom, Chaetoceros pseudocurvisetus, formed under nutrient depletion. *Marine Ecology Progress Series*, **102**(3), 245–255.

Láska, K., Witoszová, D., Prošek, P. (2012). Weather patterns of the coastal zone of

Petuniabukta, central Spitsbergen in the period 2008–2010. *Polish Polar Research*, **33** (4), 297–318.

Laybourn-Parry, J., Roberts, E.C., Bell, E.M. (2000). Protozoan growth rates in Antarctic lakes. *Polar Biology*, **23**(7), 445–451.

Lembo, V., Bordi, I., Speranza, A. (2017). Annual and semiannual cycles of mid latitude near-surface temperature and tropospheric baroclinicity: reanalysis data and AOGCM simulations. *Earth System Dynamics*, **8**(2), 295–312.

Leya, T. (2013). Snow algae: adaptation strategies to survive on snow and ice. In: J. Seckbach, A. Oren and H. Stan-Lotter (eds) *Polyextremophiles. Life Under Multiple Forms of Stress*. Springer, Dordrecht, pp. 401–423.

Lund, J. W. G. (1954). The seasonal cycle of the plankton diatom, Melosira italica (Ehr.) Kutz. subsp. subarctica O. Müll. *The Journal of Ecology*, **42**(1), 151–179.

Matsuzaki, R., Nozaki, H., Takeuchi, N., Hara, Y., Kawachi, M. (2019). Taxonomic re-examination of 'Chloromonas nivalis (Volvocales, Chlorophyceae) zygotes' from Japan and description of C. muramotoi sp. nov. *PLoS ONE*, **14**(1), e0210986.

Mazur, P. (1984). Freezing of living cells: mechanisms and implications. *American Journal of Physiology–Cell Physiology*, **247**(3), 125–142.

McLean, R.J., Pessoney, G.F. (1971). Formation and resistance of akinetes of Zygnema. In: B. C. Parker, R. M. Brown, Jr. (eds) *Contributions in Phycology*. Allen Press, Lawrence, KS, pp. 145–152.

McQuoid, M.R., Hobson, L.A. (1995). Importance of resting stages in diatom seasonal succession. *Journal of Phycology*, **31**(1), 44–50.

McQuoid, M.R., Hobson L.A. (1996). Diatom resting stages. *Journal of Phycology*, **32**(6), 889–902.

Mock, T., Valentin, K. (2004). Photosynthesis and cold acclimation: molecular evidence from a polar diatom. *Journal of Phycology*, **40** (4), 732–741.

Morgan-Kiss, R.M., Priscu, J.C., Pocock, T., Gudynaite-Savitch, L.G., Huner, N.P.A. (2006). Adaptation and acclimation of photosynthetic microorganisms to permanently cold environments. *Microbiology and Molecular Biology Reviews*, **70**(1), 222–252.

Müller, T., Bleiss, W., Martin, C.D., Rogaschewski, S., Fuhr, G. (1998). Snow algae from northwest Svalbard their identification, distribution, pigment and nutrient content. *Polar Biology*, **20**(1), 14–32.

Nagao, M., Arakawa, K., Takezawa, D., Yoshida, S., Fujikawa, S. (1999). Akinete formation in Tribonema bombycinum Derbes et Solier (Xanthophyceae) in relation to freezing tolerance. *Journal of Plant Research*, **112**, 163–174.

Nagao, M., Matsui, K., Uemura, M. (2008). Klebsormidium flaccidum, a charophycean green alga, exhibits cold acclimation that is closely associated with compatible solute accumulation and ultrastructural changes. *Plant, Cell & Environment*, **31**(3), 872–885.

Pichrtová, M., Hájek, T., Elster, J. (2014a). Osmotic stress and recovery in field populations of Zygnema sp. (Zygnematophyceae, Streptophyta) on Svalbard (High Arctic) subjected to natural desiccation. *FEMS Microbiology Ecology*, **89**(2), 270–280.

Pichrtová, M., Kulichová, J., Holzinger, A. (2014b). Nitrogen limitation and slow drying induce desiccation tolerance in conjugating green algae (Zygnematophyceae, Streptophyta) from polar habitats. *Plos ONE*, **9**, e113137.

Pichrtová, M., Arc, E., Stöggl, W., et al. (2016a). Formation of lipid bodies and changes in fatty acid composition upon pre-akinete formation in Arctic and Antarctic Zygnema (Zygnematophyceae, Streptophyta) strains. *FEMS Microbiology Ecology*, **92**, fiw096.

Pichrtová, M., Hájek, T., Elster, J. (2016b). Annual development of mat-forming conjugating green algae Zygnema spp. in hydroterrestrial habitats in the Arctic. *Polar Biology*, **39**(9), 1653–1662.

Pichrtová, M., Holzinger, A., Kulichová, J., et al. (2018). Molecular and morphological diversity of Zygnema and Zygnemopsis (Zygnematophyceae, Streptophyta) on Svalbard (High Arctic). *European Journal of Phycology*, **53**(4), 492–508.

Post, E., Forchhammer, M.C., Bret-Harte, M.S., et al. (2009). Ecological dynamics across the Arctic associated with recent climate change. *Science*, **325**(5946), 1355–1358.

Poulíčková, A., Žižka, Z., Hašler, P., Benada, O. (2007). Zygnematalean zygospores: morphological features and use in species identification. *Folia Microbiologica*, **52**(2), 135–145.

Procházková, L., Remias, D., Řezanka, T., Nedbalová, L. (2018). Chloromonas nivalis subsp. tatrae, susbp. nov. (Chlamydomonadales, Chlorophyta): re-examination of a snow alga from the High Tatra Mountains (Slovakia). *Fottea*, **18**(1), 1–18.

Raymond, J.A. (2000). Distribution and partial characterization of ice-active molecules associated with sea-ice diatoms. *Polar Biology*, **23**(10), 721–729.

Raymond, J.A. (2014). The ice-binding proteins of a snow alga, Chloromonas brevispina: probable acquisition by horizontal gene transfer. *Extremophiles*, **18**(6), 987–994.

Raymond, J.A., Janech, M.G., Fritsen, C.H. (2009). Novel ice-binding proteins from a psychrophilic Antarctic alga (Chlamydomonadaceae, Chlorophyceae). *Journal of Phycology*, **45**(1), 130–136.

Reed, R.H., Richardson, D.L., Warr, S.R.C., Stewart, W.D.P. (1984). Carbohydrate accumulation and osmotic stress in cyanobacteria. *Microbiology*, **130**, 1–4.

Remias, D. (2012). Cell structure and physiology of alpine snow and ice algae. In: C. Lütz (ed.) *Plants in Alpine Regions*. Springer, Wien, pp. 175–185.

Remias, D., Lütz-Meindl, U., Lütz, C. (2005). Photosynthesis, pigments and ultrastructure of the alpine snow alga

Chlamydomonas nivalis. *European Journal of Phycology*, **40**(3), 259–268.

Remias, D., Karsten, U., Lütz, C. (2010). Physiological and morphological processes in the Alpine snow alga Chloromonas nivalis (Chlorophyceae) during cyst formation. *Protoplasma*, **243**(1–4), 73–86.

Remias, D., Holzinger, A., Aigner, S., Lütz, C. (2011). Ecophysiology and ultrastructure of Ancylonema nordenskiöldii (Zygnematales, Streptophyta), causing brown ice on glaciers in Svalbard (high arctic). *Polar Biology*, **35**(6), 899–908.

Remias, D., Wastien, H., Lütz, C., Leya, T. (2013). Insights into the biology and phylogeny of Chloromonas polyptera (Chlorophyta), an alga causing orange snow in Maritime Antarctica. *Antarctic Science*, **25**(5), 648–656.

Rengefors, K., Karlsson, I., Hansson, L.-A. (1998). Algal cyst dormancy: a temporal escape from herbivory. *Proceedings of the Royal Society of London Series B: Biological Sciences*, **265**, 1353–1358.

Rippin, M., Becker, B., Holzinger, A. (2017). Enhanced desiccation tolerance in mature cultures of the Streptophytic green alga Zygnema circumcarinatum revealed by transcriptomics. *Plant & Cell Physiology*, **58**(12), 2067–2084.

Robinson, S.A., Wasley, J., Tobin, A.K. (2003). Living on the edge – plants and global change in continental and maritime Antarctica. *Global Change Biology*, **9**(12), 1681–1717.

Roos, J.C., Vincent, W.F. (1998). Temperature dependence of UV radiation effects on Antarctic cyanobacteria. *Journal of Phycology*, **34**(1), 118–125.

Round, F.E., Crawford, R.M., Mann, D.G. (1990). *Diatoms: Biology and Morphology of the Genera*. Cambridge University Press, Cambridge, UK.

Ryšánek, D., Elster, J., Kováčik, L., Škaloud, P. (2016). Diversity and dispersal capacities of a terrestrial algal genus Klebsormidium (Streptophyta) in polar regions. *FEMS Microbiology Ecology*, **92**(4), fiw039.

Řezanka, T., Nedbalová, L., Sigler, K., Cepák, V. (2008). Identification of astaxanthin diglucoside diesters from snow alga Chlamydomonas nivalis by liquid chromatography-atmospheric pressure chemical ionization mass spectrometry. *Phytochemistry*, **69**(2), 479–490.

Řezanka, T., Nedbalová, L., Lukavský, J., Střížek, A., Sigler, K. (2017). Pilot cultivation of the green alga Monoraphidium sp. producing a high content of polyunsaturated fatty acids in low-temperature environment. *Algal Research*, **22**, 160–165.

Šabacká, M., Elster, J. (2006). Response of cyanobacteria and algae from Antarctic wetland habitats to freezing and desiccation stress. *Polar Biology*, **30**(1), 31–37.

Saha, S.K., Rinke, A., Dethloff, K. (2006). Future winter extreme temperature and precipitation events in the Arctic. *Geophysical Research Letters*, **33**(15), L15818.

Seaburg, K.G., Parker, B.C., Wharton, Jr., R.A., Simmons Jr., G.M. (1981). Temperature-growth responses of algal isolates from Antarctic oases. *Journal of Phycology*, **17**(4), 353–360.

Sheath, R.G., Vis, M.L., Hambrook, J.A., Cole, K. M. (1996). Tundra stream macroalgae of North America: composition, distribution and physiological adaptations. *Hydrobiologia*, **336**, 67–82.

Sicko-Goad, L., Stoermer, E.F., Fahnenstiel, G. (1986). Rejuvenation of Melosira granulata (Bacillariophyceae) resting cells from the anoxic sediments of Douglas Lake, Michigan. I. Light microscopy and ^{14}C uptake. *Journal of Phycology*, **22**(1), 22–28.

Siegele, D.A., Kolter, R. (1992). Life after log. *Journal of Bacteriology*, **174**(2), 345–348.

Skácelová, K., Barták, M., Coufalík, P., Nývlt, D., Trnková, K. (2013). Biodiversity of freshwater algae and Cyanobacteria on deglaciated northern part of James RossIsland, Antarctica. A preliminary study. *Czech Polar Reports*, **3**, 93–106.

Souffreau, C., Vanormelingen, P., Verleyen, E., Sabbe, K., Vyverman, W. (2010). Tolerance of benthic diatoms from temperate aquatic and terrestrial habitats to experimental desiccation and temperature stress. *Phycologia*, **49**(4), 309–324.

Souffreau, C., Vanormelingen, P., Sabbe, K., Vyverman, W. (2013). Tolerance of resting cells of freshwater and terrestrial benthic diatoms to experimental desiccation and freezing is habitat-dependent. *Phycologia*, **52**(3), 246–255.

Spijkerman, E., Wacker, A., Weithoff, G., Leya, T. (2012). Elemental and fatty acid composition of snow algae in Arctic habitats. *Frontiers in Microbiology*, **3**, 380.

Stibal, M., Elster, J., Šabacká, M., Kaštovská, K. (2007). Seasonal and diel changes in photosynthetic activity of the snow alga Chlamydomonas nivalis (Chlorophyceae) from Svalbard determined by pulse amplitude modulation fluorometry. *FEMS Microbiology Ecology*, **59**(2), 265–273.

Svoboda, J. (2009). Evolution of plant cold hardiness and its manifestation along the latitudinal gradient in the Canadian High Arctic. In: L. Gusta, M. Wisniewski, K. Tanino (eds.) *Plant Cold Hardiness: From the Laboratory to the Field*. CAB International, pp. 140–162.

Tamaru, Y., Takani, Y., Yoshida, T., Sakamoto, T. (2005). Crucial role of extracellular polysaccharides in desiccation and freezing tolerance in the terrestrial cyanobacterium Nostoc commune. *Applied and Environmental Microbiology*, **71**, 7327–7333.

Tanghe, A., van Dijck, P., Thevelein, J.M. (2003). Determinants of freeze tolerance in microorganisms, physiological importance, and biotechnological applications. *Advances in Applied Microbiology*, **53**, 129–176.

Teoh, M.-L., Chu, W.-L., Marchant, H., Phang, S.-M. (2004). Influence of culture temperature on the growth, biochemical composition and fatty acid profiles of six Antarctic microalgae. *Journal of Applied Phycology*, **16**(6), 421–430.

Thomas, D.N., Fogg, G.E., Convey, P., et al. (2008). *The Biology of Polar Regions*. Oxford University Press, Oxford.

Trumhová, K., Holzinger, A., Obwegeser, S., Neuner, G., Pichrtová, M. (2019). The conjugating green alga Zygnema sp. (Zygnematophyceae) from the Arctic shows high frost tolerance in mature cells (pre-akinetes). *Protoplasma*, **256**, 1681–1694.

Van De Vijver, B., Beyens, L. (1999). Biogeography and ecology of freshwater diatoms in Subantarctica: a review. *Journal of Biogeography*, **26**(5), 993–1000.

Van de Vijver, B., Frenot, Y., Beyens, L. (2002). *Freshwater Diatoms from Ile de la Possession (Crozet-Archipelago, Subantarctica)*. Bibliotheca Diatomologica, **46**. J. Cramer, Berlin/Stuttgart.

Vincent, W.F. (1988). *Microbial Ecosystem of Antarctica*. Cambridge University Press, Cambridge, UK.

Vincent, W.F., James, M.R. (1996). Biodiversity in extreme aquatic environments: lakes, ponds and streams of the Ross Sea Sector, Antarctica. *Biodiversity & Conservation*, **5**(11), 1451–1471.

Vítová, M., Bišová, K., Kawano, S., Zachleder, V. (2014). Accumulation of energy reserves in algae: From cell cycles to biotechnological applications. *Biotechnological Advances*, **3**, 1204–1218.

Vrionis, H.A., Miller, R.V., Whyte, L.G. (2013). Life at the poles in the age of global warming: Part 1. *Microbe*, **8**(11), 449–453.

Welsh, D.T. (2000). Ecological significance of compatible solute accumulation by microorganisms: from single cells to global climate. *FEMS Microbiology Reviews*, **24**(3), 263–290.

Worland, M.R., Lukešová, A. (2001). The application of differential scanning calorimetry and ice nucleation spectrometry to ecophysiological studies of algae. *Nova Hedwigia Beiheft*, **123**, 571–583.

Yancey, P.H. (2005). Organic osmolytes as compatible, metabolic and counteracting cytoprotectants in high osmolarity and other stresses. *Journal of Experimental Biology*, **208**, 2819–2830.

Youssouf, M.O., Laurent, M., Xavier, C. (2016). Statistical analysis of sea surface temperature and chlorophyll-a concentration patterns in the Gulf of Tadjourah (Djibouti). *Journal of Marine Science: Research & Development*, **6**(2), 1–9.

CHAPTER SIX

Vertebrate viruses in polar ecosystems

JIŘÍ ČERNÝ[1]
Czech Academy of Sciences
JANA ELSTEROVÁ
Czech Academy of Sciences
DANIEL RŮŽEK
Czech Academy of Sciences

a n d

LIBOR GRUBHOFFER
Czech Academy of Sciences

6.1 Viruses in polar ecosystems

Viruses are non-cellular living entities. Their particles are formed by a nucleic acid surrounded by a protein capsid, which together form the so-called nucleocapsid. Nucleocapsids of some virus species are additionally enveloped by a host-cell-derived lipid envelope containing viral proteins. All three structural components (nucleic acid, capsid and envelope) are incredibly variable; for example, the viral genome can be encoded by either an RNA or DNA molecule, which can be single or double stranded, linear or circular, unsegmented or segmented, etc. (Fields et al., 2007).

Viruses are key players in global biosystems. It has been estimated that about 3.10^{30} viral particles exist in the world oceans (Suttle, 2005). Such a high abundance makes viruses key players in biogeochemical cycles of carbon and other organic elements (Zhang et al., 2014), and they play a key role in the emergence of novel genes and their transmission between different organisms (Claverie, 2006). Although viruses are known to be the infectious agents causing many diseases, numerous studies have shown that many viruses have no effect on the host and some can even be beneficial (Márquez et al., 2007; Jaenike, 2012; Kernbauer et al., 2014).

Their strictly intracellular parasitic mode of life means viruses are not dependent on external resources and conditions, as conditions within the

---

[1] Jiří Černý and Jana Elsterová contributed equally to this work.

host cells are relatively stable. This is true especially for viruses infecting homoeothermic animals, which provide the virus with energy sources, liquid water and a stable temperature. On the other hand, viruses infecting poiki-lothermic organisms in polar regions have to be adapted to the extremely low temperatures, inaccessibility of liquid water and other limitations. Viruses face this problem using proteins optimised for such extreme conditions. For example, proteins of bacteriophages infecting psychrophilic bacteria have a typical amino acid composition that allows them to keep their enzymatic activity at low temperatures (Borriss et al., 2003; Colangelo-Lillis & Deming, 2013). Activity of such proteins is lower at higher temperatures, which is thought to be the main reason why there are fewer pathogenic microorganisms in polar regions compared to tropical ecosystems.

So what are the main drivers of virus biology in polar regions, and what affects the virus's ability to be introduced and become endemic there?

First, it is the presence of suitable permissive hosts. Polar ecosystems are usually characterised by relatively low variability of species living in these areas. During the polar winter many of the organisms either migrate to the lower latitudes or are dormant, which limits possible spread of viruses. Some terrestrial organisms, mainly mammals and birds, and marine animals remain active (Blix, 2016) and therefore circulation of viruses infecting these animals (e.g. rabies virus, influenza virus, etc.) is possible even during the winter period, but rigorous data are missing. Nevertheless, during the summer period, high productivity of polar ecosystems allows presence of relatively high numbers of potential hosts. These are either organisms which survive in polar regions year-round and use the optimal conditions for exponential growth (e.g. bacteria) or organisms which migrate to polar regions only for summer from lower latitudes (e.g. migratory birds). The first group of organisms is usually colonised by viruses well adapted to their hosts and to the polar environment (Rastrojo & Alcamí, 2018). The second type of organism can serve as an efficient vector introducing new viruses into the polar regions, which can be pathogenic for local wildlife or can cumulate their effect with additional stressors such as malnutrition, environmental pollution, etc.

The second main condition allowing successful spread of viruses in polar regions is the ability of virus to be transmitted from one host to another. This is influenced by at least three important factors.

First, the host population has to be dense enough to provide sufficient contacts between infected and permissive hosts. Viruses infecting animals living in dense populations usually cause rapid spread of infection producing high numbers of new viral particles in a short time (Kaaden et al., 2002). Viruses infecting solitarily living organisms generally produce long-lasting chronic infections with usually mild but possibly progressive disease. This allows prolonged virus shedding into the environment and increases its chances of infecting a new host (Kaaden et al., 2002).

Second, the virus has to be stable enough to survive in the external environment during transmission from one host to another, which is the only point at which viruses have to face external conditions. Compared to bacteria, most viruses are rather unstable in an external environment. They can be inactivated by numerous physical, chemical and biological agents, such as UV radiation or biological degradation by bacteria (Mateu, 2013). Therefore, numerous viruses (such as HIV, rabies virus (RABV), etc.) need extremely close contact between the infected and permissive host (e.g. sexual contact, or contact of infectious saliva with new host blood) (Fargeaud et al., 1982; Tjøtta et al., 1991). On the other hand, some viruses (especially those which are transmitted by air, contaminated food, water or contaminated objects – fomites) can survive in the external environment for up to several months (Goddard & Leisewitz, 2010). Low temperatures are usually beneficial for viral particles as they protect them against chemical and biological degradation. Some viruses can even freeze without any problems, if their structural integrity is not disrupted by ice crystals. Natural cryopreservation was observed in the influenza virus. It was speculated that it may help the virus to survive long periods when an adequate number of hosts are not available, but its true role in the ecology of the influenza virus is disputable (Zhang et al., 2006).

Third, some viruses are dependent on vectors during the transmission (e.g. bloodsucking arthropods in the case of arboviruses, or rodents in the case of hantaviruses). Global climate change allows invasion of polar regions by new vector species, which also increases the risk of introduction of novel viral diseases (Culler et al., 2015; Descamps et al., 2017). A nice example is the spread of tick-borne encephalitis virus northward in Scandinavia together with its tick vector *Ixodes ricinus* (Soleng et al., 2018).

Although there are not many studies on animal viruses in polar regions (especially when compared with the number of studies on viruses from temperate and tropical zones), we can list several important reasons why animal viruses in these regions should be studied:

1. Viruses may have a serious impact on the local wildlife. This is especially true for species being stressed by other stressors (e.g. pollution, rapid changes in ecosystem), living on the borders of their ecological limits or in dense colonies. For example, phocine distemper virus (PDV) is known to cause massive mortalities in seal populations (Härkönen et al., 2006; Duignan et al., 2014). Antibodies against this virus were found also in numerous arctic and antarctic seal species (Cattet et al., 2004; Tryland et al., 2012). In 1955, Laws and Taylor reported a massive die-off in crabeater seals in the Crown Prince Gustav Channel, which was caused by

undetermined infectious disease (probably of viral origin), a description of which fits PDV (Laws & Taylor, 1957).

2. Viruses can affect domestic animals kept by humans in the Arctic. For example, salmonid alphavirus (SAV) is an important cause of death in commercially kept salmon in sub-arctic aquacultures (Jansen et al., 2017). As other examples we can list infections by foot-and-mouth disease virus, border disease virus, bovine viral diarrhoea virus and many other viruses in semidomestic populations of reindeer (das Neves et al., 2010; Larska, 2015) or rabies virus (RABV), canine distemper virus (CDV) and canine parvovirus (CPV) infections in sledge dogs (Bohm et al., 1989; Mørk & Prestrud, 2004; Gagnon et al., 2016).

3. Viruses endemic in polar regions can also pose a threat to humans. Such infections are currently rare but their probability will increase with increasing number of people living or travelling to polar regions (Bruce et al., 2016).

Moreover, polar regions (especially the Arctic) are prone to introduction of novel organisms (including viruses). Introduction of novel pathogens in an immunologically naïve population may lead to massive outbreaks, which could pose a significant threat to polar wildlife, local agri- and aquaculture, and even human health (Chan et al., 2019). A good example is the massive epidemic of the human immunodeficiency virus (HIV) in Greenland and Russia, which can be compared to the situation in those countries most affected by HIV in Africa (Bjorn-Mortensen et al., 2013). Another classic example was the introduction of the measles virus to the Faroes in 1846 after more than 60 years followed by infection of the complete island population (Panum, 2018).

6.2 Examples of viruses infecting different hosts in polar areas
6.2.1 Viruses of microorganisms and plants
Since life in polar areas is quite limited due to the extreme conditions, it tends to be dominated by microorganisms, mainly bacteria and algae. This correlates with the most abundant type of viruses in polar areas, i.e. bacteriophages. These viruses play a critical role since they affect respiration, diversity, genetic transfer, nutrient cycling, abundance and size distribution of microbial flora because phage-induced lysis returns the organic compounds to the environment where they can be used for microbial viability and physiological activity of microbiota. Detailed description of bacteriophages and their role in polar ecosystems is beyond the scope of this review; therefore, the reader is referred to some of the recent excellent reviews focusing on this problem (Säwström et al., 2008; Cavicchioli, 2015; Rastrojo & Alcamí, 2018).

Only very limited information is available about plant (Robertson, 2007; Robertson et al., 2007; Robertson & French, 2007; Maat et al., 2017), protist, fungal (Sasai et al., 2018) and invertebrate viruses in polar regions. Usually, these viral infections are connected with mild to severe disease in infected organisms, but further metagenomics studies will most probably also identify viruses that could cause asymptomatic infections.

6.2.2 Viruses of wild living, semidomestic and domestic animals

Invertebrates, especially arthropods, carry a huge diversity of RNA viruses, including potential ancestors of many vertebrate-infecting viruses (Shi et al., 2016). Although we can expect the presence of such viruses in the Arctic, only two phasmavirids have been isolated from tundra phantom midges (*Chaoborus*) in North America (Ballinger et al., 2014, 2017).

Moreover, bloodsucking arthropods such as mosquitoes, midges, lice and ticks can serve also as vectors of many vertebrate viruses (so-called arthropod-borne viruses or arboviruses). The abundance, geographical distribution and species richness of these arthropod vectors is rapidly growing in the Arctic due to the climate change (Descamps, 2013; Culler et al., 2015). Arboviruses have been reported on all continents except continental Antarctica, including all climatic zones up to 70°N in the Arctic (McLean et al., 1973; Traavik et al., 1978) and 54°S in the sub-antarctic islands (La Linn et al., 2001; Major et al., 2009).

Arboviruses in polar areas are transmitted mostly by mosquitoes or ticks, but louse-borne viruses have been observed as well (La Linn et al., 2001). The tick-borne arboviruses could be vectored by one of many tick species parasitising mostly on birds and mammals. The most intensively studied tick is *Ixodes uriae*, which is present in both the Arctic and Antarctica. This tick species was shown to transmit many arboviruses, as well as other pathogens such as *Borrelia* spirochetes. It has circumpolar distribution and exhibits the greatest range of thermal tolerance, from −30 to 40°C (Lee & Baust, 1987). Nevertheless, genetic analyses show that the *I. uriae*'s northern and southern populations do not interbreed, which may explain also differences in viruses carried by this tick as a vector in the Arctic and Antarctica (Kempf et al., 2009; Dietrich et al., 2011). Other tick species are present in the Arctic Circle region. *Ixodes ricinus* is the vector of tick-borne encephalitis virus and is present in the sub-arctic part of Scandinavia, though in limited numbers (Soleng et al., 2018). Several non-indigenous ticks have established populations in Alaska. Most of the species are associated with avian hosts (Durden et al., 2016). Numerous mosquito species from the genera *Culex* and *Aedes* transmit arboviruses in the Arctic, while no mosquitoes are present in Antarctica and Iceland. Nevertheless, another bloodsucking insect species is present in Iceland. Icelandic black midge (*Simulium vittatum*, locally called bitmý) is known to be able to vector livestock-infecting vesicular stomatitis virus (Cupp et al., 1992).

Novel arthropods, which could potentially transmit new arboviruses, are currently being introduced into the polar areas. In the continental Arctic, penetrance of new arthropod vectors can be continuous. In Antarctica or the arctic islands, they need some transport vehicles, mostly birds, to colonise new regions, but humans and domestic animals can serve as transport vehicles as well (Durden et al., 2016). As birds migrating to the polar areas trade-off the energetic demands of the migration against a lesser requirement for immunocompetence (Piersma, 1997; Piersma et al., 2001), they may also be more susceptible to the arthropod-borne pathogens. On the other hand, southward range extension was observed in the arctic mosquito species *Ochlerotatus (Aedes) nigripes*. Nowadays, it occurs more southerly than previously observed, sharing the habitat with other mosquito species vectoring important pathogens in north Ontario, Canada (Beresford, 2011; Ringrose et al., 2013), which opens the gate for possible introduction of these arboviruses to arctic mosquito species.

As many (but not all) arboviruses are potential zoonotic pathogens, we will describe examples of individual polar arboviruses in more detail in Section 6.2.3.

Several fish viruses have been described in polar regions. These viruses could have a large impact on aquaculture, such as SAV that affects farmed Atlantic salmon (*Salmo salar*). SAV was repeatedly detected in aquacultures in the northern Atlantic from Scotland and Ireland to Norway, even as far as the polar circle and beyond (Jansen et al., 2017). Various viruses were detected also in free-living fishes such as *Cyclopterus lumpus* virus detected in northern Atlantic (Skoge et al., 2018) and *Trematomus pennellii* polyomavirus detected close to McMurdo, Antarctica (Buck et al., 2016).

The migratory bird species play a great role in introducing novel virus species into the polar regions, which can then interact with other animal species. As many different migratory flyways meet in the polar region, there is an increased risk of meeting different virus strains there, which could recombine or reassert, leading to the emergence of novel, potentially more virulent variants (Ramey et al., 2016). Understanding of virus diversity in wildlife provides epidemiological and ecological information that can lead to identification of future microbial threats, not only in polar regions but globally. Together with other stresses to the polar animals, such as environmental pollution, decreased food availability and changing climate, the pathogens contribute to changes in structure and behaviour of the animal populations (Andersen et al., 2017).

The majority of bird species living in the Arctic migrate there for their breeding season. On the other hand, Antarctica with its Antarctic Polar Front, a well-known faunal barrier, represents a relatively closed ecosystem, which only a few species can migrate to. However, north of the Antarctic Circumpolar Current the fauna migration is more diverse, fluctuating from temperate zones to sub-antarctic and antarctic zones (Smeele et al., 2018).

Polar avifauna is affected by diverse viruses. The known viruses which circulate in birds of the Arctic and Antarctica are avian influenza virus, paramyxoviruses, adenoviruses, poxviruses and birnaviruses (Alexander et al., 1989; Gardner et al., 1997; Shearn-Bochsler et al., 2008; Thomazelli et al., 2010; Park et al., 2012; Hurt et al., 2014; Hartby et al., 2016; Hurt et al., 2016; Gaidet et al., 2018; Smeele et al., 2018). Many of them are transmissible to mammals and even to humans where the birds play the role of primordial reservoirs (Austin & Webster, 1993).

It is important to mention influenza virus. The ecology and migratory patterns of the avian hosts have a direct effect on distribution and diversity of avian influenza viruses. Spatially segregated aquatic birds with their different intercontinental flyways harbour distinct lineages of avian influenza viruses. During summer bird migration to the Arctic, where numerous migratory flyways cross, different otherwise spatially distinct avian influenza virus strains can meet and reassert together to produce novel strains (Hill & Runstadler, 2016). For the polar areas, avian influenza viruses of low pathogenicity are typical of bird species such as ducks, gulls, terns and waders, which are the natural reservoirs of these viruses (Olsen et al., 2006; Hurt et al., 2014). The avian influenza virus also circulates in penguins (Austin & Webster, 1993; Hurt et al., 2014; Hurt et al., 2016). Introduction of avian influenza infection to the penguins of Antarctica may be via a migrating bird, the snowy sheathbill, which has a close interaction with penguin colonies and may facilitate the movement of avian influenza virus between the mainland of South America and Antarctica (Hurt et al., 2016).

Influenza virus has also been found in marine mammals, such as seals and whales. It was shown that influenza virus caused severe pneumonia in harbour seals (Geraci et al., 1982; Krog et al., 2015). Anti-influenza antibodies were found in ringed seal in the Canadian Arctic (Nielsen et al., 2001). It is important to note that laboratory infection of seals by the influenza isolate did not cause as severe a course of infection as observed in nature, which suggests an important role for the environment and co-infections for the course of the disease caused by influenza virus (Webster et al., 1981; Callan et al., 1995). Anti-influenza antibodies were also detected in beluga whales but not in narwals in arctic Canada (Nielsen et al., 2001). Influenza virus in marine mammals is usually closely related to the influenza virus currently circulating in seabirds in the same area. Therefore, marine mammals are thought to be the dead-end host, being just occasionally infected by seabirds (Groth et al., 2014). Avian influenza antibodies were also found in arctic foxes but not in polar bears and reindeers (Van Hemert et al., 2019). Fortunately, zoonotic potential of avian influenza viruses circulating in the Arctic is low, and it seems that these strains do not cause human infections (Liberda et al., 2017), despite the ability of seal influenza virus H3N8 to infect human tissues under laboratory conditions (Hussein et al., 2016).

Apart from influenza virus, other viral infections affect marine mammals, such as those caused by viruses from the families *Anelloviridae, Poxviridae, Polyomaviridae* and *Togaviridae*, which have been found in Weddell seals (*Leptonychotes weddellii*) and southern elephant seals (*Mirounga leonina*) in Antarctica (La Linn et al., 2001; Tryland et al., 2005; Fahsbender et al., 2017; Varsani et al., 2017).

Viral infections can affect reindeer populations, which play a major role in agriculture in the Arctic, as well as other large ungulates such as musk ox. For example, Cervid herpesvirus 2 and Orf virus was detected in reindeer and musk ox populations in the North American Arctic as well as in the Eurasian Arctic (das Neves et al., 2010; Pirisinu et al., 2018). Both of these viruses can be implicated in the relatively high mortality of calves (das Neves et al., 2010). Antibodies against pestiviruses were found in reindeers in Scandinavia, where pestiviruses were eradicated in domestic cattle, showing semidomestic reindeers as possible reservoirs for this virus (Kautto et al., 2012; Larska, 2015).

Terrestrial carnivores living in the northern hemisphere are frequent hosts of many viruses. As apex predators, they affect the whole ecosystem of arctic fauna; therefore, pathogens affecting their fitness also have an impact on all other arctic species.

The current concern over the future of polar bear (*Ursus maritimus*) populations brings also the question of diseases circulating in the wildlife of the arctic ecosystem. Immunosuppression (Kirk et al., 2010) and shifts in food sources together with altered behaviour and increasing contact with human settlements puts polar bears at greater risk for viral infections. Wild-living polar bears can be infected by morbilliviruses (Follmann et al., 1996; Philippa et al., 2004; Tryland et al., 2005; Kirk et al., 2010), rhabdoviruses (Loewen et al., 1990), adenoviruses (Philippa et al., 2004) and caliciviruses (Tryland et al., 2005), but epidemiological data shows that these infections have only minor effects on population fitness (Fagre et al., 2015). The same morbillivirus, phocine distemper virus (PDV), that infects polar bears also infects seals (Heide-Joergensen et al., 1992; Thompson et al., 2002) and walruses (Duignan et al., 1994), the main prey of polar bears and reservoir for the virus. However, it is possible that the polar bear is infected by a morbillivirus that is unique to its population, since the antibodies are more reactive to canine distemper virus than to phocine distemper virus, which is widely spread in pinnipeds (Follmann et al., 1996).

Another virus from the genus *Morbillivirus*, canine distemper virus (CDV), also circulates in arctic carnivores, mainly in foxes and wolves. CDV antibodies have also been found in polar bears (Cattet et al., 2004). The virus is related to several distinct lineages. The arctic lineage was first identified in arctic canids but is present throughout Europe and North America in domestic dogs, and it has also been isolated from severely diseased wolves and other canids in the Italian Apennines (Di Sabatino et al., 2014, 2016). Moreover, arctic foxes (together with red foxes and wolves in Scandinavia and Svalbard) are also carriers of canine adenovirus (CAV). Both CDV and CAV are enzootic in the arctic fox population of Svalbard (Akerstedt et al., 2010; Balboni et al., 2019; Tryland et al., 2018). Interestingly, there are no comparative studies for seroprevalence of either CDV or CAV in the arctic fox populations of Canada and Alaska, although a recent publication describes lipid pneumonia in association with morbillivirus infection in these regions (Stimmelmayr et al., 2018). Nevertheless, even though phylogenic analysis of the hemagglutinin gene indicates that this virus strain is related to arctic CDV lineage, phylogenetic analysis of the phosphoprotein gene revealed no relationship to any other previously described CDV (Stimmelmayr et al., 2018).

6.2.3 Zoonotic viruses

Rhabdoviruses, namely rabies virus, are endemic in most parts of the Arctic. The main reservoir of the rabies virus is the arctic fox, although it also occurs in wolves in Alaska (Weiler et al., 1995; Ballard & Krausman, 1997). Nevertheless, rabies virus has a wide spectrum of potential hosts. In the Arctic, infections in polar bears, seals, reindeers and sledge dogs have been reported (Odegaard & Krogsrud, 1981; Loewen et al., 1990; Taylor et al., 1991; Macdonald et al., 2011). There are several examples of the rabies virus spreading southwards from arctic regions, and for that reason arctic rabies could serve as a permanent potential source of infection for sub-arctic areas (Mørk & Prestrud, 2004). Hence, antirabies vaccination should be recommended for all people in the Arctic who would come into contact with wild living animals (especially researchers) (Macdonald et al., 2011; Orpetveit et al., 2011).

Studies in natural host animal and native human populations show, for example, circulation of Sindbis, Inkoo, Uukuniemi, Tyuleniy, Zaliv Terpeniya, Soldado and Great Island viruses in European sub-arctic areas (Hubálek & Rudolf, 2012), although no arboviruses have been found in the European High Arctic yet (Elsterova et al., 2015; Müllerová et al., 2018). Data from the North American Arctic show the presence of several pathogens circulating in arthropod vectors and their natural hosts in north Canada and Alaska, such as Snowshoe hare viruses (SSHV), the James Canyon virus (JCV) and the Cache Valley virus (CVV) (Corbet & Downe, 1966; Deardorff et al., 2013; Carson et al., 2017).

Medically, the most important arboviruses endemic in polar regions include: Sindbis virus (SINV) and orthobunyaviruses from the California encephalitis virus serogroup such as Inkoo virus (INKV) in arctic Eurasia and SSHV in arctic North America. Sindbis virus is endemic in many countries of the Old World, causing a skin rash (Adouchief et al., 2016). In northern Scandinavia and Karelia, SINV was described as a causative agent of Pogosta disease (Finland), Ockelbo disease (Sweden) or Karelian fever (Russia) (Skogh & Espmark, 1982; Julkunen et al., 1986; Lvov et al., 1988). INKV is endemic in northern Europe. Its seroprevalence can reach up to 50% in some populations (Evander et al., 2016). INKV is associated with febrile diseases and in several cases also with encephalitis (Putkuri et al., 2016). Similar symptoms (from febrile illnesses to encephalitis) are caused by orthobunyaviruses endemic in North America (Drebot, 2015).

Due to its zoogeographical isolation, arthropod-borne viruses are less frequent in Antarctica. Nevertheless antibodies against flaviviruses have been found in polar skuas in coastal Antarctica (Miller et al., 2008), and arboviruses from the genera of *Flavivirus, Orbivirus, Phlebovirus* and *Nairovirus* were detected on the Macquarie island (Major et al., 2009). In this case, individual viruses are associated with different penguin species. Nevertheless, the relatively large phylogenetic distance of these arboviruses from known human pathogens indicates that their zoonotic potential would be very low (Major et al., 2009).

Puumala virus (PUUV) is a hantavirus endemic in northern Europe including Scandinavia, Finland and Karelia (Plyusnin et al., 1997; Razzauti et al., 2009). It is transmitted by the bank vole (*Myodes glareolus*) (Dubois et al., 2017). Humans infected by PUUV may develop a haemorrhagic fever with renal syndrome (Mustonen et al., 2017). Other hantaviruses such as Topografov virus are circulating in the Russian Arctic, being transmitted by Siberian lemmings (*Lemmus sibericus*) (Plyusnin et al., 1996; Vapalahti et al., 1999).

Examples of vertebrate viruses and zoonotic viruses potentially infecting humans which are endemic in polar regions are shown in Figure 6.1. Apart from the above-mentioned viruses, there are numerous other viruses that have been isolated in polar regions, but due to limited space, it is not possible to list them all here. Those given in Figure 6.1 are just some examples of the most important polar viruses.

6.2.4 Human-only infecting viruses

The number of infectious diseases which are endemic only in the Arctic is limited. One of the most interesting examples is Viliuisk encephalomyelitis (VE), which is an infectious disease with a very long latent stage and progressive development (McLean et al., 1997). It is typically found in the native Yakut-Evenk population in eastern Siberia (Lee et al., 2010), but it can infect Caucasians as well . Before it had been fully described, VE had multiplied over an area several times the original size. The infectious agent causing this disease is not known, but the best candidate for it is an unknown virus or prion.

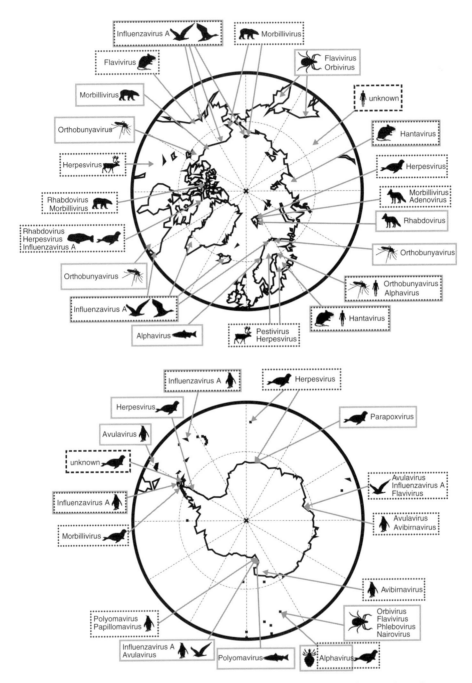

Figure 6.1 Examples of animal and zoonotic viruses endemic in the Arctic and Antarctica. Virus genera are listed next to the maps. The silhouettes indicate virus hosts. Hosts in which the virus was directly detected (e.g. by nucleic acid sequencing, virus isolation, electron microscopy, etc.) are depicted by the dashed frame. If only antivirus antibodies were detected, the frame is dotted. The figure is illustrative only and does not aim to show all viruses that have been detected in polar regions so far.

Another threat to humans in the Arctic is posed by the sexually transmitted viral diseases such as HIV and hepatitis B virus (HBV). HBV is endemic among the native populations of Greenland, Canada, Alaska and Russia (Rex et al., 2015). The genotypes B and D predominate among the circumpolar arctic populations, though other genotypes, namely A, C and F, are also prevalent. The HBV genotypes are separated geographically and also by ethnicity. Genotype B is dominant in the eastern regions of Greenland, while genotype D is found in the western part of Greenland (Rex et al., 2015). A similar distribution is found in Canada, where the B genotype prevails in the eastern far northern part of Canada and the D genotype dominates in western regions of the Canadian Arctic (Osiowy et al., 2013). The Inuit of Greenland have the highest prevalence of HBV infection among the arctic population (Børresen et al., 2015; Rex et al., 2015; Ching et al., 2016). The virus is classified as subgenotype B5 (formerly B6), which is unique to indigenous populations of Inuit and First Nations in Greenland and Alaska (Sakamoto et al., 2007; Kramvis, 2014; Bouckaert et al., 2017). This subgenotype has distinct clinical outcomes and distribution patterns, since unlike other genotypes, the geno-type B5 (B6) shows a lack of active liver disease with no biochemical and functional alterations. Furthermore, there have not been any registered cases of hepatocellular carcinoma in patients suffering from the HBV B5 (B6) genotype (Krarup et al., 2008). These findings suggest potential pathogen attenuation within the population due to host–pathogen coevolution (Bouckaert et al., 2017). Contrary to these findings, a relatively high incidence of hepatocellular carcinoma was observed in association with the F1 HBV strain among Inuits in Alaska (Gounder et al., 2016).

HIV circulating in the native population of the Arctic is mostly transmitted by heterosexual contact (Bjorn-Mortensen et al., 2013; Walker et al., 2015; Kumar, 2016; Beyrer et al., 2017). Information about HIV prevalence in the native population of Siberia is sparse, but as HIV prevalence in Russia as a whole is very high, the same can be expected for its arctic region (Beyrer et al., 2017). Another region with high rates of HIV is Greenland. HIV was introduced to Greenland in the 1990s from Denmark. Its prevalence reached a peak between 2005 and 2009, but since then the epidemic has stabilised or even declined slightly (Bjorn-Mortensen et al., 2013). HIV is also present among native inhabitants of the North American Arctic (Reilley et al., 2018), along with some other sexually transmitted pathogens, such as noroviruses, chlamydia and gonorrhoea (Walker et al., 2015; Kumar, 2016).

Human viruses circulate also at the research stations in Antarctica. Controlled experiments and observations on occasional introductions showed that if some infective virus is present, most of the crew get infected quickly. Nevertheless, due to the small population sizes, such outbreaks are self-limiting and burn out quickly (Cameron & Moore, 1968; Holmes et al., 1971).

Another problem is reactivation of the latent viruses due to an extreme environment and immunosuppression (Cameron & Moore, 1968; Holmes et al., 1971; Tingate et al., 1997; Mehta et al., 2000; Reyes et al., 2017). Many viruses, such as certain herpesviruses, are present in a latent state in most of the global population, and their reactivation can cause serious medical problems. Both of these epidemiological situations probably occur in all long-time separated populations of limited size, and therefore the virus epidemiology of antarctic stations can be used as a model for future space flights (Lugg & Shepanek, 1999; Reyes et al., 2017).

6.3 Summary and perspectives

Polar regions are important for the ecology and evolution of many viral species. First, various animals wintering on different continents migrate every summer to polar regions and can serve as efficient transport vehicles for viruses. This allows close spatial interaction of different viral lineages, which would otherwise be separated. If two such lineages infect one host, subsequent recombination or reassortment can lead to the emergence of novel viral lineages. Second, animals living in polar regions can serve as a reservoir for viruses allowing their further spread to temperate climatic zones. Third, geographically isolated regions, such as Antarctica, can serve as a genetic sink, where the ancient virus variants that have been replaced elsewhere by newer descendants can remain preserved. Finally, permanently frozen ice in polar regions can serve as an abiotic reservoir for ancient viruses.

The importance of viruses for humans as well as domestic and wild-living animals in polar regions was a neglected topic in polar biology (and medicine) for a long time. Its importance became apparent just recently in connection with rapid climate change and more intensive human activity in polar regions. It led to more intense monitoring of these viruses in polar regions (Parkinson et al., 2014; Bruce et al., 2016).

Further metagenomic (DNA viruses) and metatranscriptomic (RNA viruses) studies or studies using classical methods of virology and molecular biology will almost certainly lead to the discovery of numerous new viruses and viral sequences. Some of these sequences will have no known homologues and will fall into so-called biological dark matter. These will most probably be parts of genomes of those viruses which cannot be cultivated *in vitro*. Nevertheless, even studies on the biological function of proteins encoded by these nucleic acids could lead to many results important for molecular biology and biochemistry, defining, for example, new enzymes with unique biochemical functions.

Knowing the sequence is only one side of the coin in virology. The other is understanding the role of these viruses in the ecology of their hosts and their role in ecosystems or their pathophysiology in animal hosts or even humans. We already know that viruses infect wild-living and domestic animals and even humans in polar regions, causing serious problems in conservation biology and agri- and aquaculture as well as in human health. Describing how these viruses influence their traditional hosts in rapidly changing environment would be very important for any further attempts to protect the indigenous animal species. Viruses endemic in polar regions can also pose a threat to newly introduced organisms. This can be beneficial in the case of invading species, but undesirable in the case of humans and their animal companions. Similarly, newly introduced viruses can pose a threat to indigenous species, but this threat should be investigated for each individual host and virus species. Research on mutual interactions between viruses and their hosts will be of great interest. We can expect to not only find novel viruses causing diseases in their hosts but also some beneficial features, such as the introduction of novel important genes and phenotypes into the hosts or even using the viruses as biological weapons in the fight between indigenous and invading species for ecological niches.

6.4 Acknowledgements

We would like to thank to all members of The Laboratory of Molecular Ecology of Vectors and Pathogens, The Laboratory of Arbovirology and Centre for Polar Ecology for their kind support during our field expeditions as well as laboratory work.

This work was supported by projects 'FIT (Pharmacology, Immunotherapy, nanotoxicology)' (CZ.02.1.01/0.0/0.0/15_003/0000495), 'Czech Polar Research Infrastructure (CzechPolar2)' (LM2015078), 'ECOPOLARIS' (CZ.02.1.01/0.0/0.0/16_013/0001708), 'C4SYS: Center for Systems Biology', 'INTER-EXCELLENCE, CZ-USA: Flavivirus–host interactions on the transcriptional and translational level', and 'INTER-EXCELLENCE CZ-RF: Development of technologies for early detection of tick-borne encephalitis, based on changes in gene expression and protein production in infected antigen-presenting cells' from the Ministry of Education, Youth and Sports of the Czech Republic; by the grant RO0518 and 'Changes in distribution of ticks and tick transmitted diseases: new and neglected risks for domestic animals, livestock and humans' from the Ministry of Agriculture of the Czech Republic, and by the project 'Interactions of flaviviral genomic and subgenomic RNA with host and viral proteins' from Czech Science Found.

References

Adouchief, S., Smura, T., Sane, J., Vapalahti, O., Kurkela, S. (2016). Sindbis virus as a human pathogen: epidemiology, clinical picture and pathogenesis. *Reviews in Medical Virology*, **26**, 221–241.

Akerstedt, J., Lillehaug, A., Larsen, I.L., et al. (2010). Serosurvey for canine distemper virus, canine adenovirus, Leptospira interrogans, and Toxoplasma gondii in free-ranging canids in Scandinavia and

Svalbard. *Journal of Wildlife Diseases*, **46**, 474–480.

Alexander, D.J., Manvell, R.J., Collins, M.S., et al. (1989). Characterization of paramyxoviruses isolated from penguins in Antarctica and sub-Antarctica during 1976-1979. *Archives of Virology*, **109**, 135–143.

Andersen, J.H., Berzaghi, F., Christensen, T., et al. (2017). Potential for cumulative effects of human stressors on fish, sea birds and marine mammals in Arctic waters. *Estuarine, Coastal and Shelf Science*, **184**, 202–206.

Austin, F.J., Webster, R.G. (1993). Evidence of ortho- and paramyxoviruses in fauna from Antarctica. *Journal of Wildlife Diseases*, **29**, 568–571.

Balboni, A., Tryland, M., Mørk, T., et al. (2019). Unique genetic features of canine adenovirus type 1 (CAdV-1) infecting red foxes (Vulpes vulpes) in northern Norway and arctic foxes (Vulpes lagopus) in Svalbard. *Veterinary Research Communications*, **43**(2), 67–76.

Ballard, W.B., Krausman, P.R. (1997). Occurrence of rabies in wolves of Alaska. *Journal of Wildlife Diseases*, **33**, 242–245.

Ballinger, M.J., Bruenn, J.A., Hay, J., Czechowski, D., Taylor, D.J. (2014). Discovery and evolution of bunyavirids in arctic phantom midges and ancient bunyavirid-like sequences in insect genomes. *Journal of Virology*, **88**, 8783–8794.

Ballinger, M.J., Medeiros, A.S., Qin, J., Taylor, D.J. (2017). Unexpected differences in the population genetics of phasmavirids. *Virus Evolution*, **3**, vex015.

Beresford, D. (2011). Insect collections from Polar Bear Provincial Park, Ontario, with new records. *Journal of the Entomological Society of Ontario*, **142**, 19–27.

Beyrer, C., Wirtz, A.L., O'Hara, G., Léon, N., Kazatchkine, M. (2017). The expanding epidemic of HIV-1 in the Russian Federation. *PLoS Medicine*, **14**, e1002462.

Bjorn-Mortensen, K., Ladefoged, K., Obel, N., Helleberg, M. (2013). The HIV epidemic in Greenland: a slow spreading infection among adult heterosexual Greenlanders. *International Journal of Circumpolar Health*, **72**, 19558.

Blix, A.S. (2016). Adaptations to polar life in mammals and birds. *Journal of Experimental Biology*, **219**(Pt 8), 1093–1105.

Bohm, J., Blixenkrone-Møller, M., Lund, E. (1989). A serious outbreak of canine distemper among sled-dogs in northern Greenland. *Arctic Medical Research*, **48**, 195–203.

Børresen, M.L., Andersson, M., Wohlfahrt, J., et al. (2015). Hepatitis B prevalence and incidence in Greenland: a population-based cohort study. *American Journal of Epidemiology*, **181**, 422–430.

Borriss, M., Helmke, E., Hanschke, R., Schweder, T. (2003). Isolation and characterization of marine psychrophilic phage-host systems from Arctic sea ice. *Extremophiles*, **7**, 377–384.

Bouckaert, R., Simons, B.C., Krarup, H., Friesen, T.M., Osiowy, C. (2017). Tracing hepatitis B virus (HBV) genotype B5 (formerly B6) evolutionary history in the circumpolar Arctic through phylogeographic modelling. *PeerJ*, **5**, e3757.

Bruce, M., Zulz, T., Koch, A. (2016). Surveillance of infectious diseases in the Arctic. *Public Health*, **137**, 5–12.

Buck, C.B., Van Doorslaer, K., Peretti, A., et al. (2016). The ancient evolutionary history of polyomaviruses. *PLoS Pathogens*, **12**, e1005574.

Callan, R.J., Early, G., Kida, H., Hinshaw, V.S., 1995. The appearance of H3 influenza viruses in seals. *Journal of General Virology*, **76** (Pt 1), 199–203.

Cameron, A.S., Moore, B.W. (1968). The epidemiology of respiratory infection in an isolated Antarctic community. *Journal of Hygiene (London)*, **66**, 427–437.

Carson, P.K., Holloway, K., Dimitrova, K., et al. (2017). The seasonal timing of snowshoe hare virus transmission on the Island of Newfoundland, Canada. *Journal of Medical Entomology*, **54**, 712–718.

Cattet, M.R., Duignan, P.J., House, C.A., Aubin, D.J. (2004). Antibodies to canine distemper and phocine distemper viruses in polar bears from the Canadian arctic. *Journal of Wildlife Diseases*, **40**, 338–342.

Cavicchioli, R. (2015). Microbial ecology of Antarctic aquatic systems. *Nature Reviews Microbiology*, **13**, 691–706.

Chan, F.T., Stanislawczyk, K., Sneekes, A.C., et al. (2019). Climate change opens new frontiers for marine species in the Arctic: current trends and future invasion risks. *Global Change Biology*, **25**, 25–38.

Ching, L.K., Gounder, P.P., Bulkow, L., et al. (2016). Incidence of hepatocellular carcinoma according to hepatitis B virus genotype in Alaska Native people. *Liver International*, **36**, 1507–1515.

Claverie, J.M. (2006). Viruses take center stage in cellular evolution. *Genome Biology*, **7**, 110.

Colangelo-Lillis, J.R., Deming, J.W. (2013). Genomic analysis of cold-active Colwelliaphage 9A and psychrophilic phage-host interactions. *Extremophiles*, **17**, 99–114.

Corbet, P.S., Downe, A.E.R. (1966). Natural hosts of mosquitoes in Northern Ellesmere Island. *Arctic*, **19**, 153–161.

Culler, L.E., Ayres, M.P., Virginia, R.A. (2015). In a warmer Arctic, mosquitoes avoid increased mortality from predators by growing faster. *Proceedings of the Royal Society of London Series B: Biological Sciences*, **282**, 1–8.

Cupp, E.W., Maré, C.J., Cupp, M.S., Ramberg, F. B. (1992). Biological transmission of vesicular stomatitis virus (New Jersey) by Simulium vittatum (Diptera: Simuliidae). *Journal of Medical Entomology*, **29**, 137–140.

das Neves, C.G., Roth, S., Rimstad, E., Thiry, E., Tryland, M. (2010). Cervid herpesvirus 2 infection in reindeer: a review. *Veterinary Microbiology*, **143**, 70–80.

Deardorff, E.R., Nofchissey, R.A., Cook, J.A., et al. (2013). Powassan virus in mammals, Alaska and New Mexico, U.S.A., and Russia, 2004–2007. *Emerging Infectious Diseases*, **19**, 2012–2016.

Descamps, S. (2013). Winter temperature affects the prevalence of ticks in an Arctic seabird. *PLoS One* **8**, e65374.

Descamps, S., Aars, J., Fuglei, E., et al. (2017). Climate change impacts on wildlife in a High Arctic archipelago – Svalbard, Norway. *Global Change Biology*, **23**(2), 490–502.

Di Sabatino, D., Lorusso, A., Di Francesco, C.E., et al. (2014). Arctic lineage-canine distemper virus as a cause of death in Apennine wolves (Canis lupus) in Italy. *PLoS One*, **9**, e82356.

Di Sabatino, D., Di Francesco, G., Zaccaria, G., et al. (2016). Lethal distemper in badgers (Meles meles) following epidemic in dogs and wolves. *Infection Genetics and Evolution*, **46**, 130–137.

Dietrich, M., Gómez-Díaz, E., McCoy, K.D. (2011). Worldwide distribution and diversity of seabird ticks: implications for the ecology and epidemiology of tick-borne pathogens. *Vector Borne Zoonotic Diseases*, **11**, 453–470.

Drebot, M.A. (2015.) Emerging mosquito-borne bunyaviruses in Canada. *Canada Communicable Diseases Report*, **41**, 117–123.

Dubois, A., Galan, M., Cosson, J.F., et al. (2017). Microevolution of bank voles (Myodes glareolus) at neutral and immune-related genes during multiannual dynamic cycles: consequences for Puumala hantavirus epidemiology. *Infection Genetics and Evolution*, **49**, 318–329.

Duignan, P.J., Saliki, J.T., St Aubin, D.J., House, J. A., Geraci, J.R. (1994). Neutralizing antibodies to phocine distemper virus in Atlantic walruses (Odobenus rosmarus rosmarus) from Arctic Canada. *Journal of Wildlife Diseases*, **30**, 90–94.

Duignan, P.J., Van Bressem, M.F., Baker, J.D., et al. (2014). Phocine distemper virus: current knowledge and future directions. *Viruses*, **6**, 5093–5134.

Durden, L.A., Beckmen, K.B., Gerlach, R.F. (2016). New Records of Ticks (Acari: Ixodidae) From Dogs, Cats, Humans, and Some Wild Vertebrates in Alaska: Invasion Potential. *Journal of Medical Entomology*, **53** (6), 1391–1395.

Elsterova, J., Cerny, J., Mullerova, J., et al. (2015). Search for tick-borne pathogens in the Svalbard Archipelago and Jan Mayen. *Polar Research*, **34**, 1–7.

Evander, M., Putkuri, N., Eliasson, M., et al. (2016). Seroprevalence and risk factors of Inkoo Virus in Northern Sweden. *American Journal of Tropical Medicine and Hygiene*, **94**, 1103–1106.

Fagre, A.C., Patyk, K.A., Nol, P., et al. (2015). Review of infectious agents in polar bears (Ursus maritimus) and their long-term ecological relevance. *Ecohealth*, **12**(3), 528–539. doi:10.1007/s10393-015-1023-6. Erratum in: Ecohealth, 2015 Sep;12(3):540.

Fahsbender, E., Burns, J.M., Kim, S., et al. (2017). Diverse and highly recombinant anelloviruses associated with Weddell seals in Antarctica.*Virus Evolution*, 3, vex017.

Fargeaud, D., Bugand, M., Précausta, P., Soulebot, J.P., Tektoff, J. (1982). Thermostability of the rabies virion. Optical density measurement technique applications. *Comparative Immunology Microbiology & Infectious Diseases*, **5**, 39–47.

Fields, B.N., Knipe, D.M., Howley, P.M. (2007). In: D.M. Knipe, P.M. Howley (editors-in-chief); D. E. Griffin et al. (associate editors) *Fields' Virology*, 5th ed. Wolters Kluwer/Lippincott Williams & Wilkins, Philadelphia, PA/ London.

Follmann, E.H., Garner, G.W., Evermann, J.F., McKeirnan, A.J. (1996). Serological evidence of morbillivirus infection in polar bears (Ursus maritimus) from Alaska and Russia. *Veterinary Record*, **138**, 615–618.

Gagnon, C.A., Allard, V., Cloutier, G. (2016). Canine parvovirus type 2b is the most prevalent genomic variant strain found in parvovirus antigen positive diarrheic dog feces samples across Canada. *Canadian Veterinary Journal*, **57**, 29–31.

Gaidet, N., Leclercq, I., Batéjat, C., et al. (2018). Avian influenza virus surveillance in high arctic breeding geese, Greenland. *Avian Diseases*, **62**, 237–240.

Gardner, H., Kerry, K., Riddle, M., Brouwer, S., Gleeson, L. (1997). Poultry virus infection in Antarctic penguins. *Nature*, **387**, 245.

Geraci, J.R., St Aubin, D.J., Barker, I.K., et al. (1982). Mass mortality of harbor seals: pneumonia associated with influenza A virus. *Science*, **215**, 1129–1131.

Goddard, A., Leisewitz, A.L. (2010). Canine parvovirus. *Veterinary Clinics of North America Small Animal Practice*, **40**, 1041–1053.

Gounder, P.P., Bulkow, L.R., Snowball, M., et al. (2016). Hepatocellular carcinoma risk in Alaska native children and young adults with hepatitis B virus: retrospective cohort analysis. *Journal of Pediatrics*, **178**, 206–213.

Groth, M., Lange, J., Kanrai, P., et al. (2014). The genome of an influenza virus from a pilot whale: relation to influenza viruses of gulls and marine mammals. *Infection Genetics and Evolution*, **24**, 183–186.

Härkönen, T., Dietz, R., Reijnders, P., et al. (2006). The 1988 and 2002 phocine distemper virus epidemics in European harbour seals. *Diseases of Aquatic Organisms*, **68**, 115–130.

Hartby, C.M., Krog, J.S., Merkel, F., et al. (2016). First characterization of avian influenza viruses from Greenland 2014. *Avian Diseases*, **60**, 302–310.

Heide-Joergensen, M.P., Haerkoenen, T., Aaberg, P. (1992). Long-term effects of epizootic in harbor seals in the Kattegat-Skagerak and adjacent areas. *AMBIO A Journal of the Human Environment*, **21**, 511–516.

Hill, N.J., and Runstadler, J.A. (2016). A bird's eye view of influenza A virus transmission: challenges with characterizing both sides of a co-evolutionary dynamic. *Integrative and Comparative Biology*, **56**(2), 304–316.

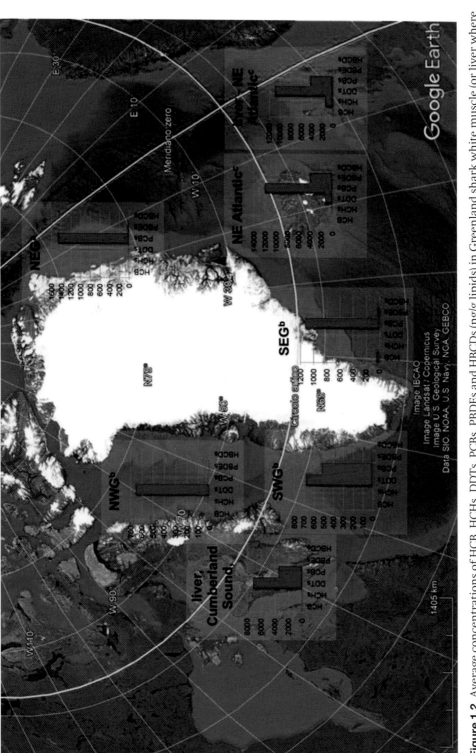

Figure 1.2 Average concentrations of HCB, HCHs, DDTs, PCBs, PBDEs and HBCDs (ng/g lipids) in Greenland shark white muscle (or liver where specified) samples from NE Greenland (NEG), NW Greenland (NWG), SW Greenland (SWG), SE Greenland (SEG), NE Atlantic and Cumberland Sound (data are from: a = Corsolini et al., 2014, 2016; b = Cotronei et al., 2017, 2018a, 2018b; c = Strid et al., 2007; d = Fisk et al., 2002). (A black and white version of this figure will appear in some formats.)

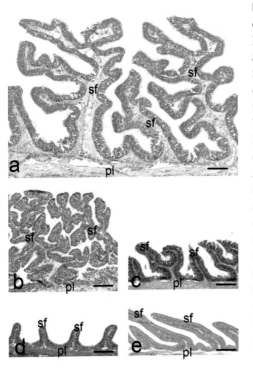

Figure 1.4 Histological sections of olfactory organs of elasmobranchs. Haematoxylineosin stain. (a) *S. microcephalus* (230-cm total length). Some secondary folds (sf) on the side of a primary lamella (pl). Almost all of the secondary folds present several branches. (b) *S. rostratus* (96-cm total length). Also in this species the secondary folds are very branched. (c) *Raja miraletus* (41-cm disc width). The secondary folds branch once or not at all. (d) *Pteroplatytrygon violacea* (107-cm disc width). The secondary folds are small, not branched and quite distant from one another. (e) *Scyliorhinus canicula* (27.5-cm total length). The secondary folds are not branched and elongated. Scale bars 200 μm. (A black and white version of this figure will appear in some formats.)

Figure 2.1 Hydrothermal vent fauna represents a high biomass in the deep-sea. The fauna is distributed according to its tolerance and requirements in the mixing zone of hydrothermal fluids and deep-sea water. In this picture, the gastropods *Ifremeria nautilei* and *Alviniconcha* sp. form dense patches. Copyright Ifremer/Chubacarc 2019. (A black and white version of this figure will appear in some formats.)

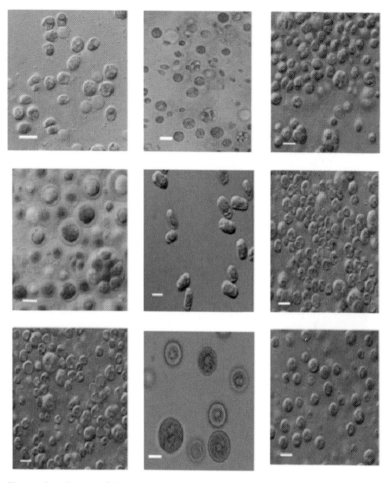

Figure 3.1 Some of the green microalgae (Chlorophyta) isolated from HBRAs in Iran. Scales, 10 μm. (A black and white version of this figure will appear in some formats.)

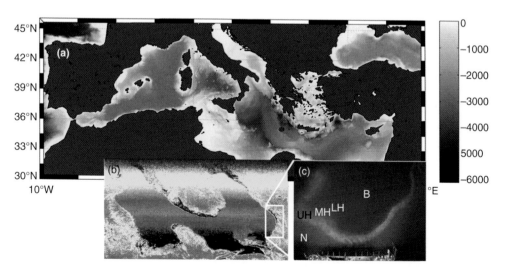

Figure 4.1 Location of the deep hypersaline anoxic basin in (a) the Mediterranean basin, showing details of the top view of the L'Atalante (b) basin and (c) halocline (panel c from Bernhard et al., 2015: UH, Upper Halocline; MH, Mid Halocline; LH, Lower Halocline; N, normoxic normal saline). (A black and white version of this figure will appear in some formats.)

Figure 4.2 Animals retrieved from the deep hypersaline anoxic L'Atalante basin. Scanning electron microscopy images of (a) the lorica and (b) the anterior edge of the lorica showing the spikes of the *Spinoloricus cinziae*; and images of Loricifera belonging to the (c) *Spinoloricus cinziae* (stained with Rose Bengal); (d) *Pliciloricus* (non-stained with Rose Bengal); (e) *Rugiloricus* containing an oocyte (non-stained with Rose Bengal); (f) *Spinoloricus* (a moulting exuvium). (A black and white version of this figure will appear in some formats.)

Figure 5.1 Stream with well acclimated (hardened) dry and frozen biomass of viable *Klebsormidium* sp. Central Svalbard, Endalen, late autumn. Photo J. Elster. (A black and white version of this figure will appear in some formats.)

Figure 5.2 Moss stream with dead *Tribonema* sp. biomass, central Svalbard, Endalen, beginning of July. Because of water erosion during spring melt, the stream started to flow in a different direction resulting in *Tribonema* biomass drying and not surviving the summer period. Photo J. Elster. (A black and white version of this figure will appear in some formats.)

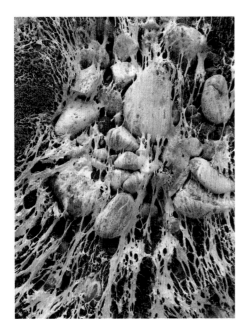

Figure 5.3 Detail of moss stream with dead *Tribonema* sp. biomass. Photo J. Elster. (A black and white version of this figure will appear in some formats.)

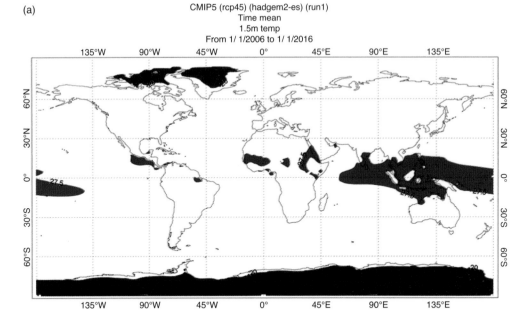

Figure 7.1 Extreme hot (red) and cold (blue) areas of the planet, under: (a) current state (2006–2016 average); (b) end-century (2090–2100) projection where radiative forcing due to human activities stabilises at ~+4.5W m² (end-century CO₂ = ~540 ppm); (c) end century (2090–2100) where radiative forcing due to human activities stabilises at ~+8.5 W m² (end-century CO₂ = ~935 ppm). Plots were all generated from the HadGEM2-ES model and are on an equirectangular projection which increases polar areas relative to other regions. Red areas are those where annual average temperatures (air at 2 m above ground) are above 27.5˚C, and blue areas are those where average temperatures are below 20˚C. Plots were generated by A. Phillips and J. Turner, British Antarctic Survey. (A black and white version of this figure will appear in some formats.)

(b)

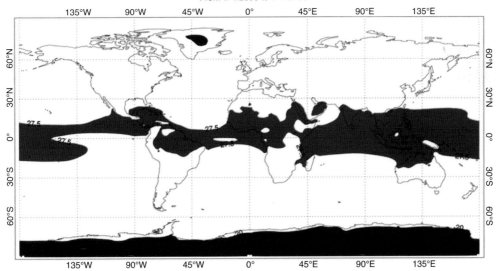

(c)

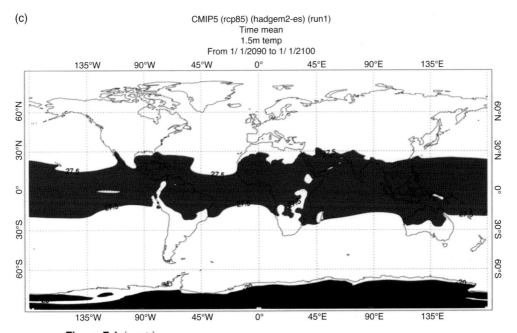

Figure 7.1 (cont.)

Figure 9.1 The Emperor penguin colony on the fast ice of the Atka Bay in the Eastern Weddell Sea, SO, during austral winter 2004. © Anna Müller. (A black and white version of this figure will appear in some formats.)

Figure 9.2 *Pleuragramma antarctica* is the only abundant midwater fish in high latitudes of the SO and, thus, plays an especially important key role in the pelagic food web. © Dieter Piepenburg, Alfred Wegener Institute. (A black and white version of this figure will appear in some formats.)

Figure 9.3 Environmental conditions on the continental shelf of the high-latitude southeastern Weddell Sea are obviously not limiting for a locally high biomass of sponges and a relatively rich associated fauna comprising among others sea cucumbers, brittle stars, bryozoans and hemichordates. © Julian Gutt and Werner Dimmler, Alfred Wegener Institute/Webseiten des Zentrum für Marine Umweltwissenschaften (MARUM). (A black and white version of this figure will appear in some formats.)

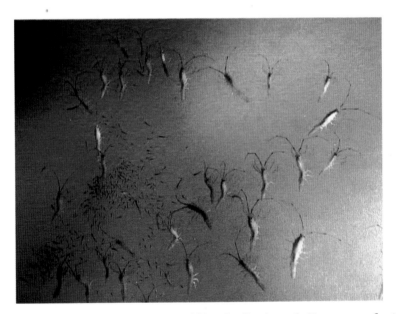

Figure 9.4 Adults and juveniles of filter-feeding isopods (*Antarcturus* cf. *spinacoronatus*) attached head-down to the underside of the floating Riiser-Larsen Iceshelf at depths of around 100 m in the Drescher Inlet, southeastern Weddell Sea in February 2016. Camera line of sight facing upwards. ROV-borne HD-video-image, © Nils Owsianowski, AWI. (A black and white version of this figure will appear in some formats.)

Figure 9.5 Benthic life below the Larsen B iceshelf, collapsed east of the Antarctic Peninsula 5 years before this photograph was taken, can be extremely poor and inhabited by some deep-sea organisms such as this multi-armed asteroid Freyella fragillisima. © Julian Gutt and Werner Dimmler, AWI/MARUM. (A black and white version of this figure will appear in some formats.)

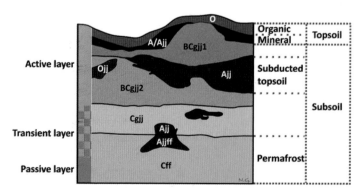

Figure 10.1 Schematic overview of the typical Cryosol influenced by cryoturbation – Turbic Cryosol. C-rich cryoturbated A horizon and O layers are shown in dark brown with code Ajj, Ojj, respectively. Horizon nomenclature according to Keys to Soil Taxonomy (Soil Survey Staff, 2010); O, organic layer; A, mineral top soil; BCg, Cg – mineral subsoil, Cff, permafrost (for details please refer to Gentsch et al., 2015). (A black and white version of this figure will appear in some formats.)

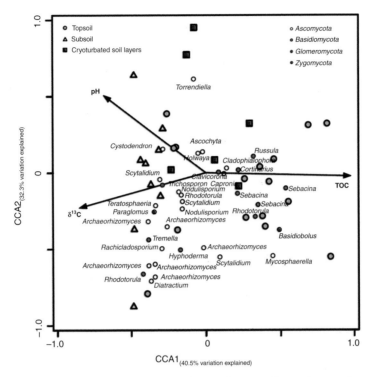

Figure 10.4 CCA ordination plots for the first two dimensions to show the relationship between fungal community structure, fungal taxa and environmental parameters. Organic topsoils: green circles; mineral subsoils, yellow triangles; cryoturbated soil layers: red squares. Correlations between environmental variables and CCA axes are represented by the length and angle of arrows (environmental factor vectors, Gittel et al., 2014a). (A black and white version of this figure will appear in some formats.)

(a)

(b)

Figure 11.1 Antarctic benthic communities photographed at shallow waters in Deception Is. (Antarctica). (a) Typical invertebrate associations in a hard-bottom sub-strate (18-m depth). (b) Common algal communities on a rocky wall (15-m depth). (A black and white version of this figure will appear in some formats.)

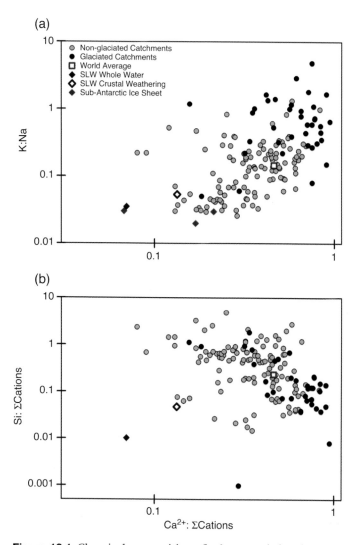

Figure 12.1 Chemical composition of sub-antarctic ice sheet water and meltwater draining glacier-covered and non-glacier-covered catchments. The two panels highlight the typical differences in glaciated versus non-glaciated catchment weathering products: (a) general enrichment of potassium and calcium in glaciated catchments; (b) the typically low concentrations of silica in glacial meltwaters relative to the typically high amounts of calcium. These typical patterns of geochemical composition in glacial meltwater are not seen when water is analysed from beneath the antarctic ice sheet (blue and red diamonds in a and b). Sub-antarctic ice sheet samples include Kamb and Bindschadler Ice Streams and a jökulhaup near Casey Station in Wilkes Land, on the East Antarctic coastline. Data provided courtesy of S. P. Anderson and figure modified from Anderson et al. (1997). (A black and white version of this figure will appear in some formats.)

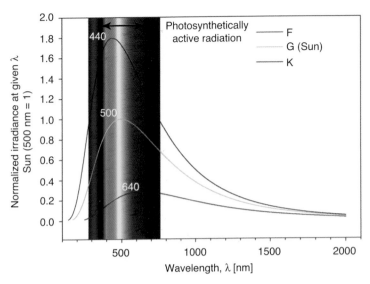

Figure 14.4 The energy flux for stars of spectral classes of relevance to astrobiology. (A black and white version of this figure will appear in some formats.)

Figure 14.5 (a) Bioweathering of sandstone rock in Svalbard, (b) detail of the endolithic community, (c) image of samples and (d) detection of endolithic community using variable chlorophyll fluorescence where the value of F_0 (minimum fluorescence in the dark) serves as a proxy of the chlorophyll a concentration. (A black and white version of this figure will appear in some formats.)

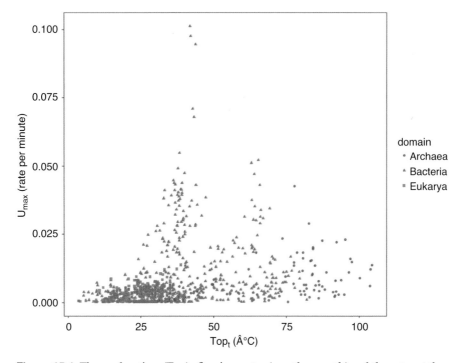

Figure 15.1 Thermal optima (T_{opt}) of various rates (mostly growth) and the rates at those optima (U_{max}) for life's major domains. Data from Corkrey et al. (2016) and Sørensen et al. (2018). Code for extracting this information from these sources is available from the author. (A black and white version of this figure will appear in some formats.)

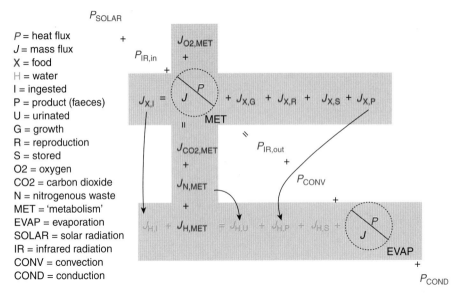

P_{SOLAR}

P = heat flux
J = mass flux
X = food
H = water
I = ingested
P = product (faeces)
U = urinated
G = growth
R = reproduction
S = stored
O2 = oxygen
CO2 = carbon dioxide
N = nitrogenous waste
MET = 'metabolism'
EVAP = evaporation
SOLAR = solar radiation
IR = infrared radiation
CONV = convection
COND = conduction

$P_{IR,in}$
$J_{O2,MET}$

$J_{X,I} = J \bigcirc P / MET + J_{X,G} + J_{X,R} + J_{X,S} + J_{X,P}$

$J_{CO2,MET}$ $P_{IR,out}$

P_{CONV}

$J_{N,MET}$

$J_{H,I} + J_{H,MET} = J_{H,U} + J_{H,P} + J_{H,S} + J \bigcirc P / EVAP$

P_{COND}

Figure 15.2 Coupled heat/mass balance equations for an organism (figure republished with permission of John Wiley & Sons - Books, from Balancing heat, water and nutrients under environmental change: a thermodynamic niche framework, Kearney et al. 2012 https://doi.org/10.1111/1365-2435.12020; permission conveyed through Copyright Clearance Center, Inc.). The diagonal equation represents the heat balance and, in a biophysical model, are calculated using the physical equations for heat exchange. The horizontal equations, within the shaded grey area, represent connections between the mass balances (food/water/gas exchange/excretion) and the heat balance, and are normally based on empirically determined relationships. (A black and white version of this figure will appear in some formats.)

Figure 15.3 The Succulent Karoo of southern Africa is an exceptionally species-rich arid region, in part owing to the predictability of its rainfall. The image here was taken near Springbok, Northern Cape, South Africa in August 2006. (A black and white version of this figure will appear in some formats.)

Holmes, M.J., Allen, T.R., Bradburne, A.F., Stott, E.J. (1971). Studies of respiratory viruses in personnel at an Antarctic base. *Journal of Hygiene (London)*, **69**, 187–199.

Hubálek, Z., Rudolf, I. (2012). Tick-borne viruses in Europe. *Parasitology Research*, **111**, 9–36.

Hurt, A.C., Vijaykrishna, D., Butler, J., et al. (2014). Detection of evolutionarily distinct avian influenza A viruses in antarctica. *MBio*, **5**, e01098–01014.

Hurt, A.C., Su, Y.C., Aban, M., et al. (2016). Evidence for the introduction, reassortment, and persistence of diverse Influenza A viruses in Antarctica. *Journal of Virology*, **90**, 9674–9682.

Hussein, I.T., Krammer, F., Ma, E., et al. (2016). New England harbor seal H3N8 influenza virus retains avian-like receptor specificity. *Science Report*, **6**, 21428.

Jaenike, J. (2012). Population genetics of beneficial heritable symbionts. *Trends in Ecology & Evolution*, **27**, 226–232.

Jansen, M.D., Bang Jensen, B., McLoughlin, M.F., et al. (2017). The epidemiology of pancreas disease in salmonid aquaculture: a summary of the current state of knowledge. *Journal of Fish Diseases*, **40**, 141–155.

Julkunen, I., Brummer-Korvenkontio, M., Hautanen, A., et al. (1986). Elevated serum immune complex levels in Pogosta disease, an acute alphavirus infection with rash and arthritis. *Journal of Clinical and Laboratory Immunology*, **21**, 77–82.

Kaaden, O.R., Eichhorn, W., Essbauer, S. (2002). Recent developments in the epidemiology of virus diseases. *Journal of Veterinary Medicine B: Infectious Diseases and Veterinary Public Health*, **49**, 3–6.

Kautto, A.H., Alenius, S., Mossing, T., et al. (2012). Pestivirus and alphaherpesvirus infections in Swedish reindeer (Rangifer tarandus tarandus L.). *Veterinary Microbiology*, **156**, 64–71.

Kempf, F., Boulinier, T., De Meeûs, T., Arnathau, C., McCoy, K.D. (2009). Recent evolution of host-associated divergence in the seabird tick Ixodes uriae. *Molecular Ecology*, **18**, 4450–4462.

Kernbauer, E., Ding, Y., Cadwell, K. (2014). An enteric virus can replace the beneficial function of commensal bacteria. *Nature*, **516**, 94–98.

Kirk, C.M., Amstrup, S., Swor, R., Holcomb, D., O'Hara, T.M. (2010). Morbillivirus and Toxoplasma exposure and association with hematological parameters for southern Beaufort Sea polar bears: potential response to infectious agents in a sentinel species. *Ecohealth*, **7**, 321–331.

Kramvis, A. (2014). Genotypes and genetic variability of hepatitis B virus. *Intervirology*, **57**, 141–150.

Krarup, H.B., Andersen, S., Madsen, P.H., et al. (2008). Benign course of long-standing hepatitis B virus infection among Greenland Inuit? *Scandinavian Journal of Gastroenterology*, **43**, 334–343.

Krog, J.S., Hansen, M.S., Holm, E., et al. (2015). Influenza A(H10N7) virus in dead harbor seals, Denmark. *Emerging Infectious Diseases*, **21**, 684–687.

Kumar, A. (2016). HIV/aids risk and prevention issues among Inuit living in Nunavut Territory of Canada. *In Vivo*, **30**, 905–916.

La Linn, M., Gardner, J., Warrilow, D., et al. (2001). Arbovirus of marine mammals: a new alphavirus isolated from the elephant seal louse, Lepidophthirus macrorhini. *Journal of Virology*, **75**, 4103–4109.

Larska, M. (2015). Pestivirus infection in reindeer (Rangifer tarandus). *Frontiers in Microbiology*, **6**, 1187.

Laws, R.M., Taylor, R.J.F. (1957). A mass dying of Crabeater Seals, Lobodon Carcinophague (Gray). *Journal of Zoology*, **129**, 315–324.

Lee, H.S., Zhdanova, S.N., Vladimirtsev, V.A., et al. (2010). Epidemiology of Viliuisk encephalomyelitis in Eastern Siberia. *Epidemiology*, **21**, 24–30.

Lee, R.E., Baust, J.G. (1987). Cold-hardiness in the antarctic tick, Ixodes-Uriae. *Physiological Zoology*, **60**, 499–506.

Liberda, E.N., Meldrum, R., Charania, N.A., Davey, R., Tsuji, L.J. (2017). Avian influenza prevalence among hunter-harvested birds in a remote Canadian First Nation community. *Rural Remote Health*, **17**, 3864.

Loewen, K., Prins, B., Philibert, H. (1990). Northwest Territories. Rabies in a polar bear. *Canadian Veterinary Journal*, **31**, 457.

Lugg, D., Shepanek, M. (1999). Space analogue studies in Antarctica. *Acta Astronaut*, **44**, 693–699.

Lvov, D.K., Vladimirtseva, E.A., Butenko, A.M., et al. (1988). Identity of Karelian fever and Ockelbo viruses determined by serum dilution-plaque reduction neutralization tests and oligonucleotide mapping. *American Journal of Tropical Medicine and Hygiene*, **39**, 607–610.

Maat, D.S., Biggs, T., Evans, C., et al. (2017). Characterization and temperature dependence of arctic Micromonas polaris viruses. *Viruses* **9**, pii: E134.

Macdonald, E., Handeland, K., Blystad, H., et al. (2011). Public health implications of an outbreak of rabies in arctic foxes and reindeer in the Svalbard archipelago, Norway, September 2011. *Eurosurveillance*, **16**, pii: 19985.

Major, L., Linn, M.L., Slade, R.W., et al. (2009). Ticks associated with macquarie island penguins carry arboviruses from four genera. *PLoS One*, **4**, e4375.

Mateu, M.G. (2013). Assembly, stability and dynamics of virus capsids. *Archives of Biochemistry and Biophysics*, **531**, 65–79.

Márquez, L.M., Redman, R.S., Rodriguez, R.J., Roossinck, M.J. (2007). A virus in a fungus in a plant: three-way symbiosis required for thermal tolerance. *Science*, **315**, 513–515.

McLean, C.A., Masters, C.L., Vladimirtsev, V.A., et al. (1997). Viliuisk encephalomyelitis: review of the spectrum of pathological changes. *Neuropathology and Applied Neurobiology*, **23**, 212–217.

McLean, D.M., Clarke, A.M., Goddard, E.J., et al. (1973). California encephalitis virus endemicity in the Yukon Territory, 1972. *Journal of Hygiene (London)*, **71**, 391–402.

Mehta, S.K., Pierson, D.L., Cooley, H., Dubow, R., Lugg, D. (2000). Epstein-Barr virus reactivation associated with diminished cell-mediated immunity in antarctic expeditioners. *Reviews in Medical Virology*, **61**, 235–240.

Miller, G., Watts, J., Shellam, G. (2008). Viral antibodies in SouthPolar Skuas around Davis Station, Antarctica. *Antarctic Science*, **20**, 455–461.

Mustonen, J., Outinen, T., Laine, O., et al. (2017). Kidney disease in Puumala hantavirus infection. *Infectious Diseases (London)*, **49**, 321–332.

Mørk, T., Prestrud, P. (2004). Arctic rabies: a review. *Acta Veterinaria Scandinavica*, **45**, 1–9.

Müllerová, J., Elsterová, J., Jiří, Č., et al. (2018). No indication of arthropod-vectored viruses in mosquitoes (Diptera: Culicidae) collected on Greenland and Svalbard. *Polar Biology*, **41**, 1581–1586.

Nielsen, O., Clavijo, A., Boughen, J.A (2001). Serologic evidence of influenza A infection in marine mammals of arctic Canada. *Journal of Wildlife Diseases*, **37**, 820–825.

Odegaard, O.A., Krogsrud, J. (1981). Rabies in Svalbard: infection diagnosed in arctic fox, reindeer and seal. *Veterinary Record*, **109**, 141–142.

Olsen, B., Munster, V.J., Wallensten, A., et al. (2006). Global patterns of influenza a virus in wild birds. *Science*, **312**, 384–388.

Orpetveit, I., Ytrehus, B., Vikoren, T., et al. (2011). Rabies in an Arctic fox on the Svalbard archipelago, Norway, January 2011. *Eurosurveillance*, **16**, 1–2.

Osiowy, C., Simons, B.C., Rempel, J.D. (2013). Distribution of viral hepatitis in indigenous populations of North America and the circumpolar Arctic. *Antiviral Therapy*, **18**, 467–473.

Panum, P.L. (2018). *Observations Made During the Epidemic of Measles on the Faroe Islands in the*

Year 1846. Franklin Classics Trade Press, Lebanon, NJ.

Park, Y.M., Kim, J.H., Gu, S.H., et al. (2012). Full genome analysis of a novel adenovirus from the South Polar skua (Catharacta maccormicki) in Antarctica. *Virology*, **422**, 144–150.

Parkinson, A.J., Evengard, B., Semenza, J.C., et al. (2014). Climate change and infectious diseases in the Arctic: establishment of a circumpolar working group. *International Journal of Circumpolar Health*, **73**.

Philippa, J.D., Leighton, F.A., Daoust, P-Y., et al. (2004). Antibodies to selected pathogens in free-ranging terrestrial carnivores and marine mammals in Canada. *Veterinary Record*, **155**, 135–140.

Piersma, T. (1997). Do global patterns of habitat use and migration strategies co-evolve with relative investments in immunocompetence due to spatial variation in parasite pressure? *Oikos*, 623–631.

Piersma, T., Mendes, L., Hennekens, J., et al. (2001). Breeding plumage honestly signals likelihood of tapeworm infestation in females of a long-distance migrating shorebird, the bar-tailed godwit. *Zoology (Jena)*, **104**, 41–48.

Pirisinu, L., Tran, L., Chiappini, B., et al. (2018). Novel type of chronic wasting disease detected in moose (Alces alces), Norway. *Emerging Infectious Diseases*, **24**, 2210–2218.

Plyusnin, A., Hörling, J., Kanerva, M., et al. (1997). Puumala hantavirus genome in patients with nephropathia epidemica: correlation of PCR positivity with HLA haplotype and link to viral sequences in local rodents. *Journal of Clinical Microbiology*, **35**, 1090–1096.

Plyusnin, A., Vapalahti, O., Lundkvist, A., Henttonen, H., Vaheri, A. (1996). Newly recognised hantavirus in Siberian lemmings. *Lancet*, **347**, 1835.

Putkuri, N., Kantele, A., Levanov, L., et al. (2016). Acute human Inkoo and Chatanga virus infections, Finland. *Emerging Infectious Diseases*, **22**, 810–817.

Ramey, A.M., Reeves, A.B., TeSlaa, J.L., et al. (2016). Evidence for common ancestry among viruses isolated from wild birds in Beringia and highly pathogenic intercontinental reassortant H5N1 and H5N2 influenza A viruses. *Infection, Genetics and Evolution*, **40**, 176–185.

Rastrojo, A., Alcamí, A. (2018). Viruses in polar lake and soil ecosystems. *Advances in Virus Research*, **101**, 39–54.

Razzauti, M., Plyusnina, A., Sironen, T., Henttonen, H., Plyusnin, A. (2009). Analysis of Puumala hantavirus in a bank vole population in northern Finland: evidence for co-circulation of two genetic lineages and frequent reassortment between strains. *Journal of General Virology*, **90**, 1923–1931.

Reilley, B., Haberling, D. L., Person, M., et al. (2018). Assessing New Diagnoses of HIV Among American Indian/Alaska Natives Served by the Indian Health Service, 2005–2014. *Public Health Reports*, **133**(2), 163–168.

Rex, K.F., Andersen, S., Krarup, H.B. (2015). Hepatitis B among Inuit: a review with focus on Greenland Inuit. *World Journal of Hepatology*, **7**, 1265–1271.

Reyes, D.P., Brinley, A.A., Blue, R.S., et al. (2017). Clinical herpes zoster in Antarctica as a model for spaceflight. *Aerospace Medicine and Human Performance*, **88**, 784–788.

Ringrose, J.L., Abraham, K.F., Beresford, D.V. (2013). New range records of mosquitoes (Diptera: Culicidae) from northern Ontario. *Journal of the Entomological Society of Ontario*, **144**, 3–14.

Robertson, N.L. (2007). Identification and characterization of a new virus in the genus Potyvirus from wild populations of Angelica lucida L. and A. genuflexa Nutt., family Apiaceae. *Archives of Virology*, **152**, 1603–1611.

Robertson, N.L., French, R. (2007). Genetic analysis of a novel Alaska barley yellow dwarf virus in the family Luteoviridae. *Archives of Virology*, **152**, 369–382.

Robertson, N.L., Côté, F., Paré, C., et al. (2007). Complete nucleotide sequence of Nootka lupine vein-clearing virus. *Virus Genes*, **35**, 807–814.

Sakamoto, T., Tanaka, Y., Simonetti, J., et al. (2007). Classification of hepatitis B virus genotype B into 2 major types based on characterization of a novel subgenotype in Arctic indigenous populations. *Journal of Infectious Diseases*, **196**, 1487–1492.

Sasai, S., Tamura, K., Tojo, M., et al. (2018). A novel non-segmented double-stranded RNA virus from an Arctic isolate of Pythium polare. *Virology*, **522**, 234–243.

Säwström, C., Lisle, J., Anesio, A.M., Priscu, J.C., Laybourn-Parry, J. (2008). Bacteriophage in polar inland waters. *Extremophiles*, **12**, 167–175.

Shearn-Bochsler, V., Green, D.E., Converse, K. A., et al. (2008). Cutaneous and diphtheritic avian poxvirus infection in a nestling Southern Giant Petrel (Macronectes giganteus) from Antarctica. *Polar Biology*, **31**, 569–573.

Shi, M., Lin, X.D., Tian, J.H., et al. (2016). Redefining the invertebrate RNA virosphere. *Nature*, **540**, 539–543.

Skoge, R.H., Brattespe, J., Økland, A.L., Plarre, H., Nylund, A. (2018). New virus of the family Flaviviridae detected in lumpfish (Cyclopterus lumpus). *Archives of Virology*, **163**, 679–685.

Skogh, M., Espmark, A. (1982). Ockelbo disease: epidemic arthritis-exanthema syndrome in Sweden caused by Sindbis-virus like agent. *Lancet*, **1**, 795–796.

Smeele, Z.E., Ainley, D.G., Varsani, A. (2018). Viruses associated with Antarctic wildlife: From serology based detection to identification of genomes using high throughput sequencing. *Virus Research*, **243**, 91–105.

Soleng, A., Edgar, K.S., Paulsen, K.M., et al. (2018). Distribution of Ixodes ricinus ticks and prevalence of tick-borne encephalitis virus among questing ticks in the Arctic Circle region of northern Norway. *Ticks and Tick Borne Diseases*, **9**, 97–103.

Stimmelmayr, R., Rotstein, D.S., Maboni, G., Person, B.T., Sanchez, S. (2018). Morbillivirus-associated lipid pneumonia in Arctic foxes. *Journal of Veterinary Diagnostic Investigation*, **30**, 933–936.

Suttle, C.A. (2005). Viruses in the sea. *Nature*, **437**, 356–361.

Taylor, M., Elkin, B., Maier, N., Bradley, M. (1991). Observation of a polar bear with rabies. *Journal of Wildlife Diseases*, **27**(2), 337–339.

Thomazelli, L.M., Araujo, J., Oliveira, D.B., et al. (2010). Newcastle disease virus in penguins from King George Island on the Antarctic region. *Veterinary Microbiology*, **146**, 155–160.

Thompson, R.M., Thompson, H., Hall, A.J. (2002). Prevalence of morbillivirus antibodies in Scottish harbour seals. *Veterinary Record*, **151**, 609–610.

Tingate, T.R., Lugg, D.J., Muller, H.K., Stowe, R. P., Pierson, D.L. (1997). Antarctic isolation: immune and viral studies. *Immunology and Cell Biology*, **75**, 275–283.

Tjøtta, E., Hungnes, O., Grinde, B. (1991). Survival of HIV-1 activity after disinfection, temperature and pH changes, or drying. *Journal of Medical Virology*, **35**, 223–227.

Traavik, T., Mehl, R., Wiger, R. (1978). California encephalitis group viruses isolated from mosquitoes collected in Southern and Arctic Norway. *Acta pathologica et microbiologica Scandinavica. Section B: Microbiology and immunology*, **86B**, 335–341.

Tryland, M., Balboni, A., Killengreen, S.T., et al. (2018). A screening for canine distemper virus, canine adenovirus and carnivore protoparvoviruses in Arctic foxes (Vulpes lagopus) and red foxes (Vulpes vulpes) from Arctic and sub-Arctic regions of Norway. *Polar Research*, **37**.

Tryland, M., Klein, J., Nordøy, E.S., Blix, A.S. (2005). Isolation and partial characterization of a parapoxvirus isolated from a skin lesion of a Weddell seal. *Virus Research*, **108**, 83–87.

Tryland, M., Nymo, I.H., Nielsen, O., et al. (2012). Serum chemistry and antibodies against pathogens in antarctic fur seals, Weddell seals, crabeater seals, and Ross seals. *Journal of Wildlife Diseases*, **48**, 632–645.

Van Hemert, C., Spivey, T.J., Uher-Koch, B.D., et al. (2019). Surwey of arctic Alaskan wildlife for influenza A antibodies: limited evidence for exposure of mammals. *Journal of Wildlife Diseases*, **55**(2), 387–398.

Vapalahti, O., Lundkvist, A., Fedorov, V., et al. (1999). Isolation and characterization of a hantavirus from Lemmus sibiricus: evidence for host switch during hantavirus evolution. *Journal of Virology*, **73**, 5586–5592.

Varsani, A., Frankfurter, G., Stainton, D., et al. (2017). Identification of a polyomavirus in Weddell seal (Leptonychotes weddellii) from the Ross Sea (Antarctica). *Archives of Virology*, **162**, 1403–1407.

Walker, F.J., Llata, E., Doshani, M., et al. (2015). HIV, chlamydia, gonorrhea, and primary and secondary syphilis among American Indians and Alaska Natives within Indian health service areas in the United States, 2007-2010. *Journal of Community Health*, **40**, 484–492.

Webster, R.G., Hinshaw, V.S., Bean, W.J., et al. (1981). Characterization of an influenza A virus from seals. *Virology*, **113**, 712–724.

Weiler, G.J., Garner, G.W., Ritter, D.G. (1995). Occurrence of rabies in a wolf population in northeastern Alaska. *Journal of Wildlife Diseases*, **31**, 79–82.

Zhang, G., Shoham, D., Gilichinsky, D., et al. (2006). Evidence of influenza a virus RNA in siberian lake ice. *Journal of Virology*, **80**, 12229–12235.

Zhang, R., Wei, W., Cai, L. (2014). The fate and biogeochemical cycling of viral elements. *Nature Reviews Microbiology*, **12**, 850–851.

PART III   Life in extreme
environments and the responses
to change: the example of polar
environments

JOSEF ELSTER

Introduction

Polar regions are defined as those areas that lie within the arctic and antarctic circles. They are vast areas, covering 16.5% of the Earth's surface and comprising 84 million km². The polar regions are of outstanding international scientific and environmental significance. Increasingly, studies of these areas are making major contributions to our understanding of global change – in terms of the Earth's ancient evolutionary history as well as with respect to contemporary climate change issues. However, the arctic and antarctic regions differ remarkably in geology and geological history and in energy transport and balance. From an anthropocentric point of view, the polar regions (both marine and terrestrial) are considered as extreme environments. However, there is high habitat diversity, and their extremeness varies remarkably.

The Arctic and Antarctic Peninsula areas are warming faster than the global average. Here, temperatures have increased substantially in the past few decades. This climate shift has attracted particular attention because of our understanding of climate/Earth interactions. The most visible and easily measurable responses to climatic warming are mechanisms related to the decrease in snow and ice cover. The Arctic and Antarctic Peninsula also show many other aspects of phase changes as the climate warms. In addition to ice and snow melting, these include permafrost thaw, soil and vegetation change, and many others including changes of the ecological parameters in habitats which are considered to be extreme. As a response to warming, the seasonal and diurnal variations of environmental patterns are changing in periodicity, amplitude, synchronicity and regularity, which initiate a number of adaptation responses. However, to understand the basis of the survival responses in the polar regions, it is essential to appreciate that these environments are subjected not only to extremes but also to rapid fluctuations across these extremes. For example, the speed at which water states (liquid to solid ice/snow) and low temperatures can change is an important ecological and physiological factor, which is related to climate change. Several polar terrestrial, freshwater and marine habitats, e.g. biological soil crusts, hypolithic and/or endolithic communities of polar desert/ semidesert, communities of sea and terrestrial ice and many others, have to shift their geographical ranges, seasonal activities, abundances and species interactions in response to ongoing climate change.

The study of extremophile responses to climate change in the polar regions is one of the most important current challenges of biological and ecological sciences. With the aid of relevant molecular, physiological and ecological properties interpreted on the whole species/community level, information could be gathered about the flexibility or vulnerability of polar extremophiles to ongoing climate changes.

CHAPTER SEVEN

Life in the extreme environments of our planet under pressure

Climate-induced threats and exploitation opportunities

MELODY S. CLARK
British Antarctic Survey
CINZIA VERDE
National Research Council (CNR) – Institute of Bioscience and BioResources (IBBR)
SILVIA FINESCHI
Institute of Heritage Science
FRANCESCO LORETO
Institute of Heritage Science
LLOYD S. PECK
British Antarctic Survey

and

GUIDO DI PRISCO
National Research Council (CNR) – Institute of Bioscience and BioResources (IBBR)

7.1 Extreme environments and ongoing climate change

7.1.1 Climate and life in extreme environments

In 2007, a report from the European Science Foundation on Investigating Life in Extreme Environments defined extreme environments as '*having one or more environmental parameters showing values permanently close to lower or upper limits known for life in its various forms*' (CAREX, 2011). These regions can be considered at the extreme of a continuum of environmental conditions that constitute limits for life as we know it. Extreme environments show enormous diversity of microorganisms, plants and animals in various marine and terrestrial extreme environments. Biomolecules may persist in extreme habitats, e.g. under low temperature, desiccation and/or low radiation conditions, which could provide well-preserved ancient biosignals.

Life in extreme environments (most notably through the deep subsurface and polar regions) represents the widest part of the Earth biosphere. Terrestrial, marine, polar or deep subsurface organisms are distributed across the whole globe and comprise a significant proportion of global biomass and biodiversity. The deep sea (more than 1 mile deep and largely unexplored) covers 60% of the planet, and drylands account for 41% of the terrestrial environment, hosting 35% of the human population (Millennium Ecosystem Assessment, 2005). The Antarctic, which offers a regional model for studying biological responses to environmental change across all scales of biological organisation, plays a crucial role in driving the climate and the ocean circulation of the whole planet (Hughes, 2000; Walther et al., 2002; Parmesan, 2006; Walther, 2010; Verde et al., 2016; Chown et al., 2017; Kristensen et al., 2018; Rintoul et al., 2018; and references therein). Therefore, the contribution of extreme environments to the Earth system should not be overlooked, and indeed deserves very serious consideration.

The constraints imposed on organisms (named 'extremophiles'; Rothschild & Mancinelli, 2001) and communities living in extreme environments have led to the emergence of specialised, often unique adaptation mechanisms, many of which have been poorly studied to date. In the course of millions of years, species have progressively become very finely adapted in order to be able to survive, often under conditions of scarce resources, for example, water, oxygen and light. Species unable to adapt to new adverse conditions underwent extinction (Bell & Collins, 2008; Chevin et al., 2013; Danovaro et al., 2017).

In general terms, habitability refers to the conditions of a planetary environment for sustaining life. With no proof of extraterrestrial life, habitability is oriented towards life on Earth, which needs a carbon-based chemistry, a solar or chemical energy source and liquid water. Therefore, valuable knowledge about the boundary conditions for habitability with respect to temperature, salinity, pressure, water activity, pH, availability of nutrients and redox potential can be obtained from studies of life in extreme environments (Fox-Powell et al., 2016).

The study of these mechanisms provides valuable new insights into biological processes, from molecular biology to physiology, ecology and evolution. New perspectives gained through the investigation of life in extreme environments allow the development and refinement of theories relevant across the breadth of biology. They provide valuable analogues for studying the possibilities for extraterrestrial life and also an understanding of those extreme conditions under which the earliest life arose on Earth. With climate change, many of these highly specialised environments are under threat. Thus, it is essential to develop an understanding, not only of how these finely tuned fragile ecosystems will change, but also the biochemical adaptations that have occurred in indigenous species which may be of use to society.

The perception of extreme environments as 'another world' is now changing. The past decade has seen remarkable advances in technologies for accessing extreme environments, e.g. autonomous underwater vehicles (AUVs), remotely operated vehicles (ROVs), *in situ* instrumentation and an impressive explosion of research in molecular and systems biology. Omics (genomics, proteomics, trancriptomics, metabolomics, etc.), in particular, give us the ability to identify and characterise previously elusive organisms, gaining new insights into the unique adaptations, metabolic capabilities and functionality of organisms from extreme environments (Gutt et al., 2018). While investigations have been undertaken for many decades in certain extremes (e.g. hot deserts, high-altitude ecosystems, polar regions), the breadth of extreme environments in which life has been found has emerged only recently. As examples, midocean ridge hydrothermal vent biota was only discovered in the late 1970s; deep subsurface ecosystems began to be recognised in the late 1980s and the first clues for life in the giant subglacial Lake Vostok (Antarctica) were only reported in the 1990s.

Interest in the biodiversity of extreme environments has developed for several reasons:

- Conditions now considered to be extreme were predominant on the young planet Earth and we therefore need to study extremophiles to understand how the planet and its biosphere evolved. Modern extreme environments are particularly associated with the Archaea, considered to include representatives of some of the earliest life forms on the evolutionary Tree of Life (Gribaldo & Brochier-Armanet, 2006). The fact that Archaea are particularly prevalent in hot springs and near hydrothermal vents has brought forth the hypothesis of a hot early Earth environment for life (Lunine, 2006). These robust organisms are also common in permanently cold environments and under a range of other extremes, so there is much to learn from this group. Archaean diversity provides an important focus for deciphering both the nature of the early Earth biosphere and that of the Tree of Life.

- Compared to present conditions, the early planet was an extreme habitat and we expect this to be true of other planets on which life is developing (see also some chapters of Part 4). Astrobiological analogues on Earth include geothermal sites, regions of low or high temperature, arid regions, regions with high or low pH, high or low salinity, low oxygen and high or low pressures, as well as remote locations such as subglacial environments, the deep subsurface and the deep sub-ocean. Several extreme environments on Earth bear similarities to extraterrestrial planetary conditions conceivable for supporting life, e.g. as analogues for Mars (surface and subsurface habitats in Antarctica, Greenland, Iceland, cold and hot deserts, the Rio Tinto, the Atacama Desert), or as analogues for the sub-ice ocean on Jupiter's moon Europa (subglacial lakes, hydrothermal vents) (Harrison et al., 2013). In the absence of direct access to extraterrestrial bodies, studies of model systems and communities from those extreme environments of our planet can provide important information in the search for life in planetary exploration.

- Extreme environments are important hotspots of microbial, metazoan and symbiotic diversity. These are habitats of organisms which have the genetic and physiological capacity to survive and reproduce under these harsh extreme conditions through which they have evolved while shaping the environment that we know today. Organisms have developed diverse mechanisms of resistance and bioenergetic pathways that allow them to colonise and thrive in extreme environments. For example, vent and seep invertebrates have taken up symbiotic bacteria that enable them to establish large amounts of biomass in the nutrient-poor deep sea (Petersen & Dubilier, 2009). Another example is that of cellular polyphosphates which occur throughout the biological world and have been proposed to be involved in very diverse functions, e.g. reservoirs for phosphate, chelators of metal ions, gene regulation, quorum sensing in bacteria, blood clotting in mammals and biofilm production (Seufferheld et al., 2008). Extreme environments will provide many more examples of such diverse and specialised gene functions.

- Advances in molecular biotechnology have provided tools to study biological response to stress and energy-metabolism pathways, helping to understand the mechanisms of microbial regulation and coordination in response to stress, which are still poorly understood, and the equivalent responses of complex multicellular life. The study of extreme environments offers opportunities to address these shortcomings. Thus, the monitoring of delicate ecosystems in extreme environments may yield early warning systems in view of global warming, pollution and other environmental changes. The potential for biotechnological exploitation is an additional important reason to study organisms populating extreme environments.

7.1.2 Stability/instability

Extreme environments can be either stable or unstable. In stable environments (e.g. polar oceans, deep biosphere), organisms live near the limits of their physiological potential and can persist for long periods of time with successive development and selection of genotypes. Such habitats are occupied by stably adapted organisms (e.g. psychrophiles, hyperthermophiles and acidophiles). In contrast, at their margins, the interfaces between different types of habitats, which can display steep gradients of environmental parameters, can be exposed to strong instabilities, leading to them being considered as extreme. Organisms can intermittently experience environmental conditions near the limits, compelling them to implement suitable strategies to survive at the individual or population scale (Merilä & Hendry, 2014). In these unstable environments, physicochemical parameters can change in a predictable (e.g. tidal, seasonal) or chaotic (e.g. change in hydrothermal vent regime) manner. Cyclical environmental patterns may differ in periodicity, amplitude, synchrony and regularity, and may require different adaptations.

Where more irregular catastrophic events occur (e.g. volcanic eruptions), the capacity of survival may be exceeded and may lead to massive mortality and extinction of local populations. Recovery of a stable ecosystem state then involves adaptation processes at the level of populations and communities that drive their resilience (Gunderson, 2000). Post eruption recovery also involves dispersal required to recolonise the destroyed site/community, and hence exposes the (successful) organisms to another set of extreme stresses.

In this review we will concentrate on the effects and consequences of climate warming using plants and polar species as examples, as increased temperatures are heavily impacting polar oceans, permafrost environments which are increasingly melting each summer, high-mountain ecosystems, and terrestrial ecosystems between desert and semiarid areas. A strategy for searching exploitable resources in extreme environments will also be proposed.

7.1.3 The speed of ongoing climate change

In recent decades, the Earth has been experiencing some major environmental changes, and is currently under pressure. Using meteorological and long-scale historical records, including ice cores and marine sediment records, there is clear evidence that the Earth's climate is changing at a faster rate than at any time. Most of this can be attributed to anthropogenic disturbance of biogeochemical cycles, notably the carbon cycle. Such change is having (and will continue to have) an impact on the biosphere, affecting habitat conditions for sustainability and survival all over the world (IPCC, 2014). In terms of understanding the effects of warming on our planet, it is essential to study *Biodiversity*, the final target of the impacts of climate change. Biodiversity encompasses the variability (within and between species) among living organisms at all levels of biological organisation (Anonymous, 2009). Driven by evolution, it is related to diversity of life at different levels: genes, individuals, species and habitats on Earth, as well as – above all – their interconnections and relationships. The term encompasses far more than a simple species count – it includes all aspects of variation and function at different levels of biological organisation, from genomic expression, through biochemistry, physiology, life history, ecology and biogeography (di Prisco et al., 2012a). Present patterns of biodiversity and distribution, in extreme environments as well as elsewhere, are a consequence of processes working on physiological, ecological and evolutionary timescales (Walther et al., 2002; Pörtner & Knust, 2007; Pörtner & Farrell 2008; Peck, 2011, 2018), which can be modified by environmental change.

Understanding the responses of organisms in extreme environments to such climate change is essential, as they contribute significantly to the functioning of the planet. They often operate at the boundaries of life, representing particularly fragile ecosystems. Even a small climate change may affect the structure and dynamics of biotic communities, as shown by examples from corals (Reaser et al., 2000). More extreme climate conditions may lead to unrecoverable loss of biodiversity in some areas. Species extinction may in turn seriously impact on higher organisms in the food chain and on biogeochemical cycles (Bellard et al., 2012). On the other hand, it may redistribute diversity according to clinal gradients and also include expansion of mesic species into previously inhospitable areas, due to amelioration of the environment ('Greening of the Earth'; Walther et al., 2002). Although invading species may outcompete some of the extreme local endemic biota, with survival implications, extreme environments will hopefully represent an exploitable resource for future sustainability. Organisms living in extreme environments are at the end of a continuum of adaptation where a wide range of unusual, novel and hence potentially useful features have been produced. Life at extremes can provide considerable opportunities in terms of biotechnological, industrial and medical applications and wealth creation, enabling us to take advantage of future sustainable exploitation of new territories and resources (Coker, 2016). They have the potential to play an essential role in societal adaptation to and living with environmental change via the exploitation of new marginal areas, untapped resources and colonisation of emerging niches.

7.2 Biogeography in extreme environments: expansion/contraction of climatic zones in a changing world

Under any predicted scenario (IPCC, 2014), climate-change effects produce two factors of habitat loss and habitat gain. These may pull in opposite directions, may operate in the same areas, and complement each other for affecting extreme environments/ecosystems. For example, the polar waters warm, expanding the milder temperate regions and associated ecosystems (Walther et al., 2002); reduced sea-ice cover and glacier melting expose new areas of sea and land, and both increase the area available for primary productivity (Clarke et al., 2007; Peck et al., 2010a); polar-sea warming, compared to land, is of the order of 1–2°C only, but this is by no means negligible, leading to changes in the food chain as many temperate species migrate to colder habitats and cold-adapted species are replaced by those adapted to warmer habitats. Increasing temperatures and decreasing water availability lead to desertification and salinisation of lands, with reduction of productivity (Altman & Hasegawa, 2012; Hooper et al., 2012), but may also increase the extension of temperate habitats.

Several scenarios have been predicted using plants as the model system, but the general principles may also be applied to other terrestrial forms (Aitken et al., 2008, among others). Most studies deal with extirpation (or local extinction) and migration of temperate species and are based on predictive species-distribution models. According to models, species respond individually to ecological forces and their range may shift in different directions, just as they have done in the past, therefore vegetation types may expand or contract (Thuiller, 2003, 2004; Bush et al., 2004; Parmesan, 2006; Hamilton & Miller, 2015; Bjorkman et al., 2016; Pellissier et al., 2016).

We may expect, therefore, that rapid climate change or extreme climatic events will alter community composition (Walther et al., 2002). Moreover, migration ability depends on the presence of suitable habitat corridors and dispersal mechanisms, and range-restricted species are often habitat specialists and poor dispersers (Parmesan, 2006; Colwell et al., 2008; Dirnböck et al., 2011; Pedersen et al., 2016). The importance of these factors is enhanced when combined with the loss and fragmentation of habitats due to anthropogenic activities (Travis, 2003).

In the present historical phase, for the first time, climate change and habitat fragmentation are also human-driven factors. How plants will respond and react is unpredictable as plants are facing unprecedented rates of anthropogenically induced climate change, which has no analogue in the past (Thomas et al., 2004; Christmas et al., 2016).

Chen et al. (2011) considered relationships between observed and expected range shifts in response to climate change for latitude and elevation for different taxonomic groups: birds, mammals, arthropods, plants, reptiles, fish and molluscs. Interestingly, most variations are reported among species within the same taxonomic group than between taxonomic groups. Moreover, for latitudinal shifts, many species shifted in the opposite direction to that expected. In their meta-analysis, Chen et al. (2011) confirmed the findings of Thuiller (2003, 2004) about the species' individual response to ecological forces: namely species may shift their range in different directions. In this way, the differential distribution capacity of species may lead to reassembling of communities and ecosystems (Hampe, 2011).

Such analyses are lacking for organisms from extreme environments, but all models forecast that organisms and ecosystems from hot areas will see

Table 7.1 *Predicted change in areas of the planet (in km²) colder than −20°C and warmer than 27.5°C using the global anthropogenic radiative forcing models for a medium-low (RCP4.5) and a high (RCP8.5) scenario over the next 100 and 300 years*

Experiment	Start year	End year	Cold areas <−20°C	% change on 2006–2015	Warm areas >27.5°C	% change on 2006–2015
RCP4.5	2006	2015	14.1		40.6	
RCP4.5	2090	2099	11.9	−16%	110.7	+273%
RCP4.5	2290	2299	11.0	−22%	125.7	+310%
RCP8.5	2006	2015	13.7		38.8	
RCP8.5	2090	2099	9.9	−28%	171.1	+441%
RCP8.5	2290	2299	6.0	−44%	265.6	+685%

Data generated by A. Phillips and J. Turner, British Antarctic Survey.

a massive range expansion while those from cold areas will see a decrease. In a warming world the areas lost first are cold (see examples in Table 7.1, Figure 7.1). This will significantly impact on biodiversity and conservation challenges in our future world, with cold-adapted species being the losers and warm-adapted organisms the winners. In marine environments, warming primarily affects the latitudes, ranges and depths at which adapted organisms are found, with biodiversity at high latitudes, the tropics and semi-enclosed seas most at risk of local extinction. High latitudes have been predicted to be the most vulnerable also *vis-á-vis* of competitive invasives (Cheung et al., 2009).

Given the frequent lack of correlation between predicted and observed range changes, more than ever we need to keep surveying species with limited habitat ranges, particularly extremophiles, as they include organisms with both the best and least capacities to withstand environmental changes. Further impacts are a predicted increase in environmental instability in the coming years in some areas, as a result of expansion and contraction of climatic zones, leading to an increased frequency of extreme weather events (Easterling et al., 2000). How these events impact extreme environments is unknown.

7.3 Strategies of adaptation to changes in the environment

In geological time, the continental drift, still under way, has produced climatic changes gradually approaching the current ones, also generating extreme conditions, as we know them to date, in large areas. The ability to survive under such extreme conditions implies that life, in the course of tens of millions of years, has overcome key constraints by evolving specialised adaptations. Studying the biology of organisms thriving in extreme environments is essential for understanding their historical capacity for developing adaptations to climate change.

(a)

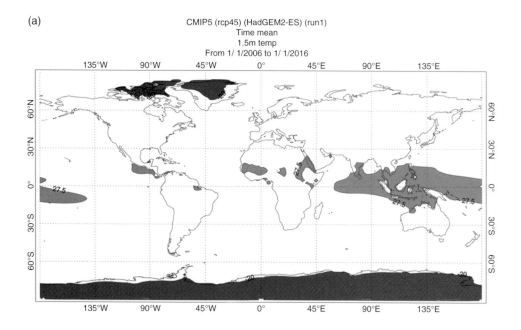

(b)

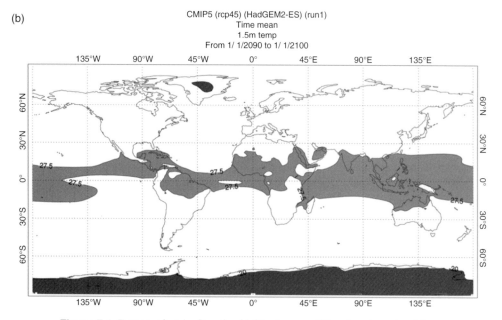

Figure 7.1 Extreme hot (red) and cold (blue) areas of the planet, under: (a) current state (2006–2016 average); (b) end-century (2090–2100) projection where radiative forcing due to human activities stabilises at ~+4.5W m^{-2} (end-century CO_2 = ~ 540 ppm); (A black and white version of this figure will appear in some formats. For the colour version, please refer to the plate section.)

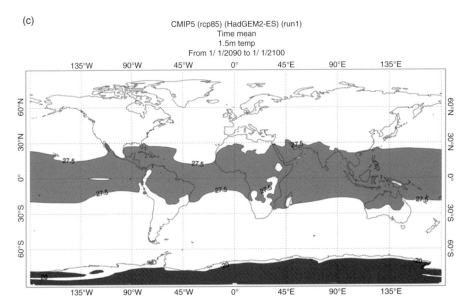

Figure 7.1 (cont.)
(c) end century (2090–2100) where radiative forcing due to human activities stabilises at
~+8.5 W m^{-2} (end-century CO_2 = ~ 935 ppm). Plots were all generated from the HadGEM2-
ES model and are on an equirectangular projection which increases polar areas relative to
other regions. Red areas are those where annual average temperatures (air at 2 m above
ground) are above 27.5°C, and blue areas are those where average temperatures are below
−20°C. Plots were generated by A. Phillips and J. Turner, British Antarctic Survey.

Global warming and associated changes may have a number of consequences. For example, altered forest productivity, desertification, changes in
ocean acidification, altered food-web dynamics, ice and permafrost melting,
impact on fishing resources, reduced abundance of habitats, shifts in species
distribution and altered disease demography (Ackerly et al., 2010; Bellard
et al., 2012; Dobrowski et al., 2013; Naudts et al., 2016) are occurring at
unprecedented speed, especially in polar regions. Consequently, investigating
impacts of fast climate change on historically adapted terrestrial and marine
organisms and their thresholds and resilience has become an urgent research
mission, in parallel with attempts to address features of newly adapted extremophiles to biotechnological, industrial and medical applications.

7.3.1 Capacity to respond to climate change
Predicting the impact of changes, even at the level of the single species, is non-
trivial. Organisms can respond to environmental threats in a range of ways,
from intracellular biochemical buffering to changes in species distribution

and ecosystem balance, depending on the scale of the change in time and magnitude (Peck, 2011).

The most important tools to develop resistance or new adaptations to climate-change impacts are phenotypic and genotypic plasticity, especially acclimation in macroorganisms, and modified genetic composition at the population level (Somero, 2010; Peck, 2018). A shift in species' potential geographical distribution may cause reorganisation of genetic diversity at the species level, because locally adapted populations may respond in different ways to environmental changes (Kremer et al., 2012).

In a scenario of globally changing conditions, species and populations unable to respond through phenotypic plasticity or genetic adaptation to climate change are at increased risk of extinction or reduced fitness (Aitken et al., 2008). Indeed, species survival is determined by both migration and adaptive potential to novel conditions within an appropriate timeframe. These features ultimately depend on the species' genetic diversity (Parmesan, 2006; Hamilton & Miller, 2015; Bjorkman et al., 2016; Pellissier et al., 2016).

The genetic make-up of populations can be altered *via* either mutation of novel genes or transfer of genes between populations (gene flow). All organisms need to have sufficient phenotypic plasticity to survive long enough to allow the mechanisms providing adaptation through altered population genetic composition to become effective. The importance of phenotypic plasticity to a given species thus depends on how rapidly the process of genetic change can be entrained (Figure 7.2), and this varies markedly with life history traits, e.g. generation time, longevity, population size and reproduction (Peck, 2011). For organisms with very short generation times, or rapid turnover times, such as bacteria or many nematode species, the need for phenotypic plasticity is small, whereas it is essential for species with generation times over 20 years, e.g. the antarctic brachiopod *Liothyrella uva* (Meidlinger et al., 1998). Species with exceptionally long generation times, e.g. the bowhead whale (*Balaena mysticetus*) at 45 years (Taylor et al., 2007), or the tropical palm *Euterpe globosa* at 101 years (Petit & Hampe, 2006), may be particularly poor at adapting to rapid environmental change *via* altered population genetic composition, and are likely to rely heavily on other characteristics. Without the phenotypic plasticity to survive at least 2–3 decades of change and reproduce, such species will have reduced chances of survival.

Other factors of importance in adapting to change are population size, dispersal capability, number of offspring at each reproductive event, type of reproduction (broadcast spawner vs. brooding strategy; Hoffman et al., 2011), mutation rates and epigenetics. It is well established that

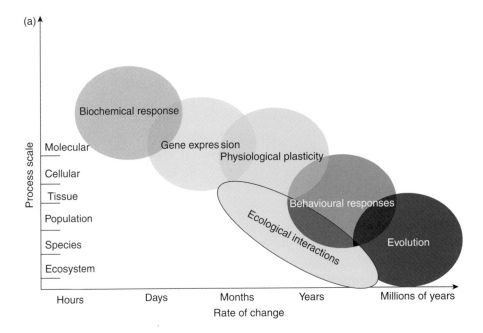

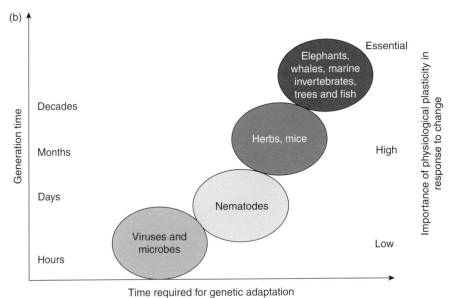

Figure 7.2 (a) Potential biological responses to environmental change and rate of change; (b) thermal responses: the importance of physiological plasticity in response to change.

epigenetic modifications can be induced by abiotic stresses and can induce changes in gene expression, for example, through cold-stress-induced DNA hypomethylation in maize roots, and salt-, cold- and toxic aluminium-induced demethylation in tobacco (Chinnusamy & Zhu, 2009). These epigenetic modifications have morphological, physiological and ecological consequences, which can be heritable across generations (namely there is a 'stress memory') and thus are potentially very relevant for adaptation to changing environments. Examples of epigenetic regulation are being identified in non-model species (e.g. Bossdorf et al., 2008; Ekblom & Galindo, 2011), and the extent to which it operates and its importance are under preliminary investigation.

Knowledge of adaptive strategies calls for further investigations to shed light on the stress caused by impacts of ongoing climate change on extremophiles and on species thresholds and resilience. Any response to stress starts with stress perception, followed (before reaching the critical 'threshold') by activation of defence and repair mechanisms that may result in hardiness to the stress. There is currently little information as to the resilience of biota when the environment changes. Organisms often take advantage of 'stress proteins', e.g. Late Embryogenesis Abundant (LEA) (Battaglia & Covarrubias, 2013) and antifreeze proteins (Duman & Olsen, 1993), chaperones, osmolytes and osmoprotectants, scavengers of reactive oxygen species (ROS) and detoxifying metabolites, which are also found in many other organisms from more amenable ecological niches (Das & Roychoudhury, 2014), as well as in some from recognised extreme environments such as Antarctica (Clark et al., 2013, 2018). Stress conditions affect not only the response of single organisms but can have even greater impact on community structure and function. The community response could be short-term, e.g. changes in net primary productivity, temporary migration of species, or long-term, e.g. permanent changes in community types (Walther, 2010).

Another important issue is the relationship of resilience with the evolutionary time scale of extreme environments. For example, the Antarctic Circumpolar Current formed about 15 million years ago, whereas the age of the Arctic Ocean is only about 1 million years. Both have similar thermal regimes, but the biotas have had different time scales to evolve and adapt, hence they may have different physiological capabilities and resiliences. Comparative investigations on the mechanisms of adaptation and the history of evolution of organisms/ecosystems thriving in the Arctic and the Antarctic will enable us to dissect out the evolution of essential cold-adapted genes from evolutionary isolation and specialisation (Wenzel et al., 2016; Gutt et al., 2018).

While physiological and biochemical studies have been carried out on many organisms over the last six decades, with the advent of pyrosequencing and synthetic biology, scientists now have access to thousands of genes and genomes, and are able to catalogue some adaptations and compare whether the same or different mechanisms operate in different species. The investigation of response mechanisms to stress using omics, complemented by experimental ecology approaches, will allow consideration of the capacity of species to adapt at individual and population scale.

All this genomic and life history information, along with a knowledge of the specialised adaptations of extremophiles, will enable us to predict future demographies. We are now able to directly access the genes and gene pathways involved (and exploit adaptations for applications), and technologies to evaluate networks of genes involved in processes are opening further avenues. With such a variety of organisms in extreme environments to choose from, the question arises as to which species are the most important to target in the immediate future.

Will climate change irreversibly change extremophile habitats to the extent that some of these organisms will become extinct (and their specialised adaptations with them)? To develop a plan for the future, we first need to understand how extremophiles evolved in past history, their specialised adaptations and the time scales involved. Adaptations have been gradually developed over thousands/millions of years to allow organisms to thrive in extreme environments, but may significantly prejudice species' abilities to respond to future change occurring over much shorter time scales, particularly in addition to reduced availability or irreversible destruction of habitat. In some instances, particularly bacteria, extreme environments can allow relatively fast changes in the genome of the species because the selection pressures and the interspecific competition are higher than in non-extreme environments (López-Maury et al., 2008; Li et al., 2014). But in other extreme environments, such as the polar oceans where growth and development are very slow and generation times long, changes in genomes are likely to be slower than elsewhere (Peck, 2018). In this respect, data are lacking in multicellular organisms.

7.3.2 Plants

The impact of global change on adaptation strategies ensuring survival of plants and terrestrial symbiotic organisms under harsh conditions needs investigation. Strategies include (a) resilience (R), i.e. being affected by the climate change-driven stress, but fully recovering after the stressful event has ended; (b) adaptation (AD); and(c) avoidance (AV). The main features involved with these strategies are as follows:

- **_Reversible alteration of photosynthetic pigments_ (R):** Plants also need to dissipate light energy when photosynthesis proceeds slowly. Poikilochlorophyllous resurrection plants survive long droughts by dismantling chlorophyll, which switches off photosynthetic electron transport (and formation of dangerous ROS) until photosynthesis is restored (Beckett et al., 2012).

- *Use of novel energy sources and pathways* **(AD)**: For example, at extreme high temperature and drought, Agave, Aloe, many epiphytes, lithophytes and xerophytes reduce water loss by nocturnal CO_2 fixation via the Crassulacean Acid Metabolism (Chaves et al., 2003).
- *Symbiotic relationships* **(AD)**: In some cases, the community can better cope with the stress as a result of symbiosis among different organisms. The physiology and biochemistry of symbiotic organisms differ from those of the individual partners in the relationship. Lichens have little control over their hydration, yet tolerate desiccation, and rehydrate similar to resurrection plants, thus surviving in the arctic tundra, hot deserts, dry rocky areas, polar regions, toxic soil surfaces and even space (Li & Wei, 2016). Extreme environments are often characterised by low levels of nutrient resources. Symbioses increase availability of soil nutrients for plants, e.g. through mycorrhiza associations (Augé, 2001).
- *Morphological/anatomical modifications* **(AD)**: Xerophytes (succulents) and halophytes carry out morphological/anatomical modifications. For example, deep rooting enables the plant to avoid surface drought and to enhance water uptake from deeper layers in extreme deserts; induction of root hairs enhances water uptake; suitable leaf orientation avoids excessive radiation, heat and freezing; leaf folding reduces light and heat. Other factors are reflective leaf hairs and/or waxes, accumulation of pigments that mask chlorophyll and act as antioxidants, low area/volume ratio, thick cuticles and/or wax deposits to decrease water loss from leaves (De Micco & Aronne, 2012).
- *Enhanced osmotic regulation* **(AD)**: Metabolites and/or water and ion transporters are used to avoid increase in cell osmotic potential from partial desiccation and salinization, e.g. in xerophytes (resurrection plants, the desert plant *Selaginella lepidophylla* and the tardigrade *Hypsibius dujardini*) and halophytes (*Atriplex*) (Blum, 2017).
- *Dormancy* **(AD-AV)**: This is connected with external stressors, but not directly induced by them (diapause), as seen in seeds, desiccation of leaves (e.g. resurrection plants), lichens, or more often involving a decrease of metabolic activity, down to undetectable levels without visible morphological cell differentiation (quiescence) (Considine & Considine, 2016). Many organisms which adopt dormancy are not extremophiles, but in their dormant state they become highly resistant to extreme challenges.
- *Migration and use of refugia* **(AV)**: Plants can 'avoid' stresses migrating into refugia or more suitable stations. Refugia do not need to be away from the site where the stress occurs. For example, endoliths survive within the rock

matrix, in deserts (Negev and Sinai), geothermal environments (Yellowstone) or the Antarctic Dry Valleys (Antony et al., 2012). For plants, climate change is a formidable challenge as migration may not occur fast enough to avoid extinction due to quick progression towards inhospitable environments (Thuiller et al., 2005).

The anticipated rates of climate change often make dispersal and colonisation more successful strategies than survival through *in situ* adaptation. Tree populations confined to the edge of a species' range are often (Yellowstone) characterised by smaller size, lower density and reduced connectivity in comparison with core populations, as a consequence of more extreme environmental conditions. Because of these features, peripheral populations are expected to have lower genetic diversity, a higher level of population subdivision and to be more prone to extinction compared with those in the central ones (Volis et al., 2016). Low genetic diversity may strongly limit the adaptive potential for range expansion/shift in response to environmental changes (Peck, 2011). These features are particularly serious for relict species and can further reduce gene flow and increase risk of maladaptation in a changing environment (Aitken & Bemmels, 2015).

In the Arctic, it is estimated that, by 2100, 40% of the tundra will be replaced by boreal forests (Walker et al., 2006). The potential impacts of climate change on tundra biodiversity at all levels have been comprehensively reviewed (Callaghan et al., 2004). The melting of arctic permafrost has other important implications. Besides releasing greenhouse gases (CO_2 and methane) to the atmosphere (Zona et al., 2016), warming also exposes arctic biota to different conditions that could threaten biodiversity. Similar warming scenarios are also true in alpine environments, with high-altitude species disappearing from lower elevated sites and becoming concentrated on the highest ridges (Klanderud & Birks, 2003). In a related study, Zhu et al. (2012) analysed tree shift in eastern USA and showed that most forest-tree populations experience range contraction rather than expansion as a consequence of climate change. It is not clear whether warming-driven tree-line shift and forest range also depends on other factors, e.g. availability of water and nutrients at higher elevations (Grace et al., 2002). Migration upwards of mountain-tree populations has negative consequences on the distribution of endemic species with restricted habitat above the tree line: the gain of new habitats for migrant species implies leaving limited suitable habitats for high-altitude endemic species, which might go extinct. Moreover, the velocity of climate change is likely to exceed the migration capacity of many endemics (Dirnböck et al., 2011; Parmesan, 2006; Pearson, 2006).

Climate change-driven migrations can cause large and sometimes catastrophic ecosystem changes. Indeed, we should also consider that the response of some species to climate change might have an indirect impact on species that depend on them, such as, for example, insects involved in plant–pollinator interactions (Bellard et al., 2012). This problem is especially clear for parasites where predictions are for potential large losses with environmental change (Carlson et al., 2017). Following climate change, phenological shifts in flowering plants and pollinator activity may lead to mismatches in plant and insect populations, and ultimately to the extinction of both populations (Bellard et al., 2012).

7.3.3 The polar marine regions

Climate predictions indicate that the organisms adapted to cold areas will see unidirectional reductions in their habitat (Table 7.1, Figure 7.1), migration will not be possible and they are the only fauna that will have the sole option of coping with living in a warmer world. The outcomes for these species, and biodiversity as a whole, will depend on whether they are metazoans or prokaryotes, and this will also indicate research priorities, based on the risk of permanent loss of capacity to count on their adaptations. Polar environments, in particular, provide several excellent case studies (Hewitt, 2000), with some urgency attached for further investigations, as temperature and pH in these regions are currently changing at more than twice the global average rates (Hansen et al., 2006; Kapsenberg et al., 2015; Verde et al., 2016).

The concentration of CO_2 in the atmosphere is being enhanced by human activities, leading to global warming. The polar waters, like other oceans, are responsible for sequestering atmospheric CO_2. They have a high capacity for the dissolved gas, but this capacity decreases with warming and will reduce the effectiveness of polar oceans in restraining CO_2 accumulation in the atmosphere. This reduction, however, does not prevent the enhanced levels of atmospheric CO_2 from making the oceans more acidic, and this is threatening many marine organisms having calcareous exoskeletons which dissolve at lower pH. Together with warming of the water, the loss of such organisms may further reduce the efficiency of polar oceans in curbing global warming by carbon sequestration (Runting et al., 2017, and references therein).

In the Antarctic ocean, which permanently experiences temperatures between −2°C and +2°C, recent decades have seen the summer temperature increase by 1°C to the west of the Antarctic Peninsula. Many organisms have very tight thermal limits (they are stenothermal), and this warming is already causing a major geographical retreat, with warmer water species invading, and could trigger a regime change in polar marine biota (Rogers, 2007; Somero, 2012). The western side of the Antarctic Peninsula experienced one of the fastest rates of climate change on the planet in the last half of the twentieth century (Meredith & King, 2005), leading to reduction of annual mean sea-ice extent. There are indications that *Pleuragramma antarctica*, a key fish species of the trophic web and whose reproduction is closely associated to sea ice, is disappearing and being replaced by myctophid fish and salps, which become new food items for predators (Moline et al., 2008; Smith et al., 2014). This event is thought to be caused by seasonal changes in sea-ice dynamics, disturbing reproduction processes. These multiple stresses are exerting a significant pressure, leading to unpredictable changes in ecosystem state and biodiversity, since even the deeper-sea regions of the ocean are impacted.

With regard to metazoans, the evolution of fishes and invertebrates of the Southern Ocean has been taking place for at least 14 million years in ice-cold waters with a high oxygen content, isolated from the rest of the ocean by the circumpolar current and the Antarctic Polar Front, and hence there is a high proportion of endemics. Unlike the Arctic, probably because of contrasting effects of the ozone hole (Andrady et al., 2016), in Antarctica the temperature is keeping constant (or even undergoes a slight decrease), and warming has only occurred in the Peninsula region. It is expected that, if and when the ozone hole disappears, warming will take place in the whole antarctic region.

A warmer world may be catastrophic for cold-adapted animals that live and reproduce in isolation. They have an aberrant heat shock response and appear particularly vulnerable to small increases in temperature (Clark & Peck, 2009; Peck et al., 2009, 2010b, 2014). Here, selection pressure has generated some novel adaptations: in the oxygen-rich waters, biosynthesis of oxygen-binding proteins has been permanently relaxed, and haemoglobin expression was lost in Channichthyidae (icefishes), a family of the dominant fish suborder Notothenioidei (Cocca et al., 1995; di Prisco et al., 2002); giant muscle fibres have evolved because diffusional constraints are relaxed (Johnston et al., 2003); fish avoid freezing by synthesising antifreeze (glyco)proteins all year round (de Vries & Cheng, 2005); and gigantism is observed in some pycnogonids and isopods (Chapelle & Peck, 1999). Warmer waters will mean reduced dissolved oxygen availability, with unknown impacts; in fact, warming is already promoting eutrophication and hypoxia, and some environments are experiencing periodic oxygen depletion. Understanding and cataloguing these adaptations is important, since it is likely that these species will be compromised in their evolutionary responses to current and future climate changes. In contrast, the arctic marine fauna comprises a relatively young assemblage characterised by species from either the Pacific or Atlantic with notably few endemics (Dunton, 1992; Mecklenburg et al., 2011). The long continental coastlines stretching from tropical latitudes up into the Arctic and the absence of a Polar Front produces no isolation (it rather encourages seasonal migration), hence marine organisms are adapted to seasonal variations of temperature and associated factors: there are no fish that lack haemoglobin, and antifreeze (glyco)proteins are often synthesised only in winter. In the Arctic, the glacier and sea-ice decrease is similar to that of the Antarctic Peninsula. It is opening new routes for sea traffic and causing displacement of many marine species (in particular those whose reproduction is associated to sea ice), with a strong impact on the food web, and strong socioeconomic outcomes (e.g. fisheries) (Notz & Stroeve, 2016; Moore & Reeves, 2018; Maksym, 2018).

Climate change has completely changed the lifestyle, entirely dependent on marine food, of the human population living in the Bering Sea area, raising both scientific and political concerns (Smetacek & Nicol, 2005). Fish stocks in the Bering and Barent Seas have fluctuated significantly in the last decades as a result of changes in fishing pressure in response to climate conditions, such as storm frequencies. The retreat of sea ice is unrelated to natural climate variability and is caused by anthropogenic impact (Johannessen & Miles, 2004; Hasselmann et al., 2003; Notz & Stroeve, 2016). By the end of the century, the Arctic Ocean might be essentially ice-free during the summer. The reductions in the extent of cover and thickness of arctic sea ice are dramatic, and potentially devastating to some species, e.g. the top predator polar bear. Marine ice algae would disappear due to loss of habitat, and this may cause a cascade effect on higher trophic levels in the food web: zooplankton feed on algae, many fish (e.g. cod) feed on zooplankton, and sea birds and mammals feed on fish.

While cold-adapted metazoans often have long life cycles and therefore will have few generations to adapt to higher temperatures, more complex, although no less important for the future, are the issues of endangered microbial species and shifts in microbial-driven biochemical changes (Verde et al., 2012, 2016). In both arctic and antarctic marine ecosystems, microorganisms dominate the gene pool and the biomass. They are integral to biogeochemical cycling and in the maintenance of proper ecosystem function. However, we may lose access to the adaptations of these short-lived species as they potentially mutate to life in a warmer world. We also do not know whether their resilience/sensitivity will translate to higher trophic levels and food webs (Wilkins et al., 2013). The gene sequencing projects currently under progress may increase our knowledge of the structure–function of polar ecosystems as a whole and help *in silico* modelling studies to predict evolutionary patterns of individual genomes and communities in response to climate changes (Verde et al., 2012; Merilä & Hendry, 2014).

These examples once more highlight the need of long-term, biologically oriented and ecosystem-based observational systems able to accommodate time- and experience-dependent species' modifications occurring through natural selection and phenotypic plasticity. However, within this monitoring system, we need to develop a catalogue of the genetic adaptations of these 'fragile' species, with an understanding of their underlying mechanisms for coping with their extreme environments (Peck, 2011).

7.4 Exploitation of extremophiles for ecosystem services

Exploring the biodiversity of extreme environments also offers opportunities in, and contributions to, the applied science realm. These include the use of 'extremozymes', compatible solutes and a broad range of other molecules in industrial applications, but also extend to issues such as sustainability where, for example, the outcomes of 'black box' ecological studies to address sustainability questions are likely to be flawed without a thorough knowledge of biological diversity and ecology in extreme environments (Dalmaso et al., 2015). Mitigation mechanisms, ecosystem dynamics and resilience, and bioremediation are among the ecological topics whose study significantly benefits from research on model extreme environments.

Extremophiles are not only valuable resources for developing novel biotechnological processes, but also offer ideal models for investigating how biomolecules are stabilised when subjected to extreme environments. Studies of life in extreme environments help to provide insights into the evolution of life and the development of distinct ecosystems. Moreover, knowledge of the physiological mechanisms underpinning how extremophiles adapt to extreme environments helps us to explore and exploit novel functionality. Such understanding ultimately feeds back into improving process efficiency, particularly exemplified in biocatalysis, which has had such a significant role in the development of industrial and environmental biotechnology. Understanding desiccation tolerance and water housekeeping in xerotolerant plants or fungi is an important basis for plant bioengineering of crop plants in the context of future increasing water shortages.

There is urgent need to evaluate the potential for the sustainable exploitation of life resources of extreme environments, to enable us to better manage future ecosystems. Improved knowledge will allow us to tailor our current exploitation of these environments via domestication and 'gardening' of natural ecosystems to ease pressure on the resources of overexploited areas. This can take many forms, from the use of extremophiles to increase agricultural production and exploitation of highly saline and arid lands, through to biotechnology applications, including a catalogue of specialised enzymes to improve the efficiency of current industrial processes, to be used in synthetic biology to replace current hydrocarbon-derived chemicals and offer novel biomedical applications (Dhamankar & Prather, 2011; Weber & Fussenegger, 2012).

Although the knowledge of living resources in extreme environments is still in its infancy, early initiatives have proven successful. Classic examples are the industrial use of Taq polymerase from the hyperthermophile *Thermus aquaticus* (Lawyer et al., 1993) in the PCR process, which revolutionised molecular biology and the use of thermophilic microorganisms to facilitate conversion of ligno-cellulose into bioethanol (e.g. Graham et al., 2011; Liszka et al., 2012).

With the predicted increased desertification of the planet and the increasing population, organisms (particularly plants) from hot dry and saline environments can be (and in fact are already) exploited to increase the greening of the planet. For example, deep-rooted trees and shrubs, as well as xerophytes and halophytes, can grow in highly dry and saline environments. Bressan et al. (2012) labelled agricultural wild species (both food and non-food) already adapted to extreme environments as 'Extremophile Energy Crops'. These plants have expressed genes, developmental processes and metabolic pathways that make them more useful to withstand extreme environments than traditional crops or genetically modified species. Further, they can grow on worthless land, thus avoiding overexploitation of valuable land for agriculture.

In coastal habitats in the tropics and subtropics which are flooded by seawater, many species of Rhizophoraceae and other families possess pneumatophores (pop roots), and their mechanisms of adaptation include limiting salt intake (by excluding salts at the root level and/or secreting salts by leaf glands) and water loss, modifying nutrient uptake from the deficient anaerobic soil, and increasing offspring survival by their floating seeds and/or vivipary. As hot areas of the planet expand, this type of adaptation is an increasingly important step towards food security. At the other end of the temperature scale, although research into antarctic metazoans is as yet limited, there certainly are species which can be exploited. For instance, icefishes are a natural mutant in which to study the human diseases of anaemia and osteopenia (Albertson et al., 2009; Maher, 2009).

Bioprospecting of microorganisms living in extreme environmental conditions has already led to the discovery of new bioactive molecules, mainly enzymes with potential commercial use, isolated in both hot and cold extreme environments. Many industrial processes currently benefit from using enzymes (e.g. proteases, amylases, cellulases, carboxymethylcellulases, xylanases) that function under extreme conditions, including the polar environments (Di Fraia et al., 2000; Cavicchioli et al., 2002; de Pascale et al., 2012; di Prisco et al., 2012b). Such enzymes are extensively used by industry in the production of pharmaceuticals, foods, beverages, confectionery and paper, as well as in textile and leather processing and wastewater treatment. Most of these enzymes are microbial in origin because they are relatively more stable than the counterparts derived from plants and animals.

Cold-adapted lipases, proteases and cellulases are key enzymes in detergent formulations due to their application in cold washing, which reduces wear-and-tear on fabrics (Gomes & Steiner, 2004). In some processes, the use of cold-adapted hydrolytic enzymes may eliminate unwanted side reactions generally taking place at high temperature, and reduce energy consumption and environmental impact (Joseph et al., 2008). These properties are important in the food and feed industries. Genetic elements that confer freeze tolerance in psychrophiles, such as cryoprotectants and antifreeze proteins, may be used to enhance cold tolerance in agricultural crops. Biological ice nucleators and antifreeze proteins have been isolated from polar extremophiles and used in stabilising foodstuffs and cosmetics. One example is their use in the production of low-fat ice cream (Leary, 2008).

In bioremediation schemes, seasonal fluctuations influence the effectiveness of enzyme preparations for pollutant degradation, and the use of both mesophilic and psychrophilic enzymes may enhance the process due to combined activity over a wider temperature range (Margesin et al., 2007). Heat-stable catalase produced by recombinant bacteria degrades excess hydrogen peroxide used in washing semiconductor boards and as a preservative in certain foods. In plant biotechnology, low-temperature enzymes are used to prepare protoplasts. Cold-adapted and heat-stable (e.g. alkaline phosphatases) proteins and/or genetic elements are used in molecular biology (Siddiqui & Cavicchioli, 2006). Halophiles have been a major source of commercially valuable molecules. For example, bacteriorhodopsin has applications in holography, optical computing and optical memory. These microorganisms have also been a source of compatible solutes used as stabilisers of biomolecules and whole cells and as stress-protection agents. Biosurfactants and exopolysaccharides from halophiles are employed in microbially enhanced oil recovery (Margesin & Schinner, 2001).

Biological ice nucleators and antifreeze proteins have been isolated from polar extremophiles and used in stabilising foodstuffs and cosmetics. Biocontrol of plant pathogens and insect pests has also been facilitated using both cold-tolerant and heat-resistant fungi. Potentially beneficial biomolecules from prokaryotes, eukaryotes and symbiotic associations still remain to be discovered from unexplored extreme environments. Bioleaching of commercially significant mineral ores has developed in recent years and has made use of cold-adapted, acid-tolerant microbes. Other extremophiles are being deployed in bioremediation of inorganic and organic environmental pollution and are being considered for the generation of alternative power, e.g. hydrogen (Marques, 2018).

Recent evidence suggests that, in some engineered bacteria, changes in a limited number of genes are sufficient to vary the optimal temperature for an essential process such as growth (Duplantis et al., 2010). In an attempt to develop a new generation of temperature-sensitive vaccines, a variant of *Francisella tularensis* subsp. *novicida*, a typical mesophile, was engineered by replacing some essential genes with homologues from cold-adapted *Colwellia psychrerythraea*. The outcome was to entirely shift the lifestyle of *Francisella* towards cold dependence. This work provided an elegant example of the biotechnological potential of psychrophiles.

Thus, the current climate change not only opens exciting biotechnological perspectives but also bears on species definition and niche concept. Protein engineering, in synergy with bioprospecting, has the capacity to revolutionise the development of extremophilic enzymes. As our understanding on how these enzymes work in extreme environments increases, we can look forward to increasingly using computational tools, such as Artificial Intelligence and deep learning, to design enzymes with new structures and functions more efficiently (Liszka et al., 2012).

7.5 Concluding remarks

7.5.1 New technology

In view of the finding of novel metabolic pathways and modes of interaction between biological systems and their environment, research on life in extreme environments is bound to encompass many areas of biological and environmental sciences.

Molecular biology is at the top of the list. Molecular tools are invaluable for studying diversity and ecosystem function and community structures, with no need for the isolation of individuals or cultivable organisms. Recent improvements in sequencing, microarray and omics have enormously extended our capacity of making such tools highly relevant to studying life in extreme environments. New techniques, e.g. pyrosequencing, can be used to analyse the biodiversity (V-tag deep sequencing) and the function (metagenomics and transcriptomics) of populations, with higher resolution and more rapidly than before. These techniques can detect rare, but often functionally important, species in a complex ecosystem. Comparative omics, in combination with flow cytometry and cell sorting, on symbionts and free-living relatives allow to understand the transition to symbiotic lifestyle. Availability of sampling devices to allow molecular tools to be used *in situ* is still limited, but will be very important in field research in extreme environments, especially under dynamic and unstable conditions.

The rapidly developing connected discipline of systems biology may become as much of a biological revolution as molecular biology has been. There is clearly a significant opportunity for extremophiles in this new science area.

The need to explore and monitor hard-to-reach environments, e.g. hydrothermal vents on midocean ridges, lakes beneath kilometre-thick ice sheets, with space-based research on the limits of life being a further example, raises serious concerns, also because financial investments to develop innovative, leading edge technologies and robust tools are quite high. There is need for infrastructures (stations, ships, platforms), automation (satellites, robots,

AUVs, ROVs), alternative power (radiation, wind power), exploration, *in situ* experimentation, simulation, securing samples, data exchange, bioinformatics, sensor technology, outreach and education.

Time scales related to climate changes and their impacts are worth consideration. Research in some areas can be addressed in relatively short time scales (3–5 years) by small collaborative groups, but many will need a decade or more to make significant progress, and may require international commitments to bring relevant infrastructure into a collaborative programme.

7.5.2 Future perspectives

While our understanding of the impact of current climate change on overall ecological interactions and responses of life in extreme environments is to date very limited, distinct adaptation mechanisms have been discovered and analysed in many extremophiles. Some of these are species-specific, others are more common across different species living in extreme environments. Our improved understanding of some of these basic mechanisms call for mobilising all intellectual resources to achieve research advancements and development activities. There are still many unknowns and there is much to investigate.

The responses of organisms to current changes in extreme environments must be studied over short-term or long-term periods, taking account of acclimation, adaptation and evolution. The phenotypic and physiological plasticity is fundamental in allowing species to adjust to current and future conditions, but the limits of plastic responses may make them unlikely to provide long-term solutions. Genetic modification is needed, through either changes in gene frequency in populations (e.g. via gene flow between populations) and/or genetic adaptation. The time needed for genetic adaptation may not be adequate for some multicellular organisms (e.g. invertebrates, fish) to respond effectively to changing conditions, notably in the polar environment.

Securing and intensifying these activities will allow sustainable exploitation and greening of new regions at the border of life and untapped resources and biomaterials, for the benefit of humanity in the changing planet. To reach this societal goal, along with expanding our knowledge of adaptations developed in geological time, it is mandatory that we succeed in forecasting and understanding the impacts of current changes, as well as the development of future adaptations, as part of investigating and predicting resilience capacity. We will thus have the chance to minimise the negative climatic impacts as much as possible, not only by having the opportunity to devise counteractions, but also by overturning such impacts, at least in part, to the advantage of society. How humans respond to climatic change, particularly to range shift of land and sea species, often has catastrophic implications. Since environmental outcomes and climate outcomes are inextricably linked, human communities should set a policy agenda that actively rewards those countries, industries and entrepreneurs who develop ecosystem-sensitive adaptation strategies (Martin & Watson, 2016).

In the next 10 years, we need to develop a detailed understanding of:

- the rates at which animals and plants can adapt to new conditions, how this is influenced by gene flow, life history traits and selection pressures;
- the range of ecological community interactions in extreme environments, and how they vary over different space and time scales;
- the impacts of unstable and warming climate on, and response of, organisms that live in extreme environments where the adaptations are easily perturbed, and how this may provide valuable information in analysing the impacts on other environments, especially at midlatitudes. In this respect, knowledge of vulnerable species from cold regions that are disappearing could provide early warning systems for species that will be vulnerable elsewhere;
- the impact of 'farming' extreme environments, namely how the human footprint (and associated overexploitation of natural resources) will change the ecosystem stability of extreme environments.

We also need to develop an extensive and updated catalogue of genes from organisms in extreme environments for exploitation by the biomedical, industrial, agricultural and synthetic biology fields. This should include screening the genetic code of species likely to disappear, and especially those species with unusual or potentially valuable adaptations. These include adaptations for low-temperature enzyme function at the poles and drought resistance in deserts. Extremophiles represent a natural toolbox for coping with a changing world, however, there is no doubt that ongoing climate changes are taking place at high speed. Although a warming planet will not affect the polar regions only, in the Arctic and in parts of the Antarctic such changes are taking place at twice the rate measured in the rest of the world, and time may be particularly pressing for the continued existence of many species.

7.6 Acknowledgements

This work was supported by the European Science Foundation, Polar and Life Science Scientific Committees, and by the European Commission FP7 Environment project: 'Coordination activities on research about life in extreme environments' (CAREX). FL was further supported by the EC-FP7 Environment project: 'Effects of Climate Change on Air Pollution and Response Strategies for European Ecosystems' (ECLAIRE). The research was carried out in the framework of the SCAR programme 'Evolution and

Biodiversity in the Antarctic' (EBA), and of the 'Integrated Land-Ecosystem-Atmosphere Process Study' (iLEAPS). MSC and LSP were funded by the Natural Environment Research Council, under the British Antarctic Survey National core funding. CV and GdP were financially supported by the Italian National Programme for Antarctic Research (PNRA). We thank A. Phillips and J. Turner (British Antarctic Survey) for producing Table 7.1 and Figure 7.1.

References

Ackerly, D.D., Loarie, S.R., Cornwell, W.K., et al. (2010). The geography of climate change: implications for conservation biogeography. *Diversity and Distribution*, **16**, 476–487; https://doi.org/10.1111/j.1472 -4642.2010.00654.x

Aitken, S.N., Bemmels, J.B. (2015). Time to get moving: assisted gene flow of forest trees. *Evolutionary Applications*, **9**, 271–290.

Aitken, S.N., Yeaman, S., Holliday, J.A., Wang, T., Curtis-McLane, S. (2008). Adaptation, migration or extirpation: climate change outcomes for tree populations. *Evolutionary Applications*, **1**(1), 95111; https://doi.org/10.1111/j.1752–4571 .2007.00013.

Albertson, R.C., Cresko, W., Detrich, H.W., Postlethwait, J.H. (2009). Evolutionary mutant models for human disease. *Trends in Genetics*, **25**, 74–81.

Altman, A., Hasegawa, P.M. (2012). Introduction to plant biotechnology 2011: basic aspects and agricultural implications. In: A. Altman, P.M. Hasegawa (eds) *Plant Biotechnology and Agriculture: Prospects for the 21st Century*. Elsevier and Academic Press, San Diego, pp. 1–586.

Andrady, A., Aucamp, P.J., Austin, A.T., Bais, A. F., Ballaré C.L. (2016). Environmental effects of ozone depletion and its interactions with climate change: progress report, 2015. *Photochemical & Photobiological Sciences*, **15**, 141–174.

Anonymous (2009). Convention on Biological Diversity. Website. www.cbd.int/

Antony, C.P., Cockell, C.S., Shouche, Y.S. (2012). Life in (and on) the rocks. *Journal of Biosciences*, **37**, 3–11.

Augé, R.M. (2001). Water relations, drought and vesicular-arbuscular mycorrhizal symbiosis. *Mychorriza*, **11**, 3–42.

Battaglia, M., Covarrubias, A.A. (2013). Late embryogenesis abundant (LEA) proteins in legumes. *Frontiers in Plant Science*, **4**; https:// doi.org/10.3389/fpls.2013.00190. eCollection 2013.

Beckett, M., Loreto, F., Velikova, V., et al. (2012). Photosynthetic limitations and volatile and non-volatile isoprenoids in the poikilochlorophyllous resurrection plant Xerophyta humilis during dehydration and rehydration. *Plant Cell Environment*, **35**, 2061–2074.

Bell, G., Collins, S. (2008). Adaptation, extinction and global change. *Evolutionary Applications*, **1**(1), 3–16; https://doi.org/10 .1111/j.1752-4571.2007.00011.x.

Bellard, C., Bertelsmeier, C., Leadley, P., Thuiller, W., Courchamp, F. (2012). Impacts of climate change on the future of biodiversity. *Ecology Letters*, **15**, 365–377; https://doi.org/10.1111/j.1461-0248 .2011.01736.x

Bjorkman, A.D., Vellend, M., Frei, E.R., Henry, G.H.R. (2016). Climate adaptation is not enough: warming does not facilitate success of southern tundra plant populations in the high Arctic. *Global Change Biology*, **23**, 1540–1551; https://doi .org/10.1111/gcb.13417.

Blum, A. (2017). Osmotic adjustment is a prime drought stress adaptive engine in support of plant production. *Plant Cell Environment*, **40**(1), 4–10; https://doi.org/10.1111/pce.12800.

Bossdorf, O., Richards, C.L., Pigliucci, M. (2008). Epigenetics for ecologists. *Ecology Letters*, **11**, 106–115.

Bressan, R.A., Reddy, M.P., Chung, S.O., et al. (2012). Stress-adapted extremophiles provide energy without interference with food production. *Food Security*, **3**, 93–105.

Bush, M.B., Silman, M.R., Urrego, D.H. (2004). 48,000 years of climate and forest change in a biodiversity hot spot. *Science*, **303**, 827–829.

Callaghan, T.V., Bjorn, L.O., Chernov, Y., et al. (2004). Biodiversity, distributions and adaptations of arctic species in the context of environmental change. *Ambio*, **33**, 404–417.

CAREX (2011). *CAREX Roadmap for Research on Life in Extreme Environments*. CAREX Publication no. 9, pp. 1–40. Available at: www.carex-eu.org/

Carlson, C.J., Burgio, K.R., Dougherty, E.R., et al. (2017). Parasite biodiversity faces extinction and redistribution in a changing climate. *Science Advances*, **3**(9), e1602422.

Cavicchioli, R., Siddiqui, K.S., Andrews, D., Sowers, K.R. (2002). Low-temperature extremophiles and their applications. *Current Opinion in Biotechnology*, **13**, 253–261.

Chapelle, G., Peck, L.S. (1999). Polar gigantism dictated by oxygen availability. *Nature*, **399**, 114–115.

Chaves, M.M., Maroco, J.P., Pereira, J.S. (2003). Understanding plant responses to drought – from genes to the whole plant. *Functional Plant Biology*, **30**, 239–264.

Chen, I.-C., Hill, J.K., Ohlemüller, R., Roy, D.B., Thomas, C.D. (2011). Rapid range shifts of species associated with high levels of climate warming. *Science*, **333**, 1024–1026.

Cheung, W.W.L., Lam, V.W.Y., Sarmiento, J.L., et al. (2009). Projecting global marine biodiversity impacts under climate change scenarios. *Fish and Fisheries*, **10**, 235–251.

Chevin, L.-M., Gallet, R., Gomulkiewicz, R., Holt, R.D., Fellous, S. (2013). Phenotypic plasticity in evolutionary rescue experiments. *Philosophical Transactions of the Royal Society B*, **368**, 20120089; https://doi.org/10.1098/rstb.2012.0089.

Chinnusamy, W., Zhu, J.-K. (2009). Epigenetic regulation of stress responses in plants. *Current Opinions in Plant Biology*, **12**, 133–139.

Chown, S.L., Brooks, C.M., Terauds, A., et al. (2017). Antarctica and the strategic plan for biodiversity. *PLoS Biology*, **15**(3), e2001656. https://doi.org/10.1371/journal.pbio.2001656.

Christmas, M.J., Breed, M.F., Lowe, A.J. (2016). Constraints to and conservation implications for climate change adaptation in plants. *Conservation Genetics*, **17**, 305–320.

Clark, M.S., Peck, L.S. (2009). HSP70 Heat shock proteins and environmental stress in Antarctic marine organisms: a mini-review. *Marine Genomics*, **2**, 11–18.

Clark, M.S., Husmann, G., Thorne, M.A.S., et al. (2013). Hypoxia impacts large adults first: consequences in a warming world. *Global Change Biology*, **19**, 2251–2263.

Clark, M.S., Thorne, M.A.S., King, M., et al. (2018). Life in the intertidal: cellular responses, methylation and epigenetics. *Functional Ecology*, **32**, 1982–1994.

Clarke, A., Griffith, H.J., Linse, K., Crame, J.A. (2007). How well do we know the Antarctic marine fauna? A preliminary study of macroecological and biogeographic patterns in Southern Ocean gastropod and bivalve molluscs. *Diversity and Distributions*, **13**(5), 620–632; https://doi.org/10.1111/j.1472-4642.2007.00380.x

Cocca, E., Ratnayake-Lecamwasam, M., Parker, S.K., et al. (1995). Genomic remnants of α-globin genes in the hemoglobinless Antarctic icefishes. *Proceedings of the National Academy of Sciences of the USA*, **92**, 1817–1821.

Coker, J.A. (2016). Extremophiles and biotechnology: current uses and prospects. *F1000 Research, Faculty Reviews*-396; https://doi

.org/10.12688/f1000research.7432.1. eCollection 2016.

Colwell, R.K., Brehm, G.,Cardelús, C.L., Gilman, A.C., Longino, J.C. (2008). Global warming, elevational range shifts, and lowland biotic attrition in the wet tropics. *Science*, **322**, 258–261.

Considine M.J., Considine, J.A. (2016). On the language and physiology of dormancy and quiescence in plants. *Journal of Experimental Botany*, **67**, 3189–3203; https://doi.org/10.1093/jxb/erw138.

Dalmaso, G.Z.L., Ferreira, D., Vermelho, A.B. (2015). Marine extremophiles: A source of hydrolases for biotechnological applications. *Marine Drugs*, **13**, 1925–1965.

Danovaro, R., Corinaldesi, C., Dell'Anno, A., Rastelli, E. (2017). Potential impact of global climate change on benthic deep-sea microbes. *FEMS Microbiology Letters*, **364**(23); https://doi.org/10.1093/femsle/fnx214..

Das, K., Roychoudhury, A. (2014). Reactive oxygen species (ROS) and response of antioxidants as ROS-scavengers during environmental stress in plants. *Frontiers in Environmental Science*; https://doi.org 10.3389/fenvs.2014.00053.

De Micco, V., Aronne, G. (2012). Morpho-anatomical traits for plant adaptation to drought. In: R. Aroca (ed.) *Plant Responses to Drought Stress*. Springer, Berlin/Heidelberg, pp. 37–62; https://doi.org/10.1007/978-3-642-32653-0_1.

de Pascale, D., De Santi, C., Fu, J., Landfald, B. (2012). The microbial diversity of Polar environments is a fertile ground for bioprospecting. *Marine Genomics*, **8**, 15–22.

De Vries, A.L., Cheng, C.-H.C. (2005). Antifreeze proteins and organismal freezing avoidance in polar fishes. In: A.P. Farrell, J. F. Steffensen (eds) *The Physiology of Polar Fishes*, **Vol. 22**. Elsevier Academic Press, San Diego, pp. 155–201.

Dhamankar, H., Prather, K.L.J. (2011). Microbial chemical factories: recent advances in pathway engineering for synthesis of value added chemicals. *Current Opinion in Structural Biology*, **21**, 488–494.

Di Fraia, R., Wilquet, V., Ciardiello, M.A., et al. (2000). NADP$^+$-dependent glutamate dehydrogenase in the Antarctic psychrotolerant bacterium Psychrobacter sp. TAD1. *European Journal of Biochemistry*, **267**, 121–131.

di Prisco, G., Cocca, E., Parker, S., Detrich, H. (2002). Tracking the evolutionary loss of hemoglobin expression by the white-blooded Antarctic icefishes. *Gene*, **295**, 185–191.

di Prisco, G., Convey,P., Gutt, J., Cowan,D., Conlan,K., Verde, C. (2012a). Understanding and protecting the world's biodiversity: The role and legacy of the SCAR programme 'Evolution and Biodiversity in the Antarctic'. *Marine Genomics*, **8**, 3–8.

di Prisco, G., Giordano, D., Russo, R., Verde, C. (2012b). The challenges of low temperature in the evolution of bacteria. In: G. di Prisco, C. Verde (eds) *Pole to Pole, Adaptation and Evolution in Marine Environments*, **Vol. 1**. A book series on the scientific achievements of environmental research during the International Polar Year (IPY). Springer, pp. 183–195.

Dirnböck, T., Essl, F., Rabitsch, W. (2011). Disproportional risk for habitat loss of high-altitude endemic species under climate change. *Global Change Biology*, **17**, 990–996.

Dobrowski, S.Z., Abatzoglou, J., Swanson, A.K., et al. (2013). The climate velocity of the contiguous United States during the 20th century. *Global Change Biology*, **19**, 241–251; https://doi.org/10.1111/gcb.12026

Duman, J.G., Olsen, T.M. (1993). Thermal hysteresis protein activity in bacteria, fungi, and phylogenetically diverse plants. *Cryobiology*, **30**, 322–328.

Dunton, K. (1992). Arctic biogeography – The paradox of the marine benthic fauna and flora. *Trends in Ecology & Evolution*, **7**, 183–189.

Duplantis, B.N., Osusky, M., Schmerk, C.L., et al. (2010). Essential genes from Arctic bacteria used to construct stable, temperature-sensitive bacterial vaccines. *Proceedings of the National Academy of Sciences of the USA*, **107**, 13456–13460.

Easterling, D.R., Meehl, G.A., Parmesan, C., et al. (2000). Climate extremes: observations, modeling, and impacts. *Science*, **289**, 2068–2074.

Ekblom, R., Galindo, J. (2011). Applications of next generation sequencing in molecular biology of non-model organisms. *Heredity*, **107**, 1–15.

Fox-Powell, M.G., Hallsworth, J.E., Cousins, C.R., Cockell, C.S. (2016). Ionic strength is a barrier to the habitability of Mars. *Astrobiology*, **16**, 427–442. https://doi.org/10.1089/ast.2015.1432.

Gomes, J., Steiner, W. (2004). The biocatalytic potential of extremophiles and extremozymes. *Food Technology and Biotechnology*, **42**, 223–235.

Grace, J., Berninger, F., Nagy, L. (2002). Impacts of climate change on the tree line. *Annals of Botany*, **90**, 537–544.

Graham, J.E., Clark, M.E., Nadler, D.C., et al. (2011). Identification and characterization of a multidomain hyperthermophilic cellulase from an archaeal enrichment. *Nature Communications*, **2**, 375; https://doi.org/10.1038/ncomms1373.

Gribaldo, S., Brochier-Armanet, C. (2006). The origin and evolution of Archaea: a state of the art. *Philosophical Transactions of the Royal Society of London B Biological Sciences*, **361**, 1007–1022.

Gunderson, L.H. (2000). Ecological resilience – in theory and application. *Annual Review of Ecology and Systematics*, **31**, 425–439.

Gutt, J., Isla, E., Bertler, A.N., et al. (2018). Cross-disciplinarity in the advance of Antarctic ecosystem research. *Marine Genomics*, **37**, 1–18; https://doi.org/10.1016/j.margen.2017.09.006.

Hamilton, J.A., Miller, J.M. (2015). Adaptive introgression as a resource for management and genetic conservation in a changing climate. *Conservation Biology*, **30**, 33–41; https://doi.org/10.1111/cobi.12574.

Hampe, A. (2011). Plants on the move: the role of seed dispersal and initial population establishment for climate-driven range expansion. *Acta Oecologica*, **37**, 666–673.

Hansen, J., Sato, M., Ruedy, R., Lo, K., Lea, D.W., Medina-Elizade, M. (2006). Global temperature change. *Proceedings of the National Academy of Science of the USA*, **103**, 14288–14293.

Harrison, J.P., Gheeraert, N., Tsigelnitskiy, D., Cockell, C.S. (2013). The limits for life under multiple extremes. *Trends in Microbiology*, **21**, 204–212. http://dx.doi.org/10.1016/j.tim2013.01.006.

Hasselmann, K., Latif, M., Hooss, G. et al. (2003). The challenge of long-term climate change. *Science*, **302**, 1923–1925.

Hewitt, G. (2000). The genetic legacy of the Quaternary ice ages. *Nature*, **405**, 907–913.

Hoffman, J.I., Clarke, A., Linse, K., Peck, L.S. (2011). Effects of brooding and broadcasting reproductive modes on the population genetic structure of two Antarctic gastropod molluscs. *Marine Biology*, **158**, 287–296.

Hooper, D.U., Adair, E.C., Cardinale, B.J., et al. (2012). A global synthesis reveals biodiversity loss as a major driver of ecosystem change. *Nature*, **486**, 105–109.

Hughes, I.I. (2000). Biological consequences of global warming: is the signal already apparent? *Trends in Ecology and Evolution*, **15**(2), 56–61.

IPCC (2014). *Climate Change 2014: Synthesis Report. Contribution of Working Groups I, II and III to the First Assessment Report of the Intergovernmental Panel on Climate Change*. Edited by Core Writing Team, R. K. Pachauri, L.A. Meyer. IPCC, Geneva, Switzerland.

Johannessen, O.M., Bengtsson, L., Miles, M.W., et al. (2004). Arctic climate change: observed and modelled temperature and sea-ice variability. *Tellus*, **56A**, 328–341.

Johnston, I.A., Fernandez, D.A., Calvo, J., et al. (2003). Reduction in muscle fibre number during the adaptive radiation of notothenioid fishes: a phylogenetic perspective. *Journal of Experimental Biology*, **206**, 2595–2609.

Joseph, B., Ramteke, P.W., Thomas, G. (2008). Cold active microbial lipases: Some hot issues and recent developments. *Biotechnology Advances*, **26**, 457–470.

Kapsenberg, L., Kelley, A.L., Shaw, E.C., et al. (2015). Near-shore Antarctic pH variability has implications for the design of ocean acidification experiments. *Scientific Reports*, **5**, 10497; https://doi.org/10.1038/srep10497.

Klanderud, K., Birks, H.J.B. (2003). Recent increases in species richness and shifts in altitudinal distributions of Norwegian mountain plants. *Holocene*, **13**, 1–6.

Kremer, A., Ronce, O., Robledo-Arnuncio, J.J., et al. (2012). Long-distance gene flow and adaptation of forest trees to rapid climate change. *Ecology Letters*, **15**, 378–392.

Kristensen, T.N., Ketola, T., Kronholm, I. (2018). Adaptation to environmental stress at different timescales. *Annals of the New York Academy of Sciences*. https://doi.org/10.1111/nyas.13974. [Epub ahead of print].

Lawyer, F.C., Stoffel, S., Saiki, R.K., et al. (1993). High-level expression, purification, and enzymatic characterization of full-length Thermus aquaticus DNA polymerase and a truncated form deficient in 5' to 3' exonuclease activity. *Genome Research*, **2**, 275–287.

Leary, D. (2008). Bioprospecting in the Arctic. United Nations University Institute of Advanced Study Report (UNU-IAS 2008 report), Nishi-ku, Yokohama, Japan, pp. 1–45.

Li, H., Wei, J.C. (2016). Functional analysis of thioredoxin from the desert lichen-forming fungus, Endocarpon pusillum Hedwig, reveals its role in stress tolerance. *Scientific Reports*, **6**, Article number 27184; https://doi.org/10.1038/srep27184.

Li, S.-J., Hua, Z.-S., Huang, L.N., et al. (2014). Microbial communities evolve faster in extreme environments. *Scientific Reports*, **4**, Article number 6205.

Liszka, M.J., Clark, M.E., Schneider, E., Clark, D. S. (2012). Nature versus nurture: developing enzymes that function under extreme conditions. *Annual Review of Chemical and Biomolecular Engineering*, **3**, 77–102.

López-Maury, L., Marguerat, S., Bähler, J. (2008). Tuning gene expression to changing environments: from rapid responses to evolutionary adaptation. *Nature Reviews Genetics*, **9**, 583–593; https://doi.org/10.1038/nrg2398.

Lunine, J.L. (2006). Physical conditions on the early Earth. *Philosophical Transactions of the Royal Society of London B Biological Sciences*, **361**, 1721–1731.

Maher, B. (2009). Evolution: biology's next top model? *Nature*, **458**, 695–698.

Maksym, T. (2018). Arctic and antarctic sea ice change: contrasts, commonalities, and causes. *Annual Review of Marine Science*, **11**; https://doi.org/10.1146/annurev-marine-010816-060610.

Margesin, R., Schinner, F. (2001). Potential of halotolerant and halophilic microorganisms for biotechnology. *Extremophiles*, **5**, 73–83.

Margesin, R., Neuner, G., Storey, B. (2007). Cold-loving microbes, plants, and animals-fundamentals and applied aspects. *Naturwissenschaften*, **94**, 77–99.

Marques, C.R. (2018). Extremophilic Microfactories: applications in Metal and Radionuclide Bioremediation. *Frontiers in Microbiology*, **9**, 1191.

Martin, T.G., Watson, J.E.M. (2016). Intact ecosystems provide best defence against climate change. *Nature Climate Change*, **6**, 122–124.

Mecklenburg, C.W., Møller, P.R., Steinke, D. (2011). Biodiversity of Arctic marine fishes: taxonomy and zoogeography. *Marine Biodiversity*, **41**, 109–140.

Meidlinger, K., Tyler, P.A., Peck, L.S. (1998). Reproductive patterns in the Antarctic brachiopod Liothyrella uva. *Marine Biology*, **132**, 153–162.

Meredith, M.P., King, J.C. (2005). Climate change in the ocean to the west of the Antarctic Peninsula during the second half of the 20th century. *Geophysics Research Letters*, **32**, L19604.

Merilä, J., Hendry, A.P. (2014). Climate change, adaptation, and phenotypic plasticity: the problem and the evidence. *Evolutionary Applications*, **7**, 1–14; https://doi.org/10.1111/eva.12137.

Millennium Ecosystem Assessment Board (2005). Dryland systems. In: U. Safriel, Z. Adeel (Lead Authors) *Ecosystems and Human Well-being: Current State and Trends*. Island Press, Washington, DC, pp. 625–656.

Moline, M.A., Karnovsky, N.J., Brown, Z., et al. (2008). High latitude changes in ice dynamics and their impact on polar marine ecosystems. *Annals of the New York Academy of Sciences*, **1134**, 267–319.

Moore, S.E., Reeves, R.R. (2018). Tracking arctic marine mammal resilience in an era of rapid ecosystem alteration. *PLoS Biology*, **16**(10); doi:10.1371/journal.pbio.2006708.

Naudts, K., Chen, Y., McGrath, M.J., et al. (2016). Europe's forest management did not mitigate climate warming. *Science*, **351**, 597.

Notz, D., Stroeve, J. (2016). Observed Arctic sea-ice loss directly follows anthropogenic CO_2 emission. *Science*, **354**, 747–750.

Parmesan, C. (2006). Ecological and evolutionary responses to recent climate change. *Annual Review of Ecology Evolution and Systematics*, **37**, 637–669.

Pearson, R.G. (2006). Climate change and the migration capacity of species. *Trends in Ecology and Evolution*, **21**, 111–113.

Peck, L.S. (2011). Organisms and responses to environmental change. *Marine Genomics*, **4**, 237–243.

Peck, L.S. (2018). Antarctic marine biodiversity: adaptations, environments and responses to change. *Oceanography and Marine Biology Annual Review*, **56**, 105–236.

Peck, L.S., Clark, M.S., Morley, S.A., Massey, A., Rossetti, H. (2009). Animal temperature limits and ecological relevance: effects of size, activity and rates of change. *Functional Ecology*, **23**, 248–253.

Peck, L.S., Barnes, D.K.A., Cook, A.J., Fleming, A.H., Clarke, A. (2010a). Negative feedback in the cold: ice retreat produces new carbon sinks in Antarctica. *Global Change Biology*, **16**, 2614–2623; https://doi.org/10.1111/j.1365-2486.2009.02071.

Peck, L.S., Morley, S.A., Clark, M.S. (2010b). Poor acclimation capacities in Antarctic marine ectotherms. *Marine Biology*, **157**, 2051–2059.

Peck, L.S., Morley, S.A., Richard, J., Clark, M.S. (2014). Acclimation and thermal tolerance in Antarctic marine ectotherms. *Journal of Experimental Biology*, **217**, 16–22.

Pedersen, M.W., Ruter, A., Schweger, C., et al. (2016). Postglacial viability and colonization in North America's ice-free corridor. *Nature*, **537**, 45–49.

Pellissier, L., Bronken Eidesen, P., Ehrich, D., et al. (2016). Past climate-driven range shifts and population genetic diversity in arctic plants. *Journal of Biogeography*, **43**, 461–470.

Petersen, J.M., Dubilier, N. (2009). Methanotrophic symbioses in marine invertebrates. *Environmental Microbiology Reports*, **1**(5), 319–335; https://doi.org/10.1111/j.1758-2229.2009.00081.x.

Petit, R.J., Hampe, A. (2006). Some evolutionary consequences of being a tree. *Annual Reviews of Ecology, Evolution and Systematics*, **37**, 187–214.

Pörtner, H.-O., Farrell, A.P. (2008). Physiology and climate change. *Science*, **322**, 690–692.

Pörtner, H.-O., Knust, R. (2007). Climate change affects marine fishes through the oxygen limitation of thermal tolerance. *Science*, **315**, 95–97.

Reaser, J.K., Pomerance, R., Thomas, P.O. (2000). Coral bleaching and global climate change:

scientific findings and policy recommendations. *Conservation Biology*, **14**, 1500–1511.

Rintoul, S.R., Chown, S.L., DeConto, R.M., et al. (2018). Choosing the future of Antarctica. *Nature*, **558**, 233–241. https://doi.org/10.1038/s41586-018-0173-4.

Rogers, A.D. (2007). Evolution and biodiversity of Antarctic organisms: a molecular perspective. *Philosophical Transactions of the Royal Society of London B Biological Sciences*, **362**, 2191–2214.

Rothschild, L.J., Mancinelli, R.L. (2001). Life in extreme environments. *Nature*, **409**, 1092–1101.

Runting, R.K., Bryan, B.A., Dee, L.E., et al. (2017). Incorporating climate change into ecosystem service assessments and decisions: a review. *Global Change Biology*, **23**, 28–41; https://doi.org/10.1111/gcb.13457.

Seufferheld, M.J., Alvarez, H.M., Farias, M.E. (2008). Role of polyphosphates in microbial adaptation to extreme environments. *Applied and Environmental Microbiology*, **74**, 5867–5874; https://doi.org/10.1128/AEM.00501-08.

Siddiqui, K.S., Cavicchioli, R. (2006). Cold-adapted enzymes. *Annual Review of Biochemistry*, **75**, 403–433.

Smetacek, V., Nicol, S. (2005). Polar ocean ecosystems in a changing world. *Nature*, **437**, 362–368.

Smith, W.O., Jr, Ainley, D.G., Arrigo, K.R., Dinniman, M.S. (2014). The oceanography and ecology of the Ross Sea *Annual Review of Marine Science*, **6**, 469–487.

Somero, G.N. (2010). The physiology of climate change: how potentials for acclimatization and genetic adaptation will determine 'winners' and 'losers'. *Journal of Experimental Biology*, **213**, 912–920.

Somero, G.N. (2012). The physiology of global change: linking patterns to mechanisms. *Annual Review of Marine Science*, **4**, 39–61.

Taylor, B.L., Chivers, S.J., Larese, J., Perrin, W.F. (2007). Generation length and percent mature estimates for IUCN assessments of cetaceans. *Administrative Report* LJ-07-01, Southwest Fisheries Science Center, 8604 La Jolla Shores Blvd., La Jolla, CA 92038, USA.

Thomas, C.D., Cameron, A., Green, R.E., et al. (2004). Extinction risk from climate change. *Nature*, **427**, 145–148.

Thuiller, W. (2003). BIOMOD: optimising predictions of species distributions and projecting potential future shifts under global change. *Global Change Biology*, **9**, 1353–1362.

Thuiller, W. (2004). Patterns and uncertainties of species' range shifts under climate change. *Global Change Biology*, **10**, 2020–2027.

Thuiller W., Lavorel S., Arau M.B., Sykes M.T., Prentice I.C. (2005). Climate change threats to plant diversity in Europe. *Proceedings of the National Academy of Sciences of the USA*, **102**, 8245–8250.

Travis, J.M.J. (2003). Climate change and habitat destruction: a deadly anthropogenic cocktail. *Proceedings of the Royal Society B – Biological Science*, **270**, 467–473.

Verde, C., di Prisco, G., Giordano, D., Russo,R., Anderson, D., Cowan, D. (2012). Antarctic psychrophiles: models for understanding the molecular basis of survival at low temperature and responses to climate change. *Biodiversity*, **13**, 349–356.

Verde, C., Giordano, D., Bellas, C.M., di Prisco, G., Anesio, A.M. (2016). Polar marine microorganisms and climate change. *Advances in Microbial Physiology*, **69**, 187–215.

Volis, S., Ormanbenkova, D., Shulgina, I. (2016). Role of selection and gene flow in population differentiation at the edge vs. interior of the species range differing in climatic conditions. *Molecular Ecology*, **25**, 1449–1464.

Walker, M.D., Wahren, C.H., Hollister, R.D., et al. (2006). Plant community responses to experimental warming across the tundra biome. *Proceedings of the National Academy of Sciences of the USA*, **103**, 1342–1346.

Walther, G.-R. (2010). Community and ecosystem responses to recent climate change. *Philosophical Transactions of the Royal Society Biological Sciences B*, **365**, 2019–2024; https://doi.org/10.1098/rstb.2010.0021.

Walther, G.-R., Post, E., Convey, P., et al. (2002). Ecological responses to recent climate change. *Nature*, **416**, 389–395.

Weber, W., Fussenegger, M. (2012). Emerging biomedical applications of synthetic biology. *Nature Reviews Genetics*, **13**, 21–35.

Wenzel, L., Gilbert, N., Goldsworthy, L., et al. (2016). Polar opposites? Marine conservation tools and experiences in the changing Arctic and Antarctic: Marine Conservation Tools and Experiences in the Arctic and Antarctic. *Aquatic Conservation Marine and Freshwater Ecosystems*, **26**(S2), 61–84; https://doi.org/10.1002/aqc.2649.

Wilkins, D., Yau, S., Williams, T.J., et al. (2013). Key microbial drivers in Antarctic aquatic environments. *FEMS Microbiology Reviews*, **37**(3), 303–303; https://doi.org/10.1111/1574-6976.12007.

Zhu, K., Woodall, C., Clark, J.S. (2012). Failure to migrate: lack of tree range expansion in response to climate change. *Global Change Biology*, **18**, 1042–1052.

Zona, D., Gioli, B., Commane, R., et al. (2016). Cold season emissions dominate the Arctic tundra methane budget. *Proceedings of the National Academy of Sciences of the USA*, **113**, 40–45; https://doi.org/10.1073/pnas.1516017113.

CHAPTER EIGHT

The ecophysiology of responding to change in polar marine benthos

LLOYD S. PECK

British Antarctic Survey

8.1 Introduction

Environments vary over all spatial scales, from subatomic to the universe and all timescales from a Plank unit to eternity. There have been very large changes in environments in the past, notably over hundreds to thousands of years during glaciations driven by Milankovitch cycles (Zachos et al., 2001). Life on Earth has survived and evolved over time to adapt to those changes and to exploit nearly all of the available environments on Earth. The very small exceptions include the insides of hydrothermal vents where temperatures often exceed 300°C and some active volcano calderas where toxic gases effectively sterilise surfaces. Surprisingly, microbial life has been isolated from deep inside polar ice sheets (Segawa et al., 2010) and from very hostile environments such as the Rio Tinto river in southwest Spain, which is red and orange in colour for over 50 km because of the high levels of iron in the water and its very low pH of around 2 (Amaral Zettler et al., 2003).

Despite the capacity of life to survive current and past variations in the environment and to colonise habitats with conditions inimical to living organisms, there is great concern that ongoing climate change on Earth has already impacted biodiversity, and there are some well cited examples of species that are now reported to be extinct due to climate change. These include the golden toad and Monteverde harlequin frog, both from Central America (IPCC, 2014). The reason for this is that on a global scale the current rate of change of several factors is faster than any that can be identified from the Earth's recent history (IPCC, 2014). Within the global envelope of change not all areas are the same and the polar regions have changed more and at more rapid rates in many aspects than the lower latitudes (IPCC, 2014; Stroeve & Notz, 2018).

8.2 Polar marine environments

The polar oceans cover a combined area of 34.4 million km², which is around 9.5% of the Earth's total. Of this, the Arctic Ocean has a continental shelf area of 6.7 million km² (Harris et al., 2014) and the Antarctic continental shelf is 4.6 million km² (Peck, 2018), and the combined total (11.3 million km²) is nearly a third of the global continental shelf. These oceans are thus not only important globally for their role in the physics of the Earth system, but provide one of the largest seabed habitats for benthic marine animals to live in.

Polar marine environments are extreme in their temperature, being characterised by having both the lowest and most stable sea temperatures on Earth (Figure 8.1). Temperatures remain below 0°C for most of the year and annual ranges are generally less than 5°C in almost all sites, but less than 0.5°C in the highest latitude sites such as McMurdo Sound (Barnes et al., 2006; Clarke & Gaston, 2006). Polar seas are also extreme in having the most intense seasonality of any marine environments on Earth with incident light varying from 24 hours of permanent light in summer to 24 hours of darkness that lasts around 5 months in the Arctic Ocean at the North Pole (www.pmel.noaa.gov/arctic-zone/gallery_np_seasons.html). This dramatic variation in light drives some of the most remarkable seasonal changes in the environment seen anywhere. Changes in sea-ice are dramatic. There is a combined arctic and antarctic winter maximum sea-ice extent of around 35 million km² (20 million km² in Antarctica, 15 million km² in the Arctic), with a minimum extent of

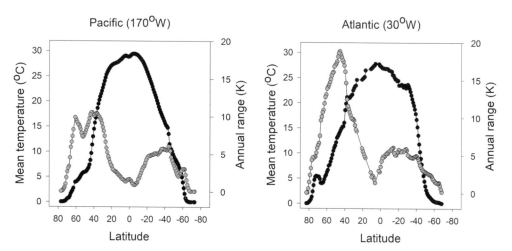

Figure 8.1 Mean annual temperature (black circles, left ordinate axes) and annual range (maximum–minimum experienced temperatures, grey circles, right ordinate axes) for transects through the Pacific and Atlantic oceans from the High Arctic to Antarctica. Modified from Clarke and Gaston (2006). Note the polar oceans are both the coldest and have the smallest annual temperature ranges across the globe.

10–15 million km^2 in summer. Thus, 20–25 million km^2 of ice forms and melts each year in the polar regions. This is an area similar to the whole of North America, or over twice the size of Europe or the USA.

The seasonal variation in light and ice drives a dramatic seasonal cycle of phytoplankton blooms which have the largest differences between summer maximum biomass and winter minimum biomass of any marine site. This is especially so in nearshore fiords where conditions can produce summer phytoplankton blooms that reach peaks in excess of 30 mg Chl m^{-3}, while winter minima are less than 0.01 mg Chl m^{-3} (Clarke et al., 1988, 2008).

Polar marine environments are dominated by ice. Some coastal areas such as the Ross, Ronne, Larsen and Getz iceshelves are permanently under ice up to hundreds of metres thick. Other areas such as large parts of the Weddell Sea and the Oates Coast are permanently covered in multi-year sea-ice up to a few metres thick. Finally, a very large area (10–15 million km^2) is open water in summer, but covered in sea-ice up to a few metres thick in winter, and the amount of time covered varies with latitude, local geography and oceanography and also interannual variation including factors such as the Antarctic Circumpolar Wave (ACW), which is a large atmospheric and oceanic set of covarying anomalies that progressively move around Antarctica in an easterly direction. The result is a wave-like pattern of sea-ice extension and contraction from the continental margin that takes 8–10 years to circumnavigate Antarctica (Cerrone et al., 2017). The sea-ice that forms in winter has powerful effects on habitats for species living beneath. Principally light levels are reduced and wind-induced mixing of the water beneath is reduced which, among other factors, allows particles to sediment out of the water column and reduces winter organic particulate levels to among the lowest on Earth.

Over 90% of species in the oceans live on the seabed (Piepenberg et al., 2010; De Broyer et al., 2014). On the polar seabed, ice is a major structuring factor through the action of icebergs and ice that grows on the seabed itself, the ice foot or anchor ice (Gutt, 2001). The ice foot and anchor ice have been reported to remove life from the seabed to depths of 33 m (Denny et al., 2011). Disturbance from icebergs in shallow polar marine sites removes over 99% of macroorganisms from any given site (Peck et al., 1999) and impacts can occur more frequently than once per year, making these possibly the most physically disturbed marine environments (Brown et al., 2004). Communities living in these habitats are held at early stages of development (Barnes, 2016) and the seabed is a patchwork of habitat age dictated by the time since last impact and the colonisation abilities of the species in the locality (Barnes & Conlan, 2012).

8.2.1 Change in polar marine environments

Climate change is occurring at different rates in different regions of the Earth. The fastest warming has, and is occurring, in the polar regions, with the

Antarctic Peninsula and parts of the Arctic heating up 3°C or more between 1950 and the end of the twentieth century. This warming has produced a dramatic loss of ice that includes both reductions in coastal glaciers and iceshelves and large losses of sea-ice. On the Antarctic Peninsula over 87% of the glaciers and iceshelves have retreated and become smaller in area since the 1950s (Cook et al., 2005), and the vast majority of the others have thinned and lost mass (Cook & Vaughan, 2010). Sea-ice patterns in Antarctica do not show clear trends of loss over large parts of the Southern Ocean, and may even have increased in extent in recent decades. This is because sea-ice formation is affected far more by strong cold winds flowing off the continent cooling surface waters and by the isolation of the Southern Ocean from lower latitude waters than by the warming of the environment (Nghiem et al., 2016). In contrast sea-ice extent in the Arctic has declined on average by over 10% per decade since 1980 (US National Snow and Ice Data Centre, 2018). This loss of ice has many consequences. There are communities that live closely associated with sea-ice, the epontic or sympagic communities. Some of these are charismatic megafauna such as polar bears in the Arctic and Crabeater and Weddell Seals in Antarctica, but a very large diversity of fish and invertebrates live associated with sea-ice including amphipods, mysid shrimps and copepods (Horner, 2018). For these species their habitat is disappearing at an alarming rate and, in the Arctic, this is highly likely the fastest habitat loss due to climate change globally.

Polar nearshore environments are also changing very rapidly in some areas. In some parts of the Arctic, warming has increased freshwater runoff from glaciers, and this has in turn increased the amounts of sediment carried into fjords (Węsławski et al., 2011). The increased sediment load smothers benthic species and reduces the biodiversity of the fjords drastically, which results in areas near the head and middle of the fjord with low diversity and then a gradient of progressively increasing diversity further out. This process does not seem to be happening in the Antarctic, where even on the Peninsula where glacier retreat is extensive, impacts of increased sedimentation have been small (Grange & Smith, 2013). This is thought to be because antarctic systems are at earlier stages in terms of climate change impacts. Future predictions are that in the Arctic this process in fiords is likely to lessen and biodiversity will increase, whereas in Antarctica there may be very large masses of freshwater runoff in some Antarctic sites with associated inundation from very large amounts of sediment deposition, especially in the West Antarctic (Węsławski et al., 2011; Sweetman et al., 2017). The consequences for the benthos in such scenarios would be severe over potentially large areas where current knowledge of biodiversity is poor; for example, much of the nearshore marine environment of the West Antarctic is unsampled in terms of its biodiversity (Griffiths et al., 2011).

As noted in the previous section, icebergs are a major structuring factor in polar nearshore environments that keep biological communities in early stages of development in shallow water less than 15 m deep. Climate change is affecting the numbers of icebergs produced by coastal glaciers. Rapid ice loss from coastal glaciers and iceshelves such as that seen on the Antarctic Peninsula and Greenland over the last 50 years (Cook & Vaughan, 2010) increases the number of icebergs in nearshore environments, and hence the frequency of scouring events (Barnes, 2016). At some stage, receding glaciers reach the grounding line and ice calving from the front no longer produces icebergs. When this happens the number of scouring events in shallow seabed will decline, and it is predicted that shallow benthic near-shore biological communities will rapidly increase biomass and sequester carbon from the system (Barnes, 2016). It is not clear when this process will reduce the number of icebergs along the Antarctic Peninsula, where large-scale glacier retreat has been seen, but it should be noted that for most coastal areas in Antarctica a significant increase in numbers of icebergs as glaciers retreat is yet to be entrained, and the Peninsula is well ahead of elsewhere in this process.

A major concern for polar marine scientists in recent years has been ocean acidification.

Human CO_2 production since the industrial revolution has increased the temperature of the Earth by absorbing radiation from the surface that would previously have passed out into space. Approximately 30% of the CO_2 produced by humans has been absorbed by the oceans. This has decreased the pH of the oceans by around 0.1 units, and most of the change has happened in the last 50 years (Caldeira & Wickett, 2003; IPCC, 2014). The concern for animals is that the overwhelming majority make their skeletons from calcium carbonate and carbonate solubility varies strongly with pH, pressure and temperature. Solubility is higher in more acidic conditions (lower pH), but also at higher pressure and lower temperature. Thus, for the same amounts dissolved per unit seawater and the same concentrations, the solubility state is lower in all three cases. All of these factors also make it harder for animals to remove calcium carbonate from seawater to make skeletons, but also their skeletons become more likely to dissolve as solubility increases. These factors also combine to make all areas of the deep sea undersaturated and the polar oceans to have lower carbonate saturation at the surface than temperate and tropical oceans (Orr et al., 2005; Fabry et al., 2009). Models predict that acidification will increase fastest in the polar regions, especially the Southern Ocean, and the most soluble form of calcium carbonate, aragonite, will become undersaturated at the sea surface, between 2040 and 2060 (McNeil & Matear, 2008; Feely et al., 2009, 2012). When seawater becomes undersaturated ($\Omega < 1$), calcium carbonate dissolves and the lower the value of Ω the more rapid is the dissolution.

8.3 Adaptations of polar marine animals

8.3.1 Historical background

Both the High Arctic and Antarctic have similar marine physical conditions of low temperatures with small variation, extreme seasonality of light, productivity and ice, and intense physical disturbance of the seabed by iceberg scour. However, the level of adaptation to these conditions differs between the two, with the Antarctic fauna demonstrating more extreme adaptations than the Arctic, which we shall see in detail later. The scale of adaptation to an environment depends on several factors, but three of the most important are environmental heterogeneity (the number of niches or niche space for new species), the time that the relevant environment has existed and the degree of isolation from other similar environments.

The polar oceans have very large environmental heterogeneity. The extent of seasonality of light changes markedly from the highest latitudes, where there is 6 months of darkness with no direct sunlight to the lowest polar latitudes (66°33′47.4″) where there are only 1 or 2 days when there is no Sun above the horizon. This alternates with 6 months of permanent light at the poles and only 1 or 2 days of permanent light at the polar circles. The temperate and tropical zones have far less variation than this. Polar seas are also heterogeneous because of the local physical effects of sea-ice melting and forming annually. This provides a surface for species to inhabit and can also affect local salinities significantly (Peck, 2018). Other factors that give great variation in the environment include the disturbance of the seabed by icebergs that varies greatly with depth, current regime and seabed topography.

The polar regions differ in their age, and hence the amount of time there has been for species to adapt to the conditions. In the Arctic, ice sheets extended to over 1 km depth and closed off the whole ocean in the MIS 6 period 140 thousand years ago (Jacoksson et al., 2016). It is also likely that in other glacial maximum periods the Arctic was either completely icebound or ostensibly so. Cold marine habitats would have existed during these extreme glacial maxima, but they would have been in deep water in the Arctic *per se* and for shallow coastal sites at much lower latitudes. Thus, the current Arctic habitats are at most around 125 thousand years old, with shallow communities having had to move hundreds, if not thousands, of kilometres from lower latitude habitats into areas newly available due to ice retreat. In contrast, cold marine environments with a significant biota have existed in some places around Antarctica for probably 15–17 million years, but current extremes were reached around 1 million years ago (Zachos et al., 2008). The extra age is one of the factors that has allowed more fine scale adaptations to arise in the Southern Ocean compared to the Arctic (Peck, 2018). Another major factor is isolation.

The two polar regions, in some respects, differ as much as is possible in the degree of isolation of their marine environments from those at lower latitudes. The Arctic is a shallow sea surrounded by continents with only small areas where water can flow in and out to other oceans, but these flows do occur (Figure 8.2). The Arctic also has long coastlines in both the Pacific and Atlantic sectors that run continuously from high polar to tropical latitudes. There is thus no barrier for coastal species to migrate across very large degrees of latitude. In contrast, the Antarctic is a large continent surrounded by the Southern Ocean that flows from west to east and completely circumnavigates the land mass. This circulation effectively stops surface waters flowing from temperate or sub-antarctic regions into high polar latitudes. When considering the isolation of the continental shelves, Antarctica is unique. The Arctic has continuous continental shelves that link North America with Asia (Figure 8.2). This continental shelf bridge means that all the continents in the world are linked by their continental shelves (Australia to Asia, Africa to Europe and Asia, Asia to North America and North America to South America), except Antarctica, which has no continental shelf link to any other continent. Antarctica is thus the most isolated continent on Earth. It is unique in not having a continuous continental shelf link with another continent.

The differences in isolation have been key to the evolution of marine species in the Arctic and Antarctic and the creation of biodiversity. During the glacial cycles that have occurred over at least the last 2.1 million years (Rohling et al., 2014) marine species have been repeatedly forced to lower latitudes as ice completely covered large areas during glacial maxima. Distributions then progressively moved to higher latitudes in interglacial periods. In the Arctic, benthic species moved up and down the continental coastlines. In the Antarctic, however, species were forced into isolated refugia during glacial maxima as there were, and are, no coastlines along which to migrate (Dayton & Oliver, 1977; Dömel, et al., 2015). This led to the concept of a biodiversity pump (Clarke & Crame, 1992, 2010), where repeated restriction of biodiversity into isolated pockets followed by expansion and proliferation into new niches has produced a unique highly diverse fauna with very high levels of endemism (species that only exist in Antarctica).

The scale of the isolation, history of the continent and unusual range of environmental conditions have resulted in the fauna producing a set of adaptations to the polar conditions that are unique in several ways. However, the level of adaptation seen is greater in Antarctica because of its greater age and isolation compared to the Arctic (Peck, 2018). The next sections focus on those more extreme adaptations.

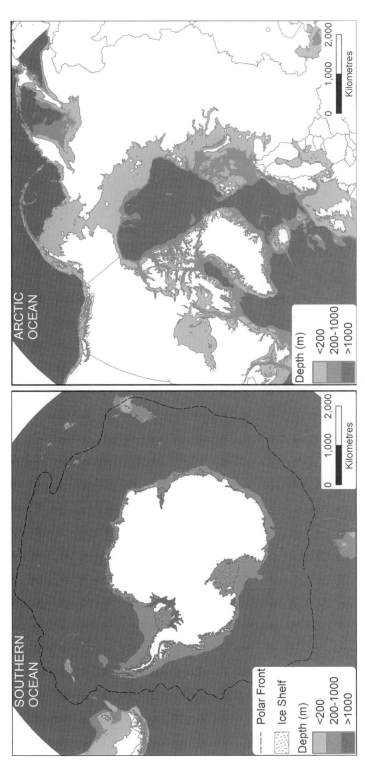

Figure 8.2 Antarctic (left pane) and Arctic (right pane) continental shelves showing depths shallower than 200 m, between 200 and 1000 m and deeper than 1000 m. This figure illustrates that in essence Antarctica is a large continent surrounded by deep oceanic waters and no continental shelf link to any other continent, whereas the Arctic is a shallow sea, over 90% bounded by land and with a shallow continental shelf bridge between Alaska in North America and Chukotka in Asia.

8.3.2 Adaptations

Life history characteristics in polar marine species have been studied for over 50 years and they were soon recognised as possessing K-selected attributes (Clarke, 1979). Arntz et al. (1994) analysed them in detail and showed they include very slow embryonic development; large, yolky eggs; low fecundity; very slow growth rates; often seasonal growth; prolonged gametogenesis; seasonal reproduction; delayed maturation; high levels of protected develop- ment (brooding); long lifetimes; low mortality rates; and large adult size (reviewed in Peck, 2002, 2018; Peck et al., 2006).

More recent research has shown that several characteristics, including embryonic and larval development, growth in juveniles and adults, the dura- tion of processes that elevate metabolism after feeding (Specific Dynamic Action of feeding, SDA) and regeneration of lost limbs in echinoderms, are all slowed beyond the expected effects of low temperature (Peck, 2016, 2018). Embryonic development is slowed by 5–10 times more than it should be compared to temperate species, and this includes when compensation for the relevant effect of temperature on biological function rates is included (Figure 8.3). Other characteristics, including resting metabolic rate and rates of performing activity, are slowed in line with the expected temperature effects. This led Peck (2016) to conclude that polar marine species have diffi- culties making well-formed proteins because all of the processes slowed beyond expectations involve either the synthesis or remodelling of significant amounts of protein, whereas other processes, e.g. those involved with energy production, are not.

In recent years data have accumulated showing that polar marine species living permanently at temperatures around or below 0°C have large problems making proteins and maintaining their body protein pool, the proteome (Peck, 2018). These data include: that the proportion of proteins that are made but not retained and are recycled is around 80% in Antarctic species compared to 25–30% in warm water species (Fraser et al., 2009); that levels of ubiquitin, the molecule that is used in cells to identify poorly formed or damaged proteins, are much higher in polar marine species than similar warmer water species (Todgham et al., 2007; Shin, et al., 2012); that the activity of the proteasome, the machinery in cells that breaks down badly formed proteins, is 2–5 times higher in polar than temperate fish (Todgham et al., 2017); and that levels of heat shock proteins (HSP), which protect cells from damaged proteins, are much higher in polar invertebrates and fish than temperate or tropical species (Place & Hofmann, 2005; Clark et al., 2008a).

A further process that appears greatly slowed in polar marine species that also may be related to problems making proteins is limb regeneration in echinoderms. Many species of starfish and brittle star regenerate limbs

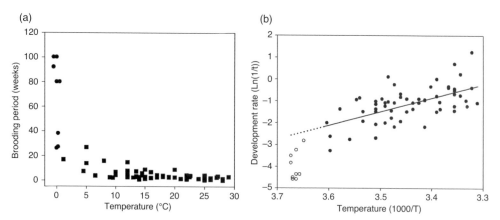

Figure 8.3 (a) Embryonic development time measured as brooding period for marine gastropod snails from the tropics to the poles. Typical times required for fertilised eggs laid in broods to emerge as juvenile snails are 1–3 weeks in the tropics (■), 2–5 weeks in temperate zones (■) and 18–102 weeks in Antarctica (●). (b) The best way of identifying the effects of temperature on a biological process is via calculating the Arrhenius relationship where the log of biological rate is plotted against the inverse of absolute temperature. In almost all cases this produces a straight line, as here for temperate and tropical species (●), but polar species (○) are all well below the line for warmer water species. Rates for polar species are slowed by a factor of 5–10 compared to expected rates from an extension of the relationship for temperate and tropical species. Figure from Peck (2018).

when they are lost. This process was studied in detail in the Antarctic brittle star *Ophionotus victoriae*. Initially, after the loss of a limb there is a lag period before regeneration starts, where the wound is healed and there is reorganisation of the adjacent tissues. In temperate species this takes between 2 and 20 days, but in the Antarctic species this lasted 150 days (Clark et al., 2007), which is a much longer delay than would be predicted by temperature effects on biological systems. Once regeneration started in *O. victoriae* it proceeded more slowly than in temperate species, but at a rate in line with the expected effects of temperature. The factor increasing the lag phase so dramatically remains unclear.

One adaptation that is probably associated with the problems that Antarctic marine species have making proteins, and possibly an adaptation to this, is the high levels of HSP permanently present in their tissues as seen above. Associated with this is their peculiarity with respect to the heat shock response (HSR). Until the early years of the twenty-first century all major multicellular organisms were thought to respond to warming beyond their regularly encountered temperature range by producing a group of proteins whose function is to help other proteins to maintain their functional state while outside their normal range for function. This class of proteins is called the heat shock proteins (HSP), and the HSR was called 'the universal heat shock response' (Gross, 2004). It is now known that HSPs are produced not only in response to temperatures outside the normal range, but also in response to a wide range of stresses, including low salinity in marine species, dehydration in insects and plants, and some pollutants (e.g. Feder & Hofmann, 1999). In a surprising contrast to the large body of previous data on the HSR, at the turn of the century Hofmann et al. (2000) showed that an Antarctic fish, *Trematomus bernacchii*, did not have the expected response to temperature stress and did not increase HSP production when heated beyond its normal range. Since then, the absence of an HSR on warming outside normal ranges has been shown to be widespread in Antarctic fish (e.g. Clark et al., 2008b; Tomanek, 2010; Beers & Jayasundara, 2014). Antarctic marine invertebrates have been shown to have an HSR in many cases (e.g. Clark et al., 2008c, 2016; González et al., 2016), although the response is very variable and often weak, e.g. in krill. Only one species of Antarctic marine invertebrate, the starfish *Odontaster validus*, has been demonstrated to lack the HSR under close scrutiny (Clark et al., 2008d; Clark & Peck, 2009). There are several possible explanations for the absence of, or reduction of, the HSR in Antarctic species (Peck, 2018). It is possible that the cellular machinery to detect an increased requirement for HSPs has been lost during the many millions of years of evolution to low constant temperature in Antarctica. Secondly it may be that the problem with synthesising and retaining functional proteins demonstrated widely in Antarctic marine species is due to low temperature unfolding or deformation and warming would reduce this problem. Thirdly it is possible that the machinery making HSPs runs as fast as it can under normal conditions because of the large constitutional need for HSPs, and hence an increase in production is limited or not possible. The latter could, again, be due to evolution to an environment where temperature variation is very small.

One of the best-known adaptations to polar marine conditions is the evolution of antifreeze in fish. All fish species living in polar waters that experience subzero temperatures produce antifreeze of one type or another. The best known of these is antifreeze in polar fish. The earliest antifreeze molecule identified was a glycoprotein (AFGP), and this was discovered in Antarctic fish in the 1960s in McMurdo Sound (DeVries & Wohlschlag, 1969). The major problem for polar fish is that water-based solutions with ionic compositions similar to their body fluids freeze between $-0.5°C$ and $-1.0°C$ (Eastman & DeVries, 1986; Peck, 2015), and polar marine environments experience temperatures below this for at least part of the year. In winter the vast majority of polar marine habitats shallower than 200 m have temperatures close to $-2°C$, the freezing point of seawater, which varies a little depending on factors such as salinity and pressure (depth). Polar fish overcome this problem by increasing the ionic concentration of their fluids and by using antifreeze proteins and glycoproteins (Figure 8.4). Most Antarctic fish lower the freezing point of their blood by 0.25–0.5°C, but they then decrease the point where they freeze by more than 1°C using antifreeze molecules (Eastman & DeVries, 1986) (Figure 8.4). This problem is especially important at temperatures close to −2°C and in the many species that live in close proximity to ice or use ice as a refuge from predation (e.g. Wiedenmann et al., 2009). In these habitats organisms are often in close contact with micro-ice crystals that readily seed the growth of ice in tissues, which makes them highly dependent on antifreeze.

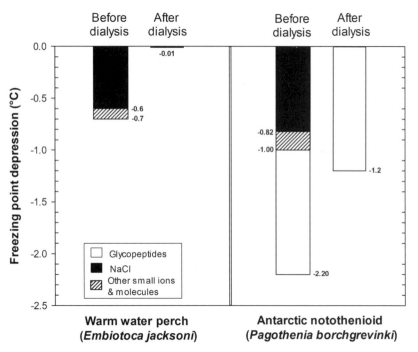

Figure 8.4 The effect of dissolved ions (including NaCl) and molecules, and glycopeptides (glycoproteins) on freezing of blood plasma from a warm-water fish (*Embiotoca jacksoni,* the perch) and *Pagothenia borchgrevinki,* an Antarctic notothenioid. The higher salt concentrations in *P. borchgrevinki* blood depresses the freezing point lower than in the perch. More than half the freezing point depression in the Antarctic fish, however, is due to antifreeze glycoproteins. The increased salt concentration in the blood of Antarctic fish lowers the freezing point by around 0.3°C compared to the temperate species. More than 1°C, however, is added to the freezing point depression by the antifreeze glycoprotein (AFGP). The effect of the AFGP on the freezing point appears small, but it is enough to make Antarctic fishes safe in the Southern Ocean where temperatures rarely fall below −2°C. Reproduced from Peck et al. (2014).

There are many different antifreeze glycoproteins in Antarctic fishes and they have been identified as being encoded in large families of polyprotein genes (Chen et al., 1997). Antifreeze glycoproteins evolved from a pancreatic trypsinogen serine protease progenitor in Antarctic fish (Chen et al., 1997). This was a crucial adaptation that allowed notothenioid fish to survive when other fish could not when the Southern Ocean cooled to freezing temperatures. Almost identical antifreeze glycoproteins in terms of structure and function have been identified in Arctic gadoid fish, but this molecule evolved from non-coding regions of DNA (C. C. Cheng, pers. comm.). This is one, if not the best, example of convergent evolution globally.

Another well-known adaptation to polar conditions is that seen only in one family of Antarctic notothenioid fishes, the crocodile icefish, or Channichthyidae. In this family 16 species do not produce haemoglobin (Hb) and do not have red blood cells; this trait was first described over 60 years ago (Ruud, 1954). These fish survive solely on oxygen carried dissolved in their blood. In red-blooded species this accounts for less than 10% of the total oxygen carried in blood (Cheng & Detrich, 2012). This appears only possible at constant low temperature where metabolic rates are 10–25 times lower than in similar fish living at tropical temperatures (Clarke & Johnston, 1999; Peck & Conway, 2000; Peck, 2018). A range of adaptations have evolved in Channichthyids that lack Hb, including large hearts, a very large circulating blood volume, large blood vessels, and higher mitochondrial densities in red muscles (Verde et al., 2012; di Prisco & Verde, 2015; Peck, 2018). The absence of Hb in Channichthyids has been described in the past as disadaptive, disadvantageous or maladaptive (e.g. Montgomery & Clements, 2000). It has also, however, been argued there may be advantages to this condition, for example in reducing problems with viscosity in circulating fluids which could be very important in higher viscosity, low-temperature conditions, especially where there may be high levels of very small ice crystals in water-based fluids (see Peck, 2018).

Several groups of polar marine species grow to very large size at low temperatures, and this has been called polar gigantism (Chapelle & Peck, 1999; Peck & Chapelle, 1999). The most commonly known theory of gigantism at the poles is Bergmann's rule, which was developed to explain large size in warm-blooded polar species and states 'within genera of warm-blooded vertebrates, species at higher latitudes tend to have larger body sizes' (Bergmann, 1847). This theory that has been developed over recent decades is that it is due to the reduction of heat loss by minimisation of the animal's surface area to volume (e.g. Blackburn et al., 1999). Very large size in ectotherms (cold-blooded species) in polar marine environments has been recognised to occur in some groups for over half a century (Thorson, 1957; De Broyer, 1977). Particularly good examples are in the sea spiders (pycnogonida) where specimens well over 50 cm across live in Antarctica, and the largest temperate species are less than 3 cm; the amphipods where species over 90 mm in length inhabit the Southern Ocean but the largest tropical species are around 10 mm; glass sponges where the Antarctic *Anoxycalyx (Scolymastra) joubini* grows to around 2 m in height and 1.5 m in diameter (Dayton & Robilliard, 1971); the errant polychaete worm *Ophioglycera eximia*, which reaches 76 cm (Hartman, 1964); and the nemertean ribbon worm *Parborlasia corrugatus*, which grows to over 2 m long and 100 g body mass, easily the heaviest globally (Davison & Franklin, 2002). The factors producing this very large size in polar seas have been debated strongly and reviewed by Moran & Woods (2012). The main factors proposed include:

1. Low temperature that reduces maintenance metabolic costs in ectotherms and allows larger body mass to be maintained at lower costs (Chapelle & Peck, 1999, 2004). Metabolic costs in ectotherms decline in line with expected Arrhenius temperature effects in marine ectotherms and are usually 10–25 times lower in polar species than similar tropical species (Peck, 2016, 2018).

2. Increasing oxygen availability at progressively lower temperatures. For amphipod crustaceans, Chapelle & Peck (1999, 2004) showed that variation in size of the largest species from tropics to poles was correlated more with oxygen availability than temperature (the relationship with temperature was curvilinear, Figure 8.5). They further demonstrated that the largest species from low salinity sites (that have higher oxygen concentrations) are larger than marine ones, again in line with ambient oxygen differences. The relationship between ambient oxygen concentration and maximum size was linear and accounted for over 98% of the variation in the data (Figure 8.5).

3. Higher concentrations of silica in the Southern Ocean making it easier for species with silica-based skeletons, such as glass sponges and radiolarians, to make larger structures (Moran & Woods, 2012). A similar, but opposite argument has been proposed for small skeleton size in many polar species with carbonate skeletons where the high solubility of calcium carbonate at low temperatures makes it more costly to remove from seawater (Arnaud, 1974; Watson et al., 2012, 2017). This argument has problems because many taxa that have relatively large skeletal requirements such as echinoderms and brachiopods are more abundant in polar regions than at lower latitude, and a competing hypothesis is that the high frequency of more fragile thinner skeletons at high latitude is due to the lack of durophagous (crushing predators) in polar seas (Aronson et al., 2007; Harper & Peck, 2016).

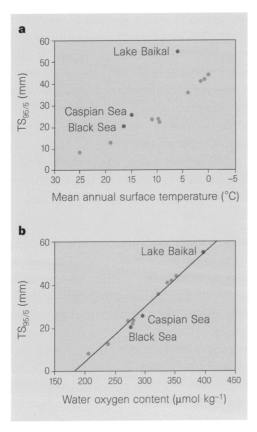

Figure 8.5 Effects of temperature and oxygen availability on amphipod maximum potential size. Data are for nine marine (light grey circles) and three reduced salinity sites (dark grey circles). The proxy for maximum size at each site is the position of the 95th percentile ($TS_{95/5}$) in the size distribution. This avoids the possibility that the largest species at any site studied is very rare and not sampled. (a) $TS_{95/5}$ plotted against mean annual water temperature (inverted scale) for each site. (b) $TS_{95/5}$ plotted against calculated dissolved oxygen content at saturation (mmol per kg), based on surface water mean temperature and salinity. Not every habitat will experience permanent high oxygen saturation, but the 100% value represents optimal conditions for attaining large size. The relationship between size and ambient oxygen in 2b is: $TS_{95/5}$= $-42.6 + 0.252\ O_2$ ($n = 12$; $r^2 = 0.98$; $F =$ 51.69; $P < 0.0001$). Reproduced from Chapelle & Peck (1999).

4. Seasonality, which is more pronounced with latitude should favour larger size as larger animals generally have lower costs for maintaining tissue combined with better abilities to store resources for winter (Blackburn et al., 1999). There are issues with this hypothesis because the very low metabolic rates mean even relatively small species can survive up to 3 years without food (Harper & Peck, 2003).

5. Increased resources and resource quality, both across geological time and space (Vermeij, 2016), and, in the context of this chapter, at high latitudes (Ho et al., 2010), where phytoplankton blooms are among the most intense on Earth. This combines with the low metabolic rates and seasonality to favour large size.

6. Competition is a factor that drives species from their optimal solutions towards the edges of their potential niches (what is physically possible). Chapelle & Peck (1999) in their analysis of large size in amphipod crustaceans only considered sites with more than 60 species present, so that competitive

interactions would drive some species to large size and close to both the niche potential for size and the limit set by the environment.

There has often been confusion in the debate about large size in some taxa in polar regions. In the above list of factors some directly drive an increase in size, whereas others either produce conditions that favour large size or that are neutral, but allow large size if other factors select for it. The two types of factors need clear separation for better understanding of what drives species to large size and what allows large species to exist. The factors that drive towards the evolution of large size in polar conditions include: (1) competition which, in any environment, drives species characteristics away from the optimum and towards the edge of the potential niche (Chapelle & Peck, 1999, 2004; Vermeij, 2016); and (2) seasonality, where periods of low or no resource mean that larger organisms with lower costs per unit mass of tissue and that usually have greater potential to sequester resources and lay down stores for periods of low supply are at an advantage. It should be noted here, as discussed above, that the very low metabolic rates of polar species likely reduce the impact of seasonal periods of low food supply. Factors that allow large size include:

1. High productivity, which provides resources for more species to exist and thence the competition needed for selection to drive species to large size. Gigantism in marine environments has often been correlated with high productivity, but this trend appears absent in terrestrial systems (Vermeij, 2016).
2. High concentrations of silica fall into this category because they reduce the cost of abstraction of this material from seawater to make skeletons, hence allowing larger size at smaller costs of skeleton construction.
3. Reduced costs of tissue maintenance at low temperatures again lower the costs of making and maintaining larger bodies, although it should be noted that there is good evidence that making proteins at temperatures around and below 0°C appears to be more difficult and costly than at warmer temperatures (Peck, 2016, 2018).
4. Variations in ambient oxygen availability, which are higher in colder waters, limit maximum attainable size. The size limitation works via diffusion limitations from external water to the animal and from circulating fluids to cells via the Fick equation (Peck & Chapelle, 1999, 2003). A key element of the work showing the link between oxygen and maximum size in amphipods was that the relationship of maximum size to temperature for marine sites was curvilinear.

Because these factors do not select for different sizes, but merely allow different sizes, careful analysis needs to be conducted to evaluate trends in size. For instance Chapelle & Peck (2003) found an increase in minimum size of female amphipods as temperature decreased and oxygen levels increased, and this was strongly correlated with changes in egg size (which are also likely affected by oxygen availability). However, no such trend was found in males indicating other factors set minimum size in males. Furthermore, Peck & Chapelle (2003) showed that amphipods at high altitude in Lake Titicaca where ambient oxygen is low are small for the temperature and match oxygen limitations predictions. They emphasised, however, that there needs to be competition to drive animals to large size and when analyses are done on sites with few competing species, size trends will not be apparent. Without the factors driving large size, such as competition, no trends will appear, and this helps to explain the problems identified by Moran & Woods (2012) when they rightly said it is surprisingly difficult 'to determine even how common polar gigantism is'.

Recently Vermeij (2016) analysed trends in the size of the largest organisms on Earth across the Phanerozoic and showed that there has been a regular increase in size of the largest over time, and hence with organism complexity. He showed that there are two types of giants – those with low metabolic rates living in either resource-poor areas on land (such as giant tortoises) or in cold aquatic systems such as the polar regions, and a second group of metabolically highly active organisms that control their body temperatures (e.g. mammals and dinosaurs). He further emphasised that the two main factors affecting size are competition and metabolic rate.

Several authors have argued that because marine giants only occur in low temperature environments, they will be more vulnerable to climate warming than smaller species (e.g. Chapelle & Peck, 1999; Daufresne et al., 2009). Moran & Woods (2012) argued contrary to this that temperature-dependent physiological systems (i.e. essentially all systems) are more coadapted to each other than they are to the size of the organism and hence that in warming conditions there should be no difference in likely survival for large species compared to smaller ones. Moran and Woods (2012) are correct that in many cases physiologies are coadapted to size and hence studies of performance against size will often not show differences between large and small species. However, when size limits are reduced by environmental change, smaller species have the capacity to remodel their physiologies, initially through phenotypic plasticity, especially developmental plasticity to match new conditions where large species do not and they either have to become smaller or become extinct due to failure in competition or inability to adapt fast enough. This is why, as Vermeij (2016) noted, large species disappear in mass extinction events, but on geological timescales they return rapidly and the overall geological trends continue. There is further evidence to suggest that within species size decreased across, at least some, mass extinction events, a phenomenon sometimes termed the Lilliput effect (Wiest et al., 2015).

8.4 How can polar marine organisms respond to change?

The mechanisms that organisms employ when responding to changing environments are dealt with in detail in Chapter 11. In terms of responding to climate change there is general consensus that there are three main mechanisms that are important in promoting survival. These are migration, flexibility of the phenotype (often called phenotypic plasticity) via acclimatisation of physiological processes and via alteration of the gene pool in populations, genetic adaptation (Somero, 2010, 2015; Peck, 2011). Specific problems for polar marine species are that when the Earth warms, cold environments disappear and there are no places to migrate to that can support cold-adapted species (Peck, 2018). This problem is the same for other cold environments such as high mountain areas. Polar benthic marine species have specific extra problems as there are no coastlines going to the highest latitudes for them to migrate along, and in Antarctica there are no coastlines from the sub-Antarctic into the Antarctic and hence migration possibilities are extremely limited. Antarctica is the most isolated marine benthic continent because it is the only one that lacks a continental bridge to another continent (Peck, 2018). In terms of capacities to respond using phenotypic plasticity and adaptation, the past evolution of species to the cold polar conditions and the adaptations they already have impact strongly on many species in their ability to resist or survive environmental change (Somero, 2010; Peck 2018).

There are many alterations to polar marine environments that are and will continue to occur with climate change. These include freshening, sea-ice loss, reductions in coastal glaciers and iceshelves and changes in turbidity at some sites (Peck, 2018). There have been two main areas, however, that research has focused on into how polar marine species can and will respond to climate-induced environmental change. These are responses to temperature change, primarily warming, and the impact of ocean acidification.

8.5 Responses to temperature change

There has been much interest in recent years about the abilities of Antarctic marine species to respond to temperature stress, which has been reviewed in detail in Peck (2018). Studies have been conducted on many different taxonomic groups, including molluscs (e.g. Peck, 1989; Pörtner et al., 1999, 2006; Peck et al., 2002, 2004; Clark et al., 2008a, c; Morley et al., 2012; Reed & Thatje, 2015), echinoderms (e.g. Smith & Peck, 1998; Clark et al., 2008d; Peck et al., 2008, 2009a; Morley et al., 2016), brachiopods (Peck, 1989, 2008), amphipods

(Young et al., 2006a,b; Clark et al., 2008d; Doyle et al., 2012; Clusella-Trullas et al., 2014; Faulkner et al., 2014; Schram et al., 2015), isopods (Whiteley et al., 1997; Robertson et al., 2001; Young et al., 2006a; Janecki et al., 2010), fish (e.g. Hardewig et al., 1999; Hofmann et al., 2000; Podrabsky & Somero, 2006; Robinson & Davison, 2008; Bilyk & DeVries, 2011; Todgham et al., 2017) and macroalgae or phytoplankton (Montes-Hugo et al., 2009; Schloss et al., 2012).

There has also been much effort put into assessing the impacts of warming using wider-scale approaches, in experiments and via field observations. These have primarily identified responses across many species or have assessed responses at higher ecological levels from the community to ecosystem or even overall biodiversity in a region (e.g. Aronson et al., 2007; Clarke et al., 2007; Peck et al., 2010, 2013, 2014; Schofield et al., 2010; Richard et al., 2012; Gutt et al., 2015; Morley et al., 2016; Clark et al., 2017).

The conclusion of nearly all of the investigations into polar marine organism responses to environmental warming is that they are less capable than species from lower latitudes. They perform poorly in experiments where they are exposed to warming, and this has been recognised for over 50 years (Somero & De Vries, 1967).

Antarctic fish generally perform better than invertebrates in laboratory warming experiments (e.g. Podrabsky & Somero, 2006; Robinson & Davison, 2008; Bilyk & DeVries, 2011). Some Antarctic marine invertebrates have possibly the worst reported capacities to cope with warming in experiments of any marine species globally. Good examples are the brittle star *Ophionotus victoriae*, which cannot survive long-term above 3°C (Peck et al., 2009a); the bivalve mollusc *Limopsis marionensis*, which turns to anaerobic metabolism between 2°C and 4°C (Pörtner et al., 1999); and the brachiopod *Liothyrella uva*, which failed at 4.5°C in long-term temperature elevation trials (Peck, 1989).

A very recent development in this field in Antarctica has been to carry out experiments *in situ* that warm the seabed and an overlying film of water 2–5 mm thick depending on water movement. Effects of raising temperatures by 1°C or 2°C above ambient were tested on natural field communities of biofouling organisms (Ashton et al., 2017). In the 9-month experiment temperature was raised, but all other environmental variables such as daylength, salinity, phytoplankton food supply, etc were not altered and were natural. All warming treatments had large impacts on community structure, with one species, the bryozoan *Fenestrulina rugula*, dominating more in elevated temperatures. Very surprisingly growth rates of all the species assessed nearly doubled for a 1°C warming. This increase in growth is more than six times higher than any normal direct effect of temperature on animal physiological processes, and some other factor needs to be invoked to explain the dramatic rise. It is possible that this unexpected result could be related to the problems associated with making proteins at low temperature discussed above. A warming of 2°C produced very different outcomes from a 1°C warming. At 2°C of warming, results between species differed markedly, with some species growing less than controls (Ashton et al., 2017) and some indicators of cell death (apoptosis) were present in screens of gene expression. This likely indicated that the least resistant species were close to their upper temperature limits.

One of the key aspects of phenotypic plasticity in relation to responding to altered environments is the ability of an organism to reset its physiological processes at some new steady state after a change, termed acclimation. Acclimation has a strong effect on the ability of a species to survive warming. In temperate species acclimation to a moderate temperature rise of 2–3°C takes from a few days to up to 2 weeks depending on the species studied. In Antarctic marine species the time required for acclimation to a similar temperature elevation is between 2 and 9 months (Peck et al., 2010, 2014; Morley et al., 2016). Antarctic marine species are thus poor in their abilities to acclimate their physiology to environmental temperature change.

In experiments evaluating the upper temperature limits of organisms the rate of warming has a very large effect. In a multispecies warming study of the marine benthic community at Rothera Point, Adelaide Island, on the Antarctic Peninsula, a temperature elevation of 1°C per day produced upper temperature limits between 8.3°C and 17.5°C above mean summer maximum depending on the species, whereas at 1°C per month these values fell to between 1°C and 6.5°C (Peck et al., 2009b; Peck, 2018) (Figure 8.6). The values obtained from the very slow rates of warming are clearly of more value in predicting future responses to climate change as they allow processes such as acclimation to be included, and rates of warming are closer to those seen in outside environments during normal ecological and seasonal cycles. Using these changes with warming rate, Richard et al. (2012) modelled the responses in thermal limits of the species involved to produce very long-term temperature limit asymptotes that should be similar to limits of organisms that experience several years of environmental change. When these long-term limits are compared to the habitat natural maximum temperatures in the sea, a measure of how much buffer there is between an organism's physiological abilities to resist warming and current temperatures is obtained. This was termed the warming allowance by Richard et al. (2012), and it is similar in concept and use to the 'warming tolerance' of Deutsch et al. (2008) for terrestrial ectotherms.

Peck et al. (2014) used the approach of Richard et al. (2012) to calculate warming allowances for communities of shallow water benthic species

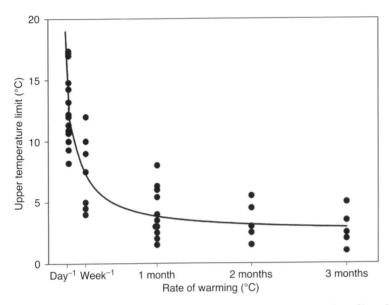

Figure 8.6 Upper temperature limit (CT_{max}) values for 14 species of benthic Antarctic marine invertebrates from Rothera Point, Adelaide Island, at different rates of warming. Upper temperature limits quoted are values above the current average summer maximum temperature (1.0°C at the study site). Reproduced from Peck (2018), updated and adapted from Peck et al. (2014).

across latitudes from the tropics to the poles. The results showed that both tropical and Antarctic communities have similar warming allowances (around 3°C above habitat mean maximum summer temperatures), but this is far less (one third to one half) than the warming allowance for species from warm temperate, temperate or cool temperate environments where the values range from 6.5°C to nearly 10°C (Figure 8.7). Data for Peru differ greatly between times outside the El Niño, where the warming allowance is around 9°C, and during an El Niño when it is 3°C and very similar to the values for the Antarctic and tropics (Figure 8.7). The large mortalities seen during El Niño years in several species on the coasts of Peru suggest that the level of warming allowance there is at or beyond that for much of the community. This in turn indicates that species in both tropical and Antarctic benthic communities are permanently living close to their physiological long-term limits, and it is only the low variation in habitat temperature that allows them to survive long-term. Polar marine species thus appear to be less capable of resisting climate warming than temperate species, but similar to tropical species. The fact that when the planet warms, hot areas increase and cold areas disappear means that polar marine species should be high on the priority list for conservation and

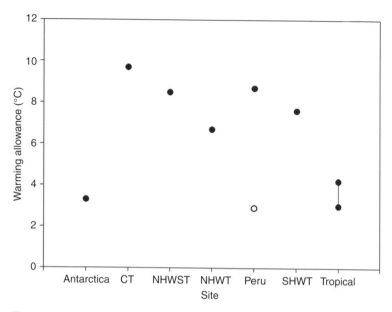

Figure 8.7 Warming allowances for multispecies assessments at seven sites from polar to tropical latitudes. Warming allowance here is the difference between long-term temperature limits in a community (CTmax) (see Figure 8.6) and the average maximum summer temperature at the site studied (°C). The sites range from Antarctica to Singapore (tropical) with abbreviations following those of Richard et al. (2012) and Peck et al. (2014): CT = cold temperate site (west coast of Scotland); NHWST = northern hemisphere warm shallow temperate site (depths less than 5 m); NHWT = northern hemisphere warm temperate site (deeper than 5 m; South of France and west coast of the USA); Peru (depths deeper than 5 m); and SHWT = southern hemisphere warm temperate (all depths deeper than 5 m). The line between the points for tropical data indicate the range of values, while the closed circle in the Peru data indicates the warming allowance outside an El Niño year, and the open symbol is for data in an El Niño. Reproduced from Peck (2018).

preservation of identity and species characteristics through mechanisms such as gene banking.

The second major way organisms have of responding to climate change levels of environmental alteration is through beneficial genetic change, or adaptation (see Chapter 11; Somero, 2010, 2015; Peck, 2011, 2018). There is a dearth of information about how fast adaptation processes can occur and hence how fast adaptation can enhance survival in a changing world. Looking at past change only tells us how rapidly a change occurred, not if this was the fastest rate at which the process could proceed. Furthermore, there are stochastic elements involved in adaptation processes. Thus, new mutations are produced by a range of mechanisms, including from damage caused by ionising radiation or oxidative damage from, for example, free radicals. In polar seas oxygen levels are higher than in warmer waters, and damage from oxygen free radicals is more of an issue (e.g. Abele et al., 2001; Clark et al., 2013). However, the most common cause of mutation that is inherited, and inheritance is of critical importance for a trait that confers resistance to change in populations and species, is via errors in reading DNA sequences during meiotic cell division (Watson et al., 2014). It should be remembered that most mutation and genetic change from this source is not beneficial or is neutral, and hence not adaptive or beneficial in a changing world.

At the population level modifications of the gene pool can come from mutation or gene flow within and between populations, and the latter is more common. Population- and species-level responses to change are also affected very strongly by life history characteristics (Peck, 2011). Especially important here are population size, number of embryos produced each time the organism reproduces, generation time and dispersal abilities (to connect different populations). Taxa with very rapid generation times, fast growth and very large populations can rely on mutation to adapt to changes in the environment. Good examples of this are viruses and microbes where nucleotide substitutions can occur rapidly, and evolutionary change, or adaptation to new conditions, can take as little as a few days to weeks (Peel & Wyndham, 1999; Duffy et al., 2008). As generation time increases, the time required for an adaptive genetic change to be entrained, on average, becomes longer, and this is also true as population sizes decline and number of offspring per reproductive event declines (Peck, 2011). At the slow end of the spectrum, species with few offspring per reproductive event and long generation times such as elephants, some Antarctic marine invertebrates such as bivalve molluscs (Butler et al., 2013), sponges (Gatti, 2002), brachiopods (Peck & Brey, 1996) and whales might need many decades, or even centuries, for this process to become effective. The reliance of such species on the resilience of the phenotype to altered environments is very high (Peck, 2011).

In polar environments information is lacking on many factors that dictate rates of adaptation to new conditions. Even data on very important characteristics such as population size, generation times, number of embryos per reproductive event, size of offspring etc. are known for only a few polar marine species, whereas these data are much more readily available for temperate and tropical species. However, the general life history characteristics of slow growth and embryonic development, long generation times and high incidence of protected development in marine species all argue for poorer abilities to adapt than species from lower latitudes (Peck, 2018).

8.6 Responses to acidification

Ocean acidification has been a major concern for science and society for over a decade. Large amounts of research have been devoted to assessing the impacts of progressively increasing levels of CO_2 in the oceans and how marine species are, or might be, responding (e.g. Doney et al. 2009), and some groups, especially corals have been shown to be strongly negatively affected by acidification (e.g. Hoegh-Guldberg, 1999; Kleypas et al., 2006; Kroeker et al., 2013).

Research in the polar regions on impacts of acidification followed a similar path to those at lower latitudes. Most early studies reported negative effects of lowered pH on marine animal physiology and survival, especially in early life history stages (e.g. Dupont et al., 2008; Watson et al., 2009). Many early studies, however, exposed animals to rapid, acute changes in pH. In a similar vein to the research on temperature, emphasis later on was given to longer-term exposures and slower rates of change of pH that allow processes such as acclimation to take effect (see Byrne, 2011 for review). As research progressed, understanding improved and experiments became more sophisticated, more longer-term exposures have been employed in laboratory studies. These have generally demonstrated much smaller impacts of acidified conditions, or no effects at all, when animals are exposed to mid- or end-century predicted conditions. There are good examples of this from studies of polar marine species in resistance in adults (e.g. Cross et al., 2015), energetics (Morley et al., 2016), reproductive characteristics (Suckling et al., 2015), and embryonic and larval development (e.g. Bailey et al., 2016). These studies demonstrate that the impacts of acidification are much less when long-term exposures are used, and negative reproductive effects are lessened greatly when parental broodstock are conditioned for long periods before spawning (e.g. Suckling et al., 2015).

One aspect of responding to change not covered by the research above is multigenerational effects. To date there have been no investigations of cross-generational responses in polar species. There have been very few studies globally that have investigated abilities to respond to acidified conditions across generations. In most of these studies adaptation and significant improvement over single generation experiments is the predominant outcome (reviewed in Sunday et al., 2014; Stillman & Paganini, 2015). Studies on bacteria have demonstrated that performance is indistinguishable from controls following a few cell divisions (e.g. Collins, 2012; Lohbeck et al., 2013). For invertebrates and fish there are very few data indeed on multigenerational responses. Where there are data, physiological performance returns to levels very close to, or indistinguishable from, controls in a few generations. Examples of this are the sea urchin *Psammechinus miliaris* (Suckling and Clark, unpublished data) and the polychaete worm *Ophryotrocha labronica* (Rodríguez-Romero et al., 2016). Some studies have, however, found evidence of long-term effects including altered retinal function, changes in neurotransmitter processes and impaired learning (e.g. Chivers et al., 2014; Roggatz et al., 2016).

Very recent research has analysed the structure and size of skeletons of marine species with shells from museum collections. An example is Cross et al. (2017) who demonstrated there was little or no change in shell structure, composition or size in the brachiopod *Calloria inconspicua* using specimens from the same population in the same bay on Stewart Island, New Zealand, every decade over the last 120 years.

Another recent development has been research involving more than one environmental stressor. These studies have usually found acidification to have a smaller impact than other factors, especially temperature (e.g. Huth & Place, 2016; Enzor et al., 2017). Furthermore, studies at the molecular level have shown significant amounts of genetic variability in response to ocean acidification (Pespeni et al., 2013), which may promote resilient populations.

The overall conclusion is that ocean acidification is not likely to be a major issue in the coming decades for most species in the oceans, even in polar regions. Most research suggests that other factors, especially temperature, are having and will have much larger impacts. There may be some species that are poorly resistant to acidification, or species where acidification tips balances as part of an overall multifactorial set of environmental insults. A good example of this from the pelagic realm is pteropods, although even here mechanisms of resilience have recently been identified (Peck et al., 2018). A challenge for the coming years for the research community in this area is to identify those species that are vulnerable and especially those that have important or critical functions in food webs and ecosystems, as declines or losses in these species could have wide-ranging consequences.

References

Abele, D., Tesch, C., Wencke, P., Pörtner, H. O. (2001). How does oxidative stress relate to thermal tolerance in the Antarctic bivalve *Yoldia eightsi*? *Antarctic Science*, **13**, 111–118.

Amaral Zettler, L.A., Messerli, M.A., Laatsch, A. D., Smith, P.J.S., Sogin, M.L. (2003). From genes to genomes: beyond biodiversity in Spain's Rio Tinto. *The Biological Bulletin*, **204**, 205–209.

Arnaud, P.M. (1974). Contribution à la bionomie marine benthique des régions antarctiques et subantarctiques. *Téthys*, **6**, 567–653.

Arntz, W.E., Brey, T., Gallardo, V.A. (1994). Antarctic zoobenthos. *Oceanography and*

Marine Biology: An Annual Review, **32**, 241–304.

Aronson, R.B., Thatje, S., Clarke, A., et al. (2007). Climate change and invisibility of the Antarctic benthos. *Annual Review of Ecology, Evolution, and Systematics*, **38**, 129–154.

Ashton, G., Morley, S.A., Barnes, D.K.A., Clark, M.S., Peck, L.S. (2017). Warming by 1°C drives species and assemblage level responses in Antarctica's marine shallows. *Current Biology*, **27**, 2698–2705.

Bailey, A., Thor, P., Browman, H.I., et al. (2016). Early life stages of the Arctic copepod Calanus glacialis are unaffected by increased seawater pCO2. *ICES Journal of Marine Science*, **74**, 996–1004.

Barnes, D.K.A. (2016). Iceberg killing fields limit huge potential for benthic blue carbon in Antarctic shallows. *Global Change Biology*, **23**, 2649–2659.

Barnes, D.K.A., Conlan, K.E. (2012). The dynamic mosaic. In: Rogers et al. (eds) *Antarctic Ecosystems: An Extreme Environment in a Changing World*. Wiley Interscience, pp. 255–290.

Barnes, D.K.A., Fuentes, V., Clarke, A., Schloss, I. R., Wallace, M.I. (2006). Spatial and temporal variation in shallow seawater temperatures around Antarctica. *Deep-Sea Research II*, **53**, 853–858.

Beers, J.M., Jayasundara, M. (2014). Antarctic notothenioid fish: what are the future consequences of 'losses' and 'gains' acquired during long-term evolution at cold and stable temperatures? *Journal of Experimental Biology*, **218**, 1834–1845.

Bergmann, C. (1847). Über die verhältnisse der wärmeökonomie der thiere zu ihrer grösse. *Göttinger Studien*, **3**, 595–708.

Bilyk, K.T., DeVries, A.L. (2011). Heat tolerance and its plasticity in Antarctic fishes. *Comparative Biochemistry and Physiology*, **158**, 382–390.

Blackburn, T.M., Gaston, K.J., Loder, N. (1999). Geographic gradients in body size: a clarification of Bergmann's Rule. *Diversity and Distributions*, **5**, 165–174.

Brown, K.M., Fraser, K.P.P., Barnes, D.K.A., Peck, L.S. (2004). Ice scour frequency dictates Antarctic shallow-water community structure. *Oecologia*, **141**, 121–129.

Butler, P.G., Wanamaker, A.D., Scourse, J.D., Richardson, C.A., Reynolds, D.J. (2013). Variability of marine climate on the North Icelandic Shelf in a 1357-year proxy archive based on growth increments in the bivalve *Arctica islandica*. *Palaeogeography, Palaeoclimatology, Palaeoecology*, **373**, 141–151.

Byrne, M. (2011). Impact of ocean warming and ocean acidification on marine invertebrate life-history stages: vulnerabilities and potential for persistence in a changing ocean. *Oceanography and Marine Biology: An Annual Review*, **49**, 1–42.

Caldeira, K., Wickett, M.E. (2003). Anthropogenic carbon and ocean pH. *Nature*, **425**, 365.

Cerrone, D., Fusco, G., Simmonds, I., Aulicino, G., Budillon, G. (2017). Dominant covarying climate signals in the Southern Ocean and Antarctic sea ice influence during the last three decades. *Journal of Climate*, **30**, 3055–3072.

Chapelle, G., Peck, L.S. (1999). Polar gigantism dictated by oxygen availability. *Nature*, **399**, 144–145.

Chapelle, G., Peck, L.S. (2004). Amphipod crustacean size spectra: new insights in the relationship between size and oxygen. *Oikos*, **106**, 167–175.

Chen, L., DeVries, A.L., Cheng, C.-H.C. (1997). Evolution of antifreeze glycoprotein gene from a trypsinogen gene in Antarctic notothenioid fish. *Proceedings of the National Academy of Sciences of the USA*, **94**, 3811–3816.

Cheng, C.-H.C., Detrich III, H.W. (2012). Molecular ecophysiology of Antarctic notothenioid fishes. In: A. Rogers et al. (eds). *Antarctic Ecosystems: An Extreme Environment in a Changing World*. Wiley Interscience, pp. 357–378.

Chivers, D.P., McCormik, M.I., Nilsson, G.E., et al. (2014). Impaired learning of predators and lower prey survival under elevated CO_2 : a consequence of neurotransmitter interference. *Global Change Biology*, **20**, 515–522.

Clark, M.S., Peck, L.S. (2009). HSP70 heat shock proteins and environmental stress in Antarctic marine organisms: a mini-review. *Marine Genomics*, **2**, 11–18.

Clark, M.S., DuPont, S., Rosetti, H., et al. (2007). Delayed arm regeneration in the Antarctic brittle star (Ophionotus victoriae). *Aquatic Biology*, **1**, 45–53.

Clark, M.S., Fraser, K.P.P., Peck, L.S. (2008a). Antarctic marine molluscs do have an HSP70 heat shock response. *Cell Stress and Chaperones*, **13**, 39–49.

Clark, M.S., Fraser, K.P.P., Burns, G., Peck, L.S. (2008b). The HSP70 heat shock response in the Antarctic fish Harpagifer antarcticus. *Polar Biology*, **31**, 171–180.

Clark, M.S, Fraser, K.P.P.F., Peck, L.S. (2008c). Antarctic marine molluscs do have an HSP70 heat shock response. *Cell Stress and Chaperones*, **13**, 39–49.

Clark, M.S., Fraser, K.P.P., Peck, L.S. (2008d). Lack of an HSP70 heat shock response in two Antarctic marine invertebrates. *Polar Biology*, **31**, 1059–1065.

Clark, M.S., Husmann, G., Thorne, M.A.S., et al. (2013). Hypoxia impacts large adults first: consequences in a warming world. *Global Change Biology*, **19**, 2251–2263.

Clark, M.S., Sommer, U., Kaur, J., et al. (2017). Biodiversity in marine invertebrate responses to acute warming revealed by a comparative multi-omics approach. *Global Change Biology*, **23**, 318–330.

Clark, M.S., Thorne, M.A., Burns, G., Peck, L.S. (2016). Age-related thermal response: the cellular resilience of juveniles. *Cell Stress and Chaperones*, **21**, 75–85.

Clarke, A. (1979). On living in cold water: K strategies in Antarctic benthos. *Marine Biology*, **55**, 111–119.

Clarke, A., Crame, J.A. (1992). The Southern Ocean benthic fauna and climate change – a historical perspective. *Philosophical Transactions of the Royal Society of London B: Biological Sciences*, **338**, 299–309.

Clarke, A., Crame, J.A. (2010). Evolutionary dynamics at high latitudes: speciation and extinction in polar marine faunas. *Philosophical Transactions of the Royal Society of London B: Biological Sciences*, **365**, 3655–3666.

Clarke, A., Gaston, K.J. (2006). Climate, energy and diversity. *Proceedings of the Royal Society of London Series B: Biological Sciences*, **273**, 2257–2266.

Clarke, A., Johnston, N. (1999). Scaling of metabolic rate and temperature in teleost fish. *Journal of Animal Ecology*, **68**, 893–905.

Clarke, A., Holmes, L.J., White, M.G. (1988). The annual cycle of temperature, chlorophyll and major nutrients at Signy Island, South Orkney Islands, 1969–1982. *British Antarctic Survey Bulletin*, **80**, 65–86.

Clarke, A., Meredith, M.P., Wallace, M.I., Brandon, M.I., Thomas, D.N. (2008). Seasonal and interannual variability in temperature, chlorophyll and macronutrients in northern Marguerite Bay, Antarctica. *Deep-Sea Research II*, **55**, 1988–2006.

Clarke, A., Murphy, E.J.M., Meredith, M.P., et al. (2007). Climate change and the marine ecosystem of the western Antarctic Peninsula. *Philosophical Transactions of the Royal Society of London B: Biological Sciences*, **362**, 149–166.

Clusella-Trullas, S., Boardman, L., Faulkner, K. T., Peck, L.S., Chown, S.L. (2014). Effects of temperature on heat-shock responses and survival of two species of marine invertebrates from sub-Antarctic Marion Island. *Antarctic Science*, **26**, 145–152.

Collins, S. (2012). Marine microbiology: evolution on acid. *Nature Geoscience*, **5**, 310–311.

Cook, A.J., Vaughan, D.G. (2010). Overview of areal changes of the ice shelves on the

Antarctic Peninsula over the past 50 years. *The Cryosphere*, **4**, 77–98.

Cook, A.J., Fox, A.J., Vaughan, D.G., Ferrigno, J. G. (2005). Retreating glacier fronts on the Antarctic Peninsula over the past half-century. *Science*, **308**, 541–544.

Cross, E.L., Peck, L.S., Harper, E.M. (2015). Ocean acidification does not impact shell growth or repair of the Antarctic brachiopod *Liothyrella uva* (Broderip, 1833). *Journal of Experimental Marine Biology and Ecology*, **462**, 29–35.

Cross, E.L., Peck, L.S., Lamare, M.D., Harper, E. M. (2016). No ocean acidification effects on shell growth and repair in the New Zealand brachiopod *Calloria inconspicua* (Sowerby, 1846). *ICES Journal of Marine Science*, **73**, 920–926.

Cross, E.L., Harper, E.M., Peck, L.S. (2017). A 120-year record of resilience to environmental change in brachiopods. *Global Change Biology*, **24**, 2262–2271.

Daufresne, M., Lengfellnera, K., Sommera, U. (2009). Global warming benefits the small in aquatic ecosystems. *Proceedings of the National Academy of Sciences of the USA*, **106**, 12788–12793.

Davison, W., Franklin, C.E. (2002). The Antarctic nemertean *Parborlasia corrugatus*: an example of an extreme oxyconformer. *Polar Biology*, **25**, 238–240.

Dayton, P.K., Oliver, J.S. (1977). Antarctic soft-bottom benthos in oligotrophic and eutrophic environments. *Science*, **197**, 55–58.

Dayton, P.K., Robilliard, G.A. (1971). Implications of pollution to the McMurdo Sound benthos. *Antarctic Journal of the United States*, **6**, 53–56.

De Broyer, C. (1977). Analysis of the gigantism and dwarfness of Antarctic and Sub-Antarctic gammaridean Amphipoda. In: G. A. Llano (ed.), *Adaptations within Antarctic ecosystems*. Proceedings of the Third SCAR Symposium on Antarctic Biology. Smithsonian Institution, Houston, pp. 327–334.

De Broyer, C., Koubbi, P., Griffiths, H.J., et al. (eds.) (2014). *Biogeographic Atlas of the Southern Ocean*. Scientific Committee on Antarctic Research, Cambridge.

Denny, M., Dorgan, K.M., Evangelista, D., et al. (2011). Anchor ice and benthic disturbance in shallow Antarctic waters: interspecific variation in initiation and propagation of ice crystals. *Biological Bulletin*, **221**(2), 155–163.

Deutsch, C. A., Tewksbury, J.J., Huey, R.B., et al. (2008). Impacts of climate warming on terrestrial ectotherms across latitude. *Proceedings of the National Academy of Sciences of the USA*, **105**, 6668–6672.

DeVries, A.L., Wohlschlag, D.E. (1969). Freezing resistance in some Antarctic fishes. *Science*, **163**, 1073–1075.

di Prisco, G., Verde, C. (2015). The Ross Sea and its rich life: research on molecular adaptive evolution of stenothermal and eurythermal Antarctic organisms and the Italian contribution. *Hydrobiologia*, **761**, 335–361.

Dömel, J.S., Convey, P., Leese, F. (2015). Genetic data support independent glacial refugia and open ocean barriers to dispersal for the Southern Ocean sea spider *Austropallene Cornigera*. *Journal of Crustacean Biology*, **35**, 480–490.

Doney, S.C., Fabry, V.J., Feely, R.A., Kleypas, J.A. (2009). Ocean acidification: the other CO_2 problem. *Annual Review of Marine Science*, **1**, 169–192.

Doyle, S.R., Momo, F.R., Brethes, J.C., Ferrera, G. A. (2012). Metabolic rate and food availability of the Antarctic amphipod *Gondogeneia antarctica* (Chevreux 1906): seasonal variation in allometric scaling and temperature dependence. *Polar Biology*, **25**, 413–424.

Duffy, S., Shackelton, L.A., Holmes, E.C. (2008). Rates of evolutionary change in viruses: patterns and determinants. *Nature Reviews Genetics*, **9**, 267–276.

Dupont, S., Havenhand, J., Thorndyke, W., Peck, L.S., Thorndyke M. (2008). CO2-driven ocean acidification radically affects larval survival and development in the brittlestar Ophiothrix fragilis. *Marine Ecology Progress Series*, **373**, 285–294.

Eastman, J.T., DeVries, A.L. (1986). Antarctic fishes. *Scientific American*, **254**, 106–114.

Enzor, L.A., Hunter, E.M., Place, S.P. (2017). The effects of elevated temperature and ocean acidification on the metabolic pathways of notothenioid fish. *Conservation Physiology*, **5**, cox019.

Fabry, V.J., McClintock, J.B., Mathis, J.T., Grebmeier, J.M. (2009). Ocean acidification at high latitudes: the bellweather. *Oceanography*, **22**, 160–171.

Faulkner, K., Clusella-Trullas, S., Peck, L.S., Chown, S. (2014). Lack of coherence in the warming responses of marine crustaceans. *Functional Ecology*, **2**, 895–903.

Feder, M.E., Hofmann, G.E. (1999). Heat shock proteins, molecular chaperones and their stress response: evolutionary and ecological physiology. *Annual Review of Physiology*, **61**, 243–282.

Feely, R.A., Doney, S.C., Cooley, S.R. (2009). Ocean acidification: present conditions and future changes in a high-CO_2 world. *Oceanography*, **22**, 36–47.

Feely, R.A., Sabine, C.L., Byrne, R.H., et al. (2012). Decadal changes in the aragonite and calcite saturation state of the Pacific Ocean. *Global Biogeochemical Cycles*, **26**, 1–15.

Fraser, L.H., Greenall, A., Carlyle, C., Turkington, R., Friedman, C.R. (2009). Adaptive phenotypic plasticity of *Pseudoroegneria spicata*: response of stomatal density, leaf area and biomass to changes in water supply and increased temperature. *Annals of Botany*, **103**(5), 769–775.

Gatti, S. (2002). The role of sponges in high-antarctic carbon and silicon cycling: a modelling approach. *Berichte zur Polarforschung/Reports on Polar Research*, **434**, 1–124.

González, K., Gaitán-Espitia, J., Font, A., Cárdenas, C.A., González-Aravena, M. (2016). Expression pattern of heat shock proteins during acute thermal stress in the Antarctic sea urchin, *Sterechinus neumayeri*. *Revista Chilena de Historia Natural*, **89**, 2.

Grange, L.J., Smith, C.R. (2013). Megafaunal communities in rapidly warming fjords along the West Antarctic Peninsula: hotspots of abundance and beta diversity. *PLoS ONE*, **8**, e77917.

Griffiths, H.J., Danis, B., Clarke, A. (2011). Quantifying Antarctic marine biodiversity: the SCAR-MarBIN data portal. *Deep Sea Research. II: Topical Studies in Oceanography*, **58**, 18–29.

Gross, M. (2004). Emergency services: a bird's eye perspective on the many different functions of stress proteins. *Current Protein & Peptide Science*, **5**, 213–223.

Gutt, J. (2001). On the direct impact of ice on marine benthic communities, a review. *Polar Biology*, **24**, 553–564.

Gutt, J., Bertler, N., Bracegirdle, T.J., et al. (2015). The Southern Ocean ecosystem under multiple climate stresses: an integrated circumpolar assessment. *Global Change Biology*, **21**, 1434–1453.

Hardewig, I., Peck, L.S., Pörtner, H.O. (1999). Thermal sensitivity of mitochondrial function in the Antarctic Notothenioid Lepidonotothen nudifrons. *Comparative Biochemistry and Physiology*, **124A**, 179–189.

Harper, E.M., Peck, L.S. (2003). Predatory behaviour and metabolic costs in the Antarctic muricid gastropod Trophon longstaffi. *Polar Biology*, **26**, 208–217.

Harper, E.M., Peck, L.S. (2016). Latitudinal and depth gradients in predation pressure. *Global Ecology and Biogeography*, **25**, 670–678.

Harris, P.T., MacMillan-Lawler, M., Rupp, J., Baker, E.K. (2014). Geomorphology of the oceans. *Marine Geology*, **352**, 4–24.

Hartman, O. (1964). *Polychaeta errantia of Antarctica*. Antarctic Research Series 3. American Geophysical Union, Washington, DC.

Ho, C.-K., Pennings, S.C., Carefoot, T.H. (2010). Is diet quality an overlooked mechanism for

Bergmann's rule? *American Naturalist*, **175**, 269–276.

Hoegh-Guldberg, O. (1999). Climate change, coral bleaching and the future of the world's coral reefs. *Marine and Freshwater Research*, **50**, 839–866.

Hofmann, G.E., Buckley, B.A., Airaksinen, S., Keen, J.E., Somero, G.N. (2000). Heat-shock protein expression is absent in the Antarctic fish *Trematomus bernacchii* (Family Nototheniidae). *Journal of Experimental Biology*, **203**, 2331–2339.

Horner, R.A. (2018). *Sea Ice Biota*. CRC Press, Boca Raton, FL.

Huth, T.P., Place, S.P. (2016). Transcriptome wide analyses reveal a sustained cellular stress response in the gill tissue of *Trematomus bernacchii* after acclimation to multiple stressors. *BMC Genomics*, **17**, 127.

IPCC; C. B.Field, V. R. Barros, D. J. Dokken, et al. (eds.) (2014). *Climate Change 2014: Impacts, Adaptation, and Vulnerability. Part A: Global and Sectoral Aspects. Contribution of Working Group II to the Fifth Assessment Report of the Intergovernmental Panel on Climate Change.* Cambridge University Press, Cambridge, UK and New York.

Jakobsson, M., Nilsson, J., Anderson, L., et al. (2016). Evidence for an ice shelf covering the central Arctic Ocean during the penultimate glaciations. *Nature Communications*, 10.1038/ncomms10365.

Janecki, T., Kidawa, A., Potocka, M. (2010). The effects of temperature and salinity on vital biological functions of the Antarctic crustacean *Serolis polita*. *Polar Biology*, **33**, 1013–1020.

Kleypas, J.A., Feely, R.A., Fabry, V.J., et al. (2006). *Impact of ocean acidification on coral reefs and other marine calcifiers: a guide for future research*. Report of a workshop held 18–20 April 2005, St. Petersburg, FL, sponsored by NSF, NOAA, and the US Geological Survey.

Kroeker, K.J., Kordas, R.L., Crim, R., et al. (2013). Impacts of ocean acidification on marine organisms: quantifying sensitivities and interaction with warming. *Global Change Biology*, **19**, 1884–1896.

Lohbeck, K.T., Riebesell, U., Collins, S., Reusch, T.B.H. (2013). Functional genetic divergence in high CO2 adapted *Emiliania huxleyi* populations. *Evolution*, **67**, 1892–1900.

McNeil, B.I., Matear, R.J. (2008). Southern Ocean acidification: a tipping point at 450-ppm atmospheric CO2. *Proceedings of the National Academy of Sciences of the USA*, **105**, 18860–18864.

Montes-Hugo, M., Doney, S.C., Ducklow, H.W., et al. (2009). Recent changes in phytoplankton communities associated with rapid regional climate change along the Western Antarctic Peninsula. *Science*, **323**, 1470–1473.

Montgomery, J., Clements, K. (2000). Disaptation and recovery in the evolution of Antarctic fishes. *Trends in Ecology and Evolution*, **15**, 267–271.

Moran, A.L., Woods, H.A. (2012). Why might they be giants? Towards an understanding of polar gigantism. *Journal of Experimental Biology*, **215**, 1995–2002.

Morley, S.A., Hirse, T., Thorne, M.A.S., Pörtner, H.O., Peck, L.S. (2012). Physiological plasticity, long term resistance or acclimation to temperature, in the Antarctic bivalve, *Laternula elliptica*. *Comparative Biochemistry and Physiology*, **162A**, 16–21.

Morley, S.A., Suckling, C.S., Clark, M.S., Cross, E.L., Peck, L.S. (2016). Long-term effects of altered pH and temperature on the feeding energetics of the Antarctic sea urchin, *Sterechinus neumayeri*. *Biodiversity*, **17**, 34–45.

Nghiem, S.V., Rigor, I.G., Clemente-Colón, P., Neumann, G., Li, P.P. (2016). Geophysical constraints on the Antarctic sea ice cover. *Remote Sensing of Environment*, **181**, 281–292.

Orr, J.C., Fabry, V.J., Aumont, O., et al. (2005). Anthropogenic ocean acidification over the twenty-first century and its impact on calcifying organisms. *Nature*, **437**, 681–686.

Peck, L.S. (1989). Temperature and basal metabolism in two Antarctic marine herbivores. *Journal of Experimental Biology*, **127**, 1–12.

Peck, L.S. (2002). Ecophysiology of Antarctic marine ectotherms: limits to life. *Polar Biology*, **25**, 31–40.

Peck, L.S. (2008). Brachiopods and climate change. *Earth and Environmental Science Transactions of The Royal Society of Edinburgh*, **98**, 451–456.

Peck, L.S. (2011). Organisms and responses to environmental change. *Marine Genomics*, **4**, 237–243.

Peck, L.S. (2015). DeVries: the Art of not freezing fish. Classics series. *Journal of Experimental Biology*, **218**, 2146–2147.

Peck, L.S. (2016). A cold limit to adaptation in the sea. *Trends in Ecology and Evolution*, **31**, 13–26.

Peck, L.S. (2018). Antarctic marine biodiversity: adaptations, environments and responses to change. *Oceanography and Marine Biology: An Annual Review*, **56**, 105–236.

Peck, L.S., Brockington, S., Brey, T. (1997). Growth and metabolism in the antarctic brachiopod *Liothyrella uva*. *Philosophical Transactions: Biological Sciences*, **352**(1355), 851–858.

Peck, L.S., Chapelle, G. (1999). Amphipod gigantism dictated by oxygen availability? *Ecology Letters*, **2**, 401–403.

Peck, L.S., Chapelle, G. (2003). Reduced oxygen at high altitude limits maximum size. *Proceedings of the Royal Society of London Series B: Biological Sciences*, **270**, S166–S167.

Peck, L.S., Conway, L.Z. (2000). The myth of metabolic cold adaptation: oxygen consumption in stenothermal Antarctic bivalves. In:E. M. Harper,J. D. Taylor, J. A. Crame (eds) *The Evolutionary Biology of the Bivalvia*. Geological Society, London, Special Publications, Vol. 177.Geological Society, London, pp. 441–445.

Peck, L.S., Brockington, S., VanHove, S., Beghyn, M. (1999). Community recovery following catastrophic iceberg impacts in Antarctica. *Marine Ecology Progress Series*, **186**, 1–8.

Peck, L.S., Clark, M.S., Morley, S.A., Massey, A., Rossetti, H. (2009b). Animal temperature limits and ecological relevance: effects of size, activity and rates of change. *Functional Ecology*, **23**, 248–253.

Peck, L. S., Convey, P., Barnes, D. K. A. (2006). Environmental constraints on life histories in Antarctic ecosystems: tempos, timings and predictability. *Biological Reviews*, **81**, 75–109.

Peck, L.S., Massey, A., Thorne, M.A.S., Clark, M. S. (2009a). Lack of acclimation in *Ophionotus victoriae*: brittle stars are not fish. *Polar Biology*, **32**, 399–402.

Peck, L.S., Morley, S.A., Clark, M.S. (2010). Poor acclimation capacities in Antarctic marine ectotherms. *Marine Biology*, **157**, 2051–2059.

Peck, L.S., Morley, S.A., Richard, J., Clark, M.S. (2014). Acclimation and thermal tolerance in Antarctic marine ectotherms. *Journal of Experimental Biology*, **217**, 16–22.

Peck, L.S., Pörtner, H.O., Hardewig, I. (2002). Metabolic demand, oxygen supply and critical temperatures in the Antarctic bivalve *Laternula elliptica*. *Physiological Biochemical Zoology*, **75**, 123–133.

Peck, L.S., Souster, T., Clark, M.S. (2013). Juveniles are more resistant to warming than adults in 4 species of Antarctic marine invertebrates. *PLoS ONE*, **8**, e66033.

Peck, L.S., Webb, K.E., Bailey, D. (2004). Extreme sensitivity of biological function to temperature in Antarctic marine species. *Functional Ecology*, **18**, 625–630.

Peck, L.S., Webb, K.E., Clark, M.S., Miller, A., Hill, T. (2008). Temperature limits to activity, feeding and metabolism in the Antarctic starfish *Odontaster validus*. *Marine Ecology Progress Series*, **381**, 181–189.

Peck, V.L., Oakes, R.L., Harper, E.M., Manno, C., Tarling, G.A. (2018). Pteropods counter mechanical damage and dissolution through extensive shell repair. *Nature Communications*, **9**, 264.

Peel, M.C., Wyndham, R.C. (1999). Selection of clc, cba, and fcb chlorobenzoate-catabolic genotypes from groundwater and surface waters adjacent to the Hyde Park, Niagara Falls, chemical landfill. *Applied and Environmental Microbiology*, **65**, 1627–1635.

Pespeni, M.H., Sanford, E., Gaylord, B., et al. (2013). Evolutionary change during experimental ocean acidification. *PNAS*, **110**(17), 6937–6942.

Piepenburg, D., Archambault, P., Ambrose, W, et al. (2010). Towards a pan-Arctic inventory of the species diversity of the macro- and megabenthic fauna of the Arctic shelf seas. *Marine Biodiversity*, **41**, 51–70.

Place, S.P., Hofmann, G.E. (2005). Constitutive expression of a stress inducible heat shock protein gene, hsp70, in phylogenetically distant Antarctic fish. *Polar Biology*, **28**, 261–267.

Podrabsky, J.E., Somero, G.N. (2006). Inducible heat tolerance in Antarctic notothenioid fishes. *Polar Biology*, **30**, 39–43.

Pörtner, H.O., Peck, L.S., Hirse, T. (2006). Hyperoxia alleviates thermal stress in the Antarctic bivalve, *Laternula elliptica*: evidence for oxygen limited thermal tolerance? *Polar Biology*, **29**, 688–693.

Pörtner, H.O., Peck, L.S. Zielinski, S., Conway, L. Z. (1999). Temperature and metabolism in the highly stenothermal bivalve mollusc *Limopsis marionensis* from the Weddell Sea, Antarctica. *Polar Biology*, **22**, 17–30.

Richard, J., Morley, S.A., Peck, L.S. (2012). Estimating long-term survival temperatures at the assemblage level in the marine environment: towards macrophysiology. *PLoS ONE*, **7**, e34655.

Robertson, R.F., El-Haj, A.J., Clarke, A., Peck, L. S., Taylor, E.W. (2001). The effects of temperature on metabolic rate and protein synthesis following a meal in the isopod *Glyptonotus antarcticus* eights (1852). *Polar Biology*, **24**, 677–686.

Robinson, E., Davison, W. (2008). The Antarctic notothenioid fish *Pagothenia borchgrevinki* is thermally flexible: acclimation changes oxygen consumption. *Polar Biology*, **31**, 317–326.

Rodríguez-Romero, A., Jarrold, M.D., Massamba-N'Siala, G., Spicer, J.I., Calosi, P. (2016). Multi-generational responses of a marine polychaete to a rapid change in seawater pCO_2. *Evolutionary Applications*, **9**, 1082–1095.

Roggatz, C.C., Lorch, M., Hardege, J.D., Benoit, D.M. (2016). Ocean acidification affects marine chemical communication by changing structure and function of peptide signalling molecules. *Global Change Biology*, **22**, 3914–3926.

Reed, A.J., Thatje, S. (2015). Long-term acclimation and potential scope for thermal resilience in Southern Ocean bivalves. *Marine Biology*, **162**, 2217–2224.

Rohling, E.J., Foster, G.L., Grant, K.M., et al. (2014). Sea-level and deep-sea-temperature variability over the past 5.3 million years. *Nature*, **508**, 477–482.

Ruud, J.T. (1954). Vertebrates without erythrocytes and blood pigment. *Nature*, **173**, 848–850.

Schloss, I.R., Abele, D., Moreau, S., et al. (2012). Response of phytoplankton dynamics to 19-year (1991–2009) climate trends in Potter Cove (Antarctica). *Journal of Marine Systems*, **92**, 53–66.

Schofield, O., Ducklow, H.W., Martinson, D.G., et al. (2010). How do polar marine ecosystems respond to rapid climate change? *Science*, **328**, 1520–1523.

Schram, J.B., McClintock, J.B., Amsler, C.D., Baker, B.J. (2015). Impacts of acute elevated seawater temperature on the feeding preferences of an Antarctic amphipod toward chemically deterrent macroalgae. *Marine Biology*, **162**, 425–433.

Segawa, T., Ushida, K., Narita, H., Kanda, H., Kohshima, S. (2010). Bacterial communities in two Antarctic ice cores analyzed by 16S rRNA gene sequencing analysis. *Polar Science*, **4**, 215–227.

Shin, S.C., Kim, S.J., Lee, J.K., et al. (2012). Transcriptomics and comparative analysis of three Antarctic notothenioid fishes. *PLoS ONE*, **16**, e43762.

Somero, G.N. (2010). The physiology of climate change: how potentials for acclimatization and genetic adaptation will determine 'winners' and 'losers'. *Journal of Experimental Biology*, **213**, 912–920.

Somero, G.N. (2015). Temporal patterning of thermal acclimation: from behaviour to membrane biophysics. *Journal of Experimental Biology*, **218**, 167–169.

Somero, G.N., De Vries A.L. (1967). Temperature tolerance of some Antarctic fishes. *Science*, **156**, 257–258.

Stanwell-Smith, D.P., Peck, L.S. (1998). Temperature and embryonic development in relation to spawning and field occurrence of larvae of 3 Antarctic echinoderms. *Biological Bulletin*, **194**, 44–52.

Stillman, J.H., Paganini, A.W. (2015). Biochemical adaptation to ocean acidification. *Journal of Experimental Biology*, **218**, 1946–1955.

Stroeve, J., Notz, D. (2018). Changing state of Arctic sea ice across all seasons. *Environmental Research Letters*, **13**,103001.

Suckling, C.C., Clark, M.S., Richard, J., et al. (2015). Adult acclimation to combined temperature and pH stressors significantly enhances reproductive outcomes compared to short-term exposures. *Journal of Animal Ecology*, **84**, 773–784.

Sunday, J.M., Calosi, P., Dupont, S., et al. (2014). Evolution in an acidifying ocean. *Trends in Ecology and Evolution*, **29**, 117–125.

Sweetman, A.K., Thurber, A.R., Smith, C.R., et al. (2017). Major impacts of climate change on deep-sea benthic ecosystems. *Elementa: Science of the Anthropocene*, **5**, 4.

Thorson, G. (1957). Bottom communities. In: Hedgpeth, J. W. (ed.) *Treatise on Marine Ecology and Paleoecology*. Geological Society of America, pp. 461–534.

Todgham, A.E., Hoaglund, E.A., Hofmann, G.E. (2007). Is cold the new hot? Elevated ubiquitin-conjugated protein levels in tissues of Antarctic fish as evidence for cold-denaturation of proteins in vivo. *Journal of Comparative Physiology B*, **177**, 857–866.

Todgham, A.E., Crombie, T.A., Hofmann, G.E. (2017). The effect of temperature adaptation on the ubiquitin-proteasome pathway in notothenioid fishes. *Journal of Experimental Biology*, **220**, 369–378.

Tomanek, L. (2010). Variation in the heat shock response and its implication for predicting the effect of global climate change on species' biogeographical distribution ranges and metabolic costs. *Journal of Experimental Biology*, **213**, 971–979.

US National Snow and Ice Data Centre (2018). Arctic sea ice extent arrives at its minimum. http://nsidc.org/arcticseaice news/2018/09/

Verde, C., Giordano, D., di Prisco, G., Andersen, Ø. (2012). The hemoglobins of polar fish: evolutionary and physiological significance of multiplicity in Arctic fish. *Biodiversity*, **13**, 228–233.

Vermeij, G.J. (2016). Gigantism and its implications for the history of life. *PLoS ONE*, **11**, e0146092.

Watson, J.D., Baker, T.A., Bell, S.P., et al. (2014). *Molecular Biology of the Gene*. Cold Spring Harbour Laboratory Press, New York.

Watson, S.-A., Southgate, P., Tyler, P.A., Peck, L. S. (2009). Early larval development of the Sydney rock oyster *Saccostrea glomerata* under near-future predictions of CO_2-driven ocean acidification. *Journal of Shellfish Research*, **28**, 431–437.

Watson, S.-A., Peck, L.S., Tyler, P.A., et al. (2012). Marine invertebrate skeleton size varies with latitude, temperature, and carbonate saturation: implications for global change and ocean acidification. *Global Change Biology*, **18**, 3026–3038.

Watson, S.-A., Peck, L.S., Morley, S.A., Munday, P.L. (2017). Latitudinal trends in shell production cost from the tropics to the poles. *Scientific Reports*, **3**, e1701362.

Węsławski, J.M., Kendall, M.A, Włodarska-Kowalczuk, M., et al. (2011). Climate change effects on Arctic fjord and coastal macrobenthic diversity – observations and predictions. *Marine Biodiversity*, **41**, 71–85.

Whiteley, N.M., Taylor, E.W., El Haj, A.J. (1997). Seasonal and latitudinal adaptation to temperature in crustaceans. *Journal of Thermal Biology*, **22**, 419–427.

Wiedenmann, J., Cresswell, K.A., Mangel, M. (2009). Connecting recruitment of Antarctic krill and sea-ice. *Limnology and Oceanography*, **54**, 799–811.

Wiest, L.A., Buynevich, I.G., Grandstaff, D.E., et al. (2015). Trace fossil evidence suggests widespread dwarfism in response to the end-Cretaceous mass extinction: Braggs, Alabama and Brazos River, Texas. *Palaeogeography, Palaeoclimatology, Palaeoecology*, **417**, 405–411.

Young, J.S., Peck, L.S., Matheson, T. (2006a). The effects of temperature on walking in temperate and Antarctic crustaceans. *Polar Biology*, **29**, 978–987.

Young, J.S., Peck, L.S., Matheson, T. (2006b). The effects of temperature on peripheral neuronal function in eurythermal and stenothermal crustaceans. *Journal of Experimental Biology*, **209**, 1976–1987.

Zachos, J., Pagani, M., Sloan, L., Thomas, E., Billups, K. (2001). Trends, rhythms, and aberrations in global climate 65 Ma to present. *Science*, **292**, 686–693.

Zachos, J.C., Dickens, G.R., Zeebe, R.E. (2008). An early Cenozoic perspective on greenhouse warming and carbon-cycle dynamics. *Nature*, **451**, 279–283.

CHAPTER NINE

The Southern Ocean: an extreme environment or just home of unique ecosystems?

JULIAN GUTT

Alfred Wegener Institute, Helmholtz Centre for Polar and Marine Research Bremerhaven

a n d

GERHARD DIECKMANN

Alfred Wegener Institute, Helmholtz Centre for Polar and Marine Research Bremerhaven

9.1 Introduction

The CAREX roadmap defined an extreme environment as having 'parameters showing values permanently close to lower or upper limits known for life in its various forms' (CAREX, 2011). In the Southern Ocean (SO) physical–chemical conditions shape life to a greater extent than in other marine ecosystems, due to for example the pronounced seasonality and freezing temperatures, particularly in sea-ice. The latter is the habitat where penguins and seals live for limited periods of their life cycle, in extreme cases at temperatures below $-40°C$ (Barbraud & Weimerskirch, 2001; www.coolantarctica.com/Antarctica%20fact%20file/wildlife/Emperor-penguins.php). Obviously, such conditions do not prevent species from attaining high biomass and abundances at least locally. In addition, at temperatures close to freezing, communities on the seabed and within sea-ice can attain high functional as well as structural diversity (Griffiths, 2010; Thomas & Dieckmann, 2010; Jacob et al., 2011). Such diversity hotspots can occur almost everywhere in the SO. Similarly, cold-spots, in terms of biomass and biological performance also exist all over the continent.

We apply two approaches to answer the question whether the SO is an extreme environment or 'simply' home to unique ecosystems:

1. Using an absolute definition of extreme conditions, where ecologically relevant environmental variables, e.g. temperature close to freezing, are assumed to determine the 'lower' polar margin of species and their assemblages. This can allow the verification or rejection of the historic view of an extreme environment.
2. Alternatively, highly degraded SO communities are monitored to judge whether they are good indicators of environmental conditions or extreme

events, which permanently or abruptly limit their development. This assessment is mostly applicable to entire species communities and not to almost uncountable exceptions of single species, which are not in accordance with general trends.

9.2 Environmental conditions assumed to be limiting for biological processes, which shape biodiversity, biomass and ecological functions

9.2.1 Sea-Ice

9.2.1.1 *Environment*

Sea-ice is the frozen ocean surface, which begins to freeze in autumn and mostly melts in spring. It is characterised by steep gradients in ecologically relevant environmental parameters, such as radiation, temperature, salinity and nutrients (Thomas & Dieckmann, 2002). Temporal shifts in these variables depend on the exposure to atmospheric and hydrodynamic processes at the corresponding interfaces. Sea-ice has multiple forms depending on formation processes, the 2- to 5-m-thick multi-year ice, the 1- to 2-m-thick first-year ice, frazil (slush) ice consisting of tiny platelets and sometimes also anchor ice, which forms on the seabed down to 30 m. Throughout its entire vertical extent, sea-ice serves as a habitat for a diverse range of organisms.

9.2.1.2 *Life on and within sea-ice, as well as on the under side*

Life on the ice surface is limited to a small number of species. Emperor penguins and Weddell, Ross and Crabeater seals especially are adapted to spend a considerable proportion of their life cycle on the ice, where they rest and reproduce (Figure 9.1). The fact that only 20–25% of the Emperor penguin chicks, which hatch under harshest conditions in winter, survive the first summer confirms the life-threatening conditions under which they live, but to which on the other hand they are well adapted (www.coolantarctica.com /Antarctica%20fact%20file/wildlife/Emperor-penguins.php and www.whoi.edu /news-release/melting-sea-ice-threatens-emperor-penguins–study-finds).

Organisms that live within sea-ice comprise micro zooplankton and algae (originally derived from phytoplankton), and other tiny organisms, which are incorporated into the sea ice when it forms during winter. The small dimensions of the brine channels, where the ice organisms live, limits their body size to a maximum of a few millimetres. The community composition differs from that in the water from which they derive, since both size and adaptation to high salinities up to 100‰ and extremely low temperatures select for the life within sea ice (Thomas & Dieckmann, 2002).

Algae at the underside of the sea-ice live under low-light conditions but compared to free-floating phytoplankton, they don't sink down to water layers below the compensation depth, where environmental conditions do not allow

Figure 9.1 The Emperor penguin colony on the fast ice of the Atka Bay in the Eastern Weddell Sea, SO, during austral winter 2004.© Anna Müller. (A black and white version of this figure will appear in some formats. For the colour version, please refer to the plate section.)

sustained net primary production (NPP). Nutrient supply is guaranteed by permanent currents, which provide constant replenishment contrary to the situation with phytoplankton, which drifts with the waterbody or lives in the sea-ice, where nutrients may become depleted. Ice algae are an important food source for krill, which play a key role in the SO pelagic food web, but also for amphipods and other zooplankton. At the end of the winter, algal biomass in the approximately 2-m-thick ice, most of which developed at the subsurface, can reach that of the summer bloom in the entire water column (Thomas & Dieckmann, 2010).

9.2.1.3 Conclusion
Biodiversity on the sea-ice surface comprises only a handful of warm-blooded animals. Thus, despite the locally high abundances, this environment must be considered extreme and not conducive to supporting a complex ecosystem in itself. Surprising adaptations, especially the reproductive cycle, of the Emperor penguins, however, demonstrate that this species is doing well in an environment that can be considered deadly for most other multicellular species on Earth. Life in the ice is not rich in species, but relatively rich and interesting with regard to life forms and interactions in the context of space limitation. Thus, with the exception of high salinity and low nutrients at the

end of the winter, this environment cannot be considered to mark the limit for a relatively diverse and rich pelagic life. The sea-ice habitat is comparable to coarse sediments in warmer oceans, where the meiofauna lives in the interstitial water between sand grains. As a consequence, environmental conditions can be considered extremely favourable for NPP, especially with respect to the available volume of water; however, diversity of life at the subsurface of the sea-ice remains low.

9.2.2 Ice-cold waters

9.2.2.1 Environment

Temperatures in the SO close to the continent, near the ocean surface and at the seafloor may drop to near freezing (−1.8°C) during winter, while relatively warm water between +4°C and 0°C occurs offshore further North at the Polar Front, West of the Antarctic Peninsula, in the warm deep water and during summer in a thin surface layer (Pardo et al., 2012). Low temperatures generally slow down biochemical processes and reduce metabolic rates, growth, reproduction, motility and biological interactions. At ambient seawater temperatures close to freezing, ice crystals may form in the bodies of cold-blooded vertebrates and invertebrates.

9.2.2.2 Life at temperatures close to freezing

In the pelagic system, especially on the continental shelf, NPP can be high and attain rates of up to >100 g C m^{-2} a^{-1} (Arrigo et al., 2008) despite low temperatures. Krill, the main consumer of the phytoplankton, concentrates in dense swarms of >200 n m^2 at temperatures between approximately +1°C and −1.5°C (Flores et al., 2012). These biological processes, of which some begin with a microbial loop, provide the basis of the food web, which ends with top predators such as seals and whales but which also includes seabirds, predatory fish and salps, the latter at slightly higher temperatures. Whether governed by the temperature or not, only one pelagic fish species, *Pleuragramma antarctica*, has become successful in terms of biomass, in the midwater region at temperatures below 0°C (Figure 9.2). Thus, this single species serves as an important food source for predators. However, ecological equivalents in temperate latitudes such as herring, sprat, sardines and anchovies have also not evolved to a high species diversity.

Most benthic organisms inhabiting the continental shelf live at temperatures between −1.0 and −1.8°C (Clarke et al., 2009). The species investigated do not seem to be specifically adapted to low temperature since they have a low metabolism (Peck, 2005). However, exceptions such as fast-growing ascidians are also known to occur (Gutt et al., 2013) and sponges cover a wide range from fast-growing species or life stages to extremely slow-growing life forms (Dayton, 1979). Obviously, different evolutionary pathways have led to their

Figure 9.2 *Pleuragramma antarctica* is the only abundant midwater fish in high latitudes of the SO and, thus, plays an especially important key role in the pelagic food web.© Dieter Piepenburg, Alfred Wegener Institute. (A black and white version of this figure will appear in some formats. For the colour version, please refer to the plate section.)

success in terms of a high diversity and (wet weight) biomass, which reaches a world record in the Weddell Sea (Figure 9.3; Gerdes, 1992). Sponges occur in similar concentrations in the Ross Sea, providing evidence that temperature does not necessarily limit the development of very high benthic biomass.

Antarctic fish are usually adapted to low temperature due to a higher metabolism compared to relatives living in warmer environments, when they are exposed to low temperatures (Somero, 1991). In addition, notothenioid fish have reduced their red blood cells and blood pigment at the expense of a lower oxygen transport through the blood vessels. However, due to the increased blood viscosity, less energy is required to pump the blood through the body (di Prisco et al., 1998). These fish have also developed antifreeze proteins in their blood to avoid the formation of ice crystals (see Chapter 8), and some species apply a 'sit-observe-and-hide' behaviour to save energy while observing potential prey and predators from an elevated position (Gutt & Ekau, 1996).

9.2.2.3 *Conclusion*
In evolutionary time scales, the cooling of the SO has led to the absence of a few systematic groups such as sharks, flat fish and lobsters. The cooling in combination with its partial isolation, however, has also led to an efficient

Figure 9.3 Environmental conditions on the continental shelf of the high-latitude southeastern Weddell Sea are obviously not limiting for a locally high biomass of sponges and a relatively rich associated fauna comprising among others sea cucumbers, brittle stars, bryozoans and hemichordates.© Julian Gutt and Werner Dimmler, Alfred Wegener Institute/Webseiten des Zentrum für Marine Umweltwissenschaften (MARUM). (A black and white version of this figure will appear in some formats. For the colour version, please refer to the plate section.)

radiation of species. In terms of ecological productivity, the pelagic system of the SO with krill in the centre of the food web in some larger areas is considered not to be limited by extreme temperatures. In addition to maintaining an autochthonous community, including an assumed 50% of the world's seal abundance, it attracts several whale species, which reproduce in warmer waters but use the SO as an important feeding ground. Also, as interesting as the lower temperature limit, is the species tolerance range. Many ectotherms in the SO are known to tolerate a narrow temperature range with an upper limit of 2° to 10°C above their normal ambient temperature of around −1.3°C (Peck, 2005). This can be considered to be extreme, especially if compared with species in temperate latitudes, e.g. black mussels, which can tolerate ranges at least from below 0° to 25°C.

9.3 Extreme seasonality

9.2.3.1 *Environment*

The position of the SO between 60°S and 80°S in combination with the Earth's inclination and its rotation around the Sun causes a change from 24 hours daylight to 24 hours darkness in its southern part during the course of a year. This can be considered to be extreme in a global comparison and, in combination with the low temperature, this seasonality is the main driver of polar ecosystems. However, the seasonal radiation pattern in the Arctic Ocean region is even more extreme, since it covers the entire area from sub-arctic latitudes to the North Pole. A third driver, governed by radiation and temperature is the seasonal advance and retreat of the sea-ice. These physical processes are accurately predictable at a larger regional scale. However, this is superimposed by non-predictable local effects with obvious consequences for the NPP (see e.g. Cape et al., 2014).

9.2.3.2 *Life in total darkness and under permanent sunlight*

NPP in the SO open water peaks at levels >100 g C m^{-2} a^{-1}, however, only for a limited period of a few weeks and very locally. Its low spatial and temporal predictability might be buffered for consumers since they have access to additional food sources. In winter, krill feeds at the subsurface of the sea-ice and when microalgae melt out of the ice in spring they provide food for other pelagic and benthic species over an extended period. Benthic filter feeders can also benefit from resuspension of phytodetritus deposited at the seafloor and extend the feeding period into autumn. Additional strategies to uncouple from the seasonal radiation and unpredictable NPP are reproductive cycles triggered by food availability and not by the seasonal light cycle (Gutt et al., 1992) or to feed on higher levels of the food chain including cannibalism. Invertebrates can also feed on heterotrophic bacteria or benefit from dissolved organic matter. The slow growth of predatory fish larvae is obviously determined by the short period of food availability in summer, rather than the low temperature (Clarke & North, 1991).

9.2.3.3 *Conclusion*

The strongly altering seasonal radiation alone cannot be considered limiting for most of the life in the SO. If radiation did have a direct effect, biological gradients would be coupled strongly to latitudinal gradients, e.g. in biomass and diversity, which are barely detectable (Piepenburg et al., 2002). For antarctic krill it is known that deviations from a predictable seasonality in sea-ice extent is unfavourable for juveniles and can lead to the loss of a complete year class (Flores et al., 2012). On the one hand, such a mismatch between environmental conditions and demands, also of juvenile fish, is also known from oceans with a less distinct seasonality and can generally be seen to be 'normal' in marine systems. On the other hand, the shorter the period of food availability, in the case of krill, related to sea-ice extent, the higher their vulnerability to unexpected environmental changes and the easier they are pushed towards their survival limits. Especially in the case of the benthos, theories exist that it was well adapted to the low NPP during glacial periods when a solid ice cover similar to today's winter conditions existed. As a consequence, during the actual interglacial their food, phytodetritus, might exist in excess and at least some species can easily starve and survive even 1 year without open water. Conditions during the glacial periods might have caused a selection towards only those species that are capable of surviving under conditions of a low food supply. However, it is a 'normal' phenomenon that all species are somehow adapted to their environment and, consequently, not all species occur everywhere on Earth.

9.2.4 Low nutrient and food availability

9.2.4.1 *Environment*

Microalgae in the SO, which form the basis of the marine food web, live under high nutrient–low chlorophyll conditions (HNLC). This means that macronutrients are available in excess but micronutrients limit NPP. Higher trophic levels can also experience a paucity of their specific food, partly explained by the seasonality described above. Oligotrophic conditions are found beneath ice shelves. These are discussed below in the Section 9.3.

9.2.4.2 *Micronutrient limitation and food paucity*

Despite the limitation of micronutrients, NPP, in a global comparison, attains intermediate annual levels, locally even high values for a limited period (Arrigo et al., 2008) and thus supports the enormous krill biomass of >100 million t (Flores et al., 2012). In addition, the coastal pelagic system obviously leaves surplus energy, which enables the accumulation of high benthic biomass on the continental shelf albeit that metabolic rates do not seem to be very energy demanding. Space and not food availability might be the limiting factor in the rich sponge grounds as well as in ascidian and filter-feeding clam aggregations. An open question is, however, why glass sponges of the same species grow to 2 m in height in shallow water in McMurdo Sound and East Antarctica, while individuals of the same species mostly remain smaller than 1 m at a 200-m depth in the Weddell Sea (Gutt, 2000). The latter can be associated to poorer food conditions, but the local density of the sponges argues against a trophic limitation (Figure 9.3). A theory about the evolution of the rich antarctic epibenthos assumes that the lack of predators, e.g. catfish or large crabs, being able to crush hard-shelled or other indigestible prey species, enabled the development of the rich epibenthic communities (Smith et al., 2012). In this case, the limitations for one group, the predators, allowed a successful development of the sessile epifauna.

9.2.4.3 *Conclusion*

Despite the HNLC conditions, the annual NPP in the SO provides sufficient energy for high biomass at higher trophic levels, for example of krill. It remains unclear to which degree krill is limiting for the next step in the trophic web since the large krill consumers have been depleted by whaling. The same question can be applied to the regionally disastrous effect of bottom trawling on demersal fish, mostly of the genus *Notothenia*, to an impending overfishing of the antarctic toothfish, *Dissostichus antarcticus*. It is also still unclear whether the antarctic benthos in particular follows the trend of a deep-sea food limited system, which, however, supports a high biodiversity (Smith et al., 2008). The high benthic biomass, however, contradicts this. In general, the HNLC conditions do not seem to have an effect on biodiversity, since macronutrient limitation is more pronounced in the SO than in the Arctic, but diversity trends show the opposite.

9.3 Poor communities (cold-spots) indicating extreme environments
9.3.1 Life on the sea-ice surface and at the iceshelf subsurface

Despite observations of thousands of Emperor penguins on the fast sea-ice even during winter (Figure 9.1) and scattered seal occurrences on ice floes in summer, the sea-ice surface is generally poorly inhabited. This applies especially to the number of species, which is not much more than a handful. The pressure to use the sea-ice surface as a habitat comes on the one hand from the incomplete adaptation of penguins and seals to the marine environment. They still have to give birth or lay eggs, also to recover and some of them to mate outside the water. On the other hand, they are so much adapted to the high-latitude marine conditions that for most of these species land is no longer a suitable substitute when they have to leave the water. An advantage to living for a certain period of time on the ice is to avoid competition with less polar adapted related species and avoid predators such as Orcas, and this habitat is much larger than the almost one-dimensional non-glaciated coastal strip around part of the continent. However, the adaptations in their ecology, behaviour and physiology to this extreme habitat has had its price. Weddell seals have to constantly maintain a hole in the ice, which they do with their teeth, to haul out onto the ice and vice versa, which might limit their life span. Emperor penguins have developed a behaviour to reduce heat loss within their colony especially in winter and at temperatures down to −40°C. However, their offspring generally experience a low survival. In addition, in spring they must walk up to 300 km on the fast ice to its margin (www .coolantarctica.com/Antarctica%20fact%20file/wildlife/Emperor-penguins.php) to access their marine food source after breeding and then return with food to feed their chicks. All these are indications that these species live at the lower limit of any higher organisms, albeit quite successfully.

The subsurface of the ice shelves is another cold-spot in species richness and abundance and, thus, traditionally not considered as a habitat. A few decades ago, however, the fish species *Trematomus nicolai* and juveniles of *Pagothenia borchgrevinki* were observed clinging to the vertical edge of an iceberg and to the margin of the iceshelf (Gutt, 2002). Recently the sea anemone *Edwardsiella andrillae* was found to live with the rear part of the body buried in the iceshelf (Daly et al., 2013), and aggregations of isopods were observed to attach themselves to the subsurface of the iceshelf with their hind legs (Figure 9.4; Bornemann et al., 2016).

9.3.1.1 Conclusion

Without exception, the sea-ice surface is an extreme environment, following the CAREX definition. It is more hostile than true deserts due to its seasonal dynamics in its extent and the extreme variability in ecologically relevant environmental factors. Traditionally, considered even less habitable was the subsurface of the ice shelves. However, three very different animals use this

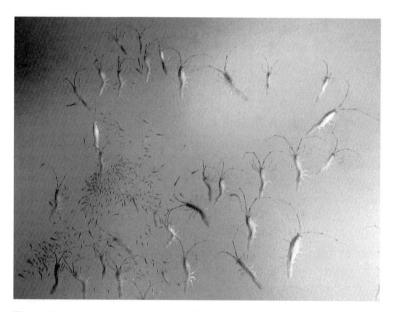

Figure 9.4 Adults and juveniles of filter-feeding isopods (*Antarcturus* cf. *spinacoronatus*) attached head-down to the underside of the floating Riiser-Larsen Iceshelf at depths of around 100 m in the Drescher Inlet, southeastern Weddell Sea in February 2016. Camera line of sight facing upwards. ROV-borne HD-video-image,© Nils Owsianowski, AWI. (A black and white version of this figure will appear in some formats. For the colour version, please refer to the plate section.)

habitat, which is only poorly investigated. It still remains totally unclear how many other life forms are also adapted to this extreme environment.

9.3.2 Low biomass in the water column

Tiny organisms cause turbidity in the oceans, which is an indicator for the amount of life at the basis of the food chain. This is especially significant in the SO because non-organic turbidity due to terrestrial runoff is only of minor relevance. Due to these specific environmental conditions and the extremely low biomass a world record in transparency was measured in austral winter/ spring 1986 in the Weddell Sea (Gieskes et al., 1987).

9.3.2.1 Conclusion
The above-described paucity of microorganisms in the open water reflects the extreme conditions, which limit most of the pelagic life in winter. The options to survive elsewhere, in deep-water layers for krill eggs and larvae, or hide below or inside the sea-ice for phyto- and zooplankton organisms demand specific adaptations. Similar to the sea-ice surface, the pelagic system during winter can be compared to a desert, which, however, flourishes regularly in spring. Conditions that do not allow population growth of microalgae in winter are also known from all marine systems experiencing a clear season-ality. The above-described world record for the lack of microorganisms, how-ever, was reached in the SO.

9.3.3 Sediments are poor in life like deserts

As in the open water and on the sea-ice, the distribution of life at the seafloor is patchy (De Broyer et al., 2014). Imaging methods show areas almost devoid of any life with only few major functional groups becoming visible. A significant impact on the ecosystem such as iceberg scouring, which locally eradicates all life (Gutt & Piepenburg, 2003), can be categorised into a relative or absolute disturbance. In the first case the disturbance is part of the 'normal' ecosystem functioning, in the second it is not. Iceberg disturbance is a process that the SO benthic system has experienced for millions of years. As a consequence, such destructive events can be seen as a local catastrophe but not as a limiting factor for entire benthic commu-nities. The recolonisation of the devastated areas can even increase benthic diversity (Gutt & Piepenburg, 2003). Also, the sea-ice impacts benthic life, but only in a narrow fringe around the continent, where it is not glaciated by inland ice down to approximately 30 m depth (Barnes, 1995). Due to the high frequency of this disturbance, this habitat can be considered extreme. Only few animals are able to avoid the high mortality, for example limpets living in crevices. Similarly, bare spots on the sediment originate from the formation and uplift of anchor ice in which benthic organisms are trapped.

Figure 9.5 Benthic life below the Larsen B iceshelf, collapsed east of the Antarctic Peninsula 5 years before this photograph was taken, can be extremely poor and inhabited by some deep-sea organisms such as this multi-armed asteroid Freyella fragillisima.© Julian Gutt and Werner Dimmler, AWI/MARUM. (A black and white version of this figure will appear in some formats. For the colour version, please refer to the plate section.)

In contrast to the spatially limited ice impact, benthic hotspots can fade out at distances of a few metres to kilometres towards cold-spots. Only speculation exists about the mechanisms behind such obvious patchiness. One reason can be differences in sediment characteristics in combination with demands of early life stages, such as microfilms or a specific grain size, which might match or mismatch with the demands of the organisms. In addition, grazing deposit feeders such as motile holothurians can remove early life stages of other benthic organisms from the seabed, which can have a long-lasting effect. Perhaps the permanently poorest life in any marine system is found under Antarctic ice shelves (Figure 9.5; Bruchhausen et al., 1979). The explanation that food conditions especially lead to the impoverishment is convincing, but it is hard to survey this habitat and understand ecological details. A closer view shows that areas under the ice shelves can locally also have a relatively rich life where the current transports organic matter under the ice, and life can be extremely poor where a current almost without any organic matter flows from under the iceshelf towards the open ocean.

9.3.3.1 Conclusion

Iceberg scouring and anchor ice impacts are examples of extreme events that shape an important part of the SO ecosystem. From a whole ecosystem perspective this is a normal phenomenon comparable to forest fires on land or hurricanes in coral reefs. In addition, largely unknown long-term conditions might exist that limit the SO benthos and lead to high mortality of sponges (Dayton et al., 2016). Extremely low provision of food under the iceshelf, especially far beyond its margins, leads to extremely poor communities considered to exist at the lower limit of life with respect to biodiversity, biomass and community metabolism.

9.4 Overarching conclusions

The SO ecosystems are generally characterised by a 'normal' diversity, a similar food web structure as exists in other marine ecosystems and an ecological complexity comparable to that of other oceans. This 'normality' includes limiting factors that other ecosystems on Earth also experience. In the SO some of these factors and their combination, however, are unique. As a consequence, life in the SO is also unique, expressed for example by the high number of endemic species, distinctive trophic pathways in the pelagic realm, life within sea-ice, a diverse recolonisation of the seafloor after iceberg scouring, and distinctive epibenthic composition. Such ecological complexity developed under physicochemical conditions not considered to be extreme and not considered to limit the performance of most organisms and their interactions. However, major exceptions exist. These are the habitats on the sea-ice surface, under ice shelves and at its subsurface. The only zones that may be devoid of almost any life for a limited period are fresh iceberg scours and the pelagic realm in winter. Independently of the difficulties to apply CAREX definitions to entire communities, we can conclude that the state at which the above-mentioned poor assemblages or populations exist is close to generally deathly conditions. If for example blood freezes in fish bodies, they would die were they not protected by specific adaptation. If food supply almost ceases, e.g. under ice shelves also a simple food web would collapse, and if water freezes to a solid compound without brine channels, organisms would be unable to live in it. As a consequence, most of the above-mentioned poor or limited communities must be considered to be especially vulnerable to environmental changes. After even only minor environmental changes, e.g. in temperature leading to ice disintegration, cold-spots could turn to 'normal' conditions and hotspots to cold-spots. These are good reasons why the study of SO ecosystems at the limit of life is not only of academic interest but may also help to understand the future development of marine ecosystems.

9.5 Outlook

The unique SO ecosystems are connected to those in all other oceans and contribute to global biodiversity as well as globally relevant ecosystem services, primarily the production of oxygen and uptake of CO_2. As a consequence, all assessments with this background are only complete if the relevant information from the SO is included. Single studies on the SO's ecosystem structure and functioning are frequent and in the past decades have contributed to an increase in knowledge on hotspots such as krill swarms and sponge communities or on the biochemical/physiological adaptation of fishes to low temperatures. What is still lacking is a comprehensive view, where marine communities or ecosystems as a whole are limited by environmental constraints. The problem in such an approach is that it is a matter of the scientific perspective whether an obviously impoverished community is a modification of a richer community existing at its lower limit of existence or whether it is just a different community being poorer in some parameters than related communities. Such approaches exist for single species and thresholds are not difficult to define. It is generally not impossible to make similar assessments for multispecies complexes, but a generally accepted concept does not exist. Apart from such an academic discussion, further investigations of 'extreme' habitats in the sense of cold-spots seems to be at least as interesting to study as the generally more attractive hotspots. Gradients from high to lower abundances can easily be explained by food limitation, grazing effects, behaviour or dispersal problems. When, however, animals are extremely rare, e.g. under the ice shelves, one could argue that at the end of this gradient, food supply is not sufficient to support any (e.g. sponge) specimens. Nevertheless, extremely low abundances have been observed that are far below a level where competition for food would have an effect. To discover the true reasons for such extremely low abundances, which might be related to the early life history of the organisms, could lead to a better understanding of the functioning of richer communities, of which most have only a handful of abundant species, while the majority is also rare. Such ecological studies demand more advanced and focused interdisciplinary approaches and coordinated surveys. As a result, spatially explicit dynamic ecosystem models could provide the basis for currently lacking comprehensive risk maps to be used in a nature conservation context.

9.6 Acknowledgements

This chapter is a product of the Scientific Research Programme Antarctic Thresholds – Ecosystem Resilience and Adaptation (AnT-ERA) of the Scientific Committee on Antarctic Research (SCAR).

References

Arrigo, K.R., van Dijken, G.L., Bushinsky, S. (2008). Primary production in the Southern Ocean, 1997–2006. *Journal of Geophysical Research Oceans*, **113**, C08004; doi:10.1029/2007JC004551

Barbraud, C., Weimerskirch, H. (2001). Emperor penguins and climate change. *Nature*, **411**, 183–186.

Barnes, D.K.A. (1995). Sublittoral epifaunal communities at Signy Island, Antarctica. I. The ice-foot zone. *Marine Biology*, **121**, 565–572.

Bornemann, H., Held, C, Nachtsheim, D., et al. (2016). Seal research at the Drescher Inlet (SEADI). In: M. Schröder (ed.) The Expedition PS96 of the Research Vessel POLARSTERN to the southern Weddell Sea in 2015/16. Reports on Polar and Marine Research. Alfred-Wegener-Institute, Bremerhaven, pp. 116–29; doi:10.2312/BzPM_0700_2016

Bruchhausen, P.M., Raymond, J.A., Jacobs, S.S., et al. (1979). Fish, crustaceans, and the sea floor under the Ross Ice Shelf. *Science*, **203**, 449–451.

Cape, M.R., Vernet, M., Kahru, M., Spreen, G.M. (2014). Polynya dynamics drive primary production in the Larsen A and B embayments following ice shelf collapse. *Journal of Geophysical Research: Oceans*, **119**, 572–594; doi:10.1002/2013JC009441.

CAREX (2011). *CAREX Roadmap for Research on Life in Extreme Environments.* CAREX publication no. 9. Carex Project Office, Strasbourg.

Clarke, A., North, A.W. (1991). Is the growth of polar fish limited by temperature? In: G. Di Prisco, B. Maresca, B. Tota (eds) *Biology of Antarctic Fish*. Springer-Verlag, Berlin, pp. 54–69; doi:10.1007/978-3-642-76217-8

Clarke, A., Griffiths, H., Barnes, D.K.A., Meredith, M.P., Grant, S.M. (2009). Spatial variation in seabed temperatures in the Southern Ocean: implications for benthic ecology and biogeography. *Journal of Geophysical Research*, **114**, G03003; doi:10.1029/2008JG000886

Daly, M., Rack, F., Zook, R. (2013). Edwardsiella andrillae, a new species of sea anemone from Antarctic ice. *PLoS ONE*, **8**(12), e83476.

Dayton, P.K. (1979). Observations of growth, dispersal and population dynamics of some sponges in Mc Murdo Sound, Antarctica. *Colloques internationaux du C.N.R.S. Biologie des Spongaires*, **291**, 271–282.

Dayton, P., Jarrell, S., Kim, S., et al. (2016). Surprising episodic recruitment and growth of Antarctic sponges: implications for ecological resilience. *Journal of Experimental Marine Biology and Ecology*, **482**, 38–55.

De Broyer, C., Koubbi, P., Griffiths, H.J., et al. (2014). *Biogeographic Atlas of the Southern Ocean*. SCAR, Cambridge.

di Prisco, G., Pisano, E., Clark A. (1998). *Fishes of Antarctica. A Biological Overview*. Springer, Milan.

Flores, H., Atkinson, A., Kawaguchi, S., et al. (2012). Impact of climate change on Antarctic krill. *Marine Ecology Progress Series*, **458**, 1–19.

Gerdes, D. (1992). Quantitative investigations on macrobenthos communities of the southeastern Weddell Sea shelf based on multibox corer samples. *Polar Biology*, **12**, 291–301.

Gieskes, W.W.C., Veth, C., Woehrmann, A., Graefe, M. (1987). Secchi disc visibility world record shattered. *EOS*, **68**, 123.

Griffiths, H.J. (2010). Antarctic marine biodiversity – what do we know about the distribution of life in the Southern Ocean? *PLoS ONE*, **5**(8), e11683; doi:10.1371/journal.pone.0011683

Gutt, J., Gerdes, D., Klages, M. (1992). Seasonality and spatial variability in the reproduction of two Antarctic holothurians (Echinodermata). *Polar Biology*, **11**, 533–544.

Gutt, J., Ekau, W. (1996). Habitat partitioning of dominant high Antarctic demersal fish in the Weddell Sea and Lazarev Sea. *Journal of*

Experimental Marine Biology and Ecology, **206**, 25–37.

Gutt, J. (2000). Some 'driving forces' structuring communities of the sublittoral antarctic macrobenthos. *Antarctic Science*, **12**, 297–313.

Gutt, J. (2002). The Antarctic iceshelf: an extreme habitat for notothenioid fish. *Polar Biology*, **25**, 320–322.

Gutt, J., Piepenburg, D. (2003). Scale-dependent impact on diversity of Antarctic benthos caused by grounding of icebergs. *Marine Ecology Progress Series*, **253**, 77–83.

Gutt, J., Cape, M., Dimmler, W., et al. (2013). Shifts in Antarctic megabenthic structure after ice-shelf disintegration in the Larsen area east of the Antarctic Peninsula. *Polar Biology*, **36**, 895–906.

Jacob, U., Thierry, A., Brose, U., et al. (2011). The role of body size in complex food webs: a cold case. In A. Belgrano, J. Reiss (eds) *Advances in Ecological Research*, Vol. **45**. Elsevier, Amsterdam, the Netherlands, pp. 181–223.

Pardo, P.C., Pérez, F.F., Velo, A., Gilcoto, M. (2012). Water masses distribution in the Southern Ocean: improvement of an extended OMP (eOMP) analysis. *Progress in Oceanography*, **103**, 92–105; doi.org/10.1016/j.pocean.2012.06.002

Peck, L.S. (2005). Prospect for survival in the Southern Ocean: vulnerability of benthic species to temperature change. *Antarctic Science*, **17**(4), 497–507; doi.10.1017/S0954102005002920

Piepenburg, D., Schmid, M.K., Gerdes, D. (2002). The benthos off King George Island (South Shetland Islands, Antarctica): further evidence for a lack of a latitudinal biomass cline in the Southern Ocean. *Polar Biology*, **25**, 146–158.

Smith, C.R., De Leo, F.C., Bernardino, A.F., Sweetman, A.K., Martinez Arbizu, P. (2008). Abyssal food limitation, ecosystem structure and climate change. *Trends in Ecology and Evolution*, **23**(9), 518–529; doi:10.1016/j.tree.2008.05.002

Smith, C.R., Grange, L.J., Honig, D.L., et al. (2012). A large population of king crabs in Palmer Deep on the west Antarctic Peninsula shelf and potential invasive impacts. *Proceedings of the Royal Society B Biological Sciences*, **279**, 1077–1026; doi:10.1098/rspb.2011.1496

Somero, G.N. (1991). Biochemical mechanisms of cold adaptation and stenothermality in Antarctic fish. In: G. Di Prisco, B. Maresca, B. Tota (eds) *Biology of Antarctic Fish*. Springer-Verlag, Berlin, pp. 232–247; doi:10.1007/978-3-642-76217-8

Thomas, D.N., Dieckmann, G.S. (2002). Antarctic sea-ice – a habitat for extremophiles. *Science*, **295**, 641–644.

Thomas, D.N., Dieckmann, G.S. (2010). *Sea Ice*. Wiley Blackwell Publishing, Oxford.

CHAPTER TEN

Microorganisms in cryoturbated organic matter of Arctic permafrost soils

JIŘÍ BÁRTA

University of South Bohemia

10.1 Characteristics and distribution of Cryosols in the Arctic

Cryosols (permafrost-affected soils) cover more than 90% of the continuous permafrost zone in the Arctic (Tarnocai & Bockheim, 2011). They represent the dominant soil in the arctic and sub-arctic regions in Canada, Alaska and Russia but also occur in boreal and alpine regions. They can be classified into the static or organic Cryosols which develop on mineral or organic deposits, respectively. A specific class is Turbic Cryosols, which are characterised by cryoturbation (i.e. by mixing of soil layers due to seasonal freezing and thawing). For example, in Canada the Turbic Cryosols cover approx. 1.8 million km^2 representing the majority of Cryosols (72% of all Cryosols and 25% of the soil area; Tarnocai, 2018).

Thawing of Cryosols is likely the most important process that may translocate carbon (C) from terrestrial ecosystems to the atmosphere by the activity of soil microbial communities as a response to global warming, thus initiating a positive feedback to climate change (Schuur et al., 2008; Schuur & Abbott, 2011). The most recent database on northern circumpolar soil organic carbon (SOC) assumes a surface permafrost OC pool (0–3 m) of 1035 Pg, and the majority of it is stored in a frozen state underneath the active layer (Hugelius et al., 2016), accounting for half of OCs in soils worldwide (2050 Pg; Jobbágy & Jackson, 2000). Warming in northern ecosystems increases the thickness of the active layer (Harden et al., 2012) and reduces the areal extent of permafrost. Schmidt (2011) suggested that about 130–160 Pg C will be released from permafrost soils within a century, with CH_4 comprising 2.3% of the overall C emission resulting in a 35–48% higher warming potential. The understanding of the composition and functioning of microbial communities in Turbic Cryosols is crucial for future prediction of SOC vulnerability.

The most typical feature of Turbic Cryosols are the unique cryogenic processes. The presence of permafrost creates a very steep temperature gradient in the soil during the summer. The unfrozen water in the active soil layer (i.e. the unfrozen soil above the permafrost) migrates along this thermal gradient into the frozen system, feeding the ice lenses (Tarnocai, 2006). Once the soil water freezes, it increases its volume which leads to differential frost heave and cryostatic pressure. This results mainly in sorting and moving of coarse fragments in the soil profile and cryoturbation. By these processes part of the organic rich topsoil horizons (mainly O and A horizons) can be pushed down into the soil profile creating C-enriched organic pockets in the deeper mineral soil horizons (Figure 10.1; Gentsch et al., 2015). Cryoturbation, therefore, leads to irregular or broken soil horizons and creates uneven distribution of the soil organic matter (SOM). Radiocarbon dating of these cryoturbated C-enriched organic pockets ranges from hundreds to several thousand years, and the occurrence of both very old and relatively recent organic matter at various depths suggests that C sequestration in these soils has been occurring for thousands of years (Tarnocai & Bockheim, 2011). The decomposition (recalculated per gram of C) of organic matter in these C-enriched organic pockets is much slower compared to that of surrounding soils. Recent studies suggest that this is mainly caused by unfavourable abiotic conditions where low temperature and anaerobiosis are the most important (Kaiser et al., 2007).

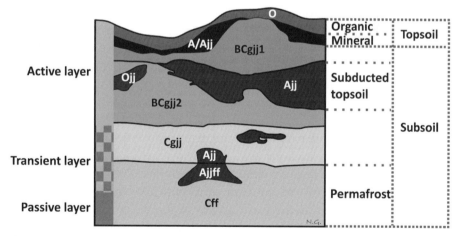

Figure 10.1 Schematic overview of the typical Cryosol influenced by cryoturbation – Turbic Cryosol. C-rich cryoturbated A horizon and O layers are shown in dark brown with code Ajj, Ojj, respectively. Horizon nomenclature according to Keys to Soil Taxonomy (Soil Survey Staff, 2010); O, organic layer; A, mineral top soil; BCg, Cg – mineral subsoil, Cff, permafrost (for details please refer to Gentsch et al., 2015). (A black and white version of this figure will appear in some formats. For the colour version, please refer to the plate section.)

10.2 Vulnerability of soil organic matter in Cryosols

Cryosols are warming rapidly, and current scenarios predict an increase of up to 4.8°C by 2100 (RCP8.5 scenario, Climate Change 2014 Synthesis Report, IPCC 2014). Higher subsoil temperatures will lead to permafrost thaw, prolonged frost-free periods, and increased soil active layer thickness and oxygen levels by soil drainage (Lawrence et al., 2015). It can promote the availability of large permafrost C pools including cryoturbated C-rich pockets for microbial decomposition. The microbial decomposition of SOM is more temperature sensitive than primary production (Davidson & Janssens, 2006), and this may generate greenhouse gas emissions from permafrost areas and create a positive feedback to climate warming (Schuur et al., 2008; Koven et al., 2011; Schuur & Abbott, 2011; Ping et al., 2015). Besides temperature, moisture and oxygen availability, the structure, functional capabilities of microbial communities and their accessibility to SOM are the crucial factors in SOM vulnerability to decomposition in Turbic Cryosols (Schmidt, 2011; Dungait et al., 2012).

Recent studies have focused on the active layer or specifically on underlying permafrost and characterisation of microbial community structure with relation to soil chemical characteristics (Yergeau et al., 2010; Mackelprang et al., 2011; Tveit et al., 2013; Gittel et al., 2014a, 2014b; Čapek et al., 2015; Gentsch et al., 2015; Wild et al., 2015, 2016, 2018; Dao et al., 2018; Šantrůčková et al., 2018). To understand the vulnerability of SOM in Turbic Cryosols including the specific cryoturbated C-rich pockets research needs to be focused on both physicochemical properties of SOM (i.e. capability to form organo-mineral associations, chemical complexity of SOM) and structure and functioning of microbial communities (Gittel et al., 2014a). For example, the information on the distribution of C in different soil fractions and aggregate sizes can help to understand how accessible the resources for microbes are (i.e. the amount of C in heavy, light and mobile fractions) (Gentsch et al., 2015). There is a need to describe and better understand the degradation capacities of the microbial communities, and interactions (e.g. symbiotic relationships vs. grazing) between different domains including bacteria, Archaea and fungi.

Vulnerability of SOM is the hot topic in recent Cryosol research with the microbial communities as the main drivers of the decomposition/stabilisation processes. Similar to temperate soils, both saprotrophic microorganisms play a crucial role in SOM transformations in Cryosols. Saprotrophic basidiomycetes are thought to be the dominant producers of extracellular enzymes catalysing efficient breakdown of biopolymers including lignin and other polyphenolic compounds to low-molecular-weight dissolved organic matter (Bailey et al., 2002; Baldrian & Valášková, 2008). When available resources are depleted, their dense hyphal network is crucial for efficient decomposition of the remaining recalcitrant lignin and lignin-like compounds. Specific taxa with different lifestyle strategies have distinct influence on OM vulnerability or stability. Mycorrhizal fungi can significantly contribute to SOM stability by the creation of the dense hyphal net which can stabilise OM in soil aggregates through the secretion of glue-like compounds (Talbot et al., 2008). Therefore, the different proportions and activity of the saprotrophic and mycorrhizal taxa influence the balance between the vulnerability and stability of OM in Cryosols (Kochkina et al., 2012; Gittel et al., 2014a; Hu et al., 2014; Coolen & Orsi, 2015; Frey et al., 2016; Yong-Liang et al., 2017).

10.3 Microbial communities in Cryosols

Determination of abundance and activity of bacterial communities in arctic soils is crucial for predicting their significance as key players in organic matter SOM turnover. Microorganisms can survive in the permafrost for decades without reproduction. They invest most of the energy into the cell maintenance functions like sustaining the proton motive force, osmoregulation or DNA repair (Bakermans et al., 2003).

10.3.1 Adaptation of microorganisms to the extreme conditions

Both prokaryotic (bacteria and Archaea) and eukaryotic (fungi) microorganisms inhabiting the Cryosols are extremophiles able to survive at low temperatures, repeating freezing/thawing of the soil, or long periods of anoxia. They developed physiological mechanisms that protect them and allow them to reproduce in these extreme conditions. One of the main mechanisms of adaptation to low temperature is the induced synthesis of cold shock proteins (Csp). Csp are small nucleic acid-binding proteins of approximately 70 amino acids in length (Czapski & Trun, 2014) destabilising the secondary structures in RNA so that the single-stranded state of RNA is maintained. This enables efficient transcription and translation at low temperatures (Phadtare, 2004). Many bacteria and fungi produce Csp as a response to rapid temperature downshift (cold shock).

Fungi can produce specific intracellular compounds like trehalose, which helps them to stabilise membranes during the drought and low temperatures. Intracellular accumulation of trehalose was detected in fungal hyphae of genera *Hebeloma*, *Humicola* and *Mortierella* when exposed to the low temperatures (Tibbett et al., 2002; Weinstein et al., 2007). Another important mechanism is the modification of the cytoplasmic membranes by increasing the proportion of unsaturated bonds in membrane lipids which helps microbes to maintain membrane fluidity and nutrient transport into the cell (Soina et al., 1995; Weinstein et al., 2007). Basidiomycetes are able to produce extracellular antifreeze proteins (AFPs) which keep the close surrounding of the cell in a fluid state (Robinson, 2001; Hoshino et al., 2009). Similar to Gram-positive bacteria, fungi can also survive in the form of spores in hostile conditions and germinate only when conditions are favourable.

10.3.2 Structure and activity of prokaryotic communities

All major bacterial and archaeal phyla have been described in Cryosols including Proteobacteria, Acidobacteria, Actinobacteria, Bacteriodetes, Firmicutes and Chloroflexi (Wilhelm et al., 2011; Jansson & Taş, 2014; Mackelprang et al., 2017, 2011; Mondav et al., 2017). Their proportions, however, differ not only between geographical localities but also along the depth in the soil profile. Similarly to temperate soils, the prokaryotic communities in Cryosols are the most active at the surface soil layers, in the rhizosphere. Unlike the temperate soils the rhizosphere layer in non-degraded permafrost Cryosols is usually very thin because of shallow rooting depth of local plants (Iversen et al., 2015). This layer promotes mostly copiotrophic organisms, fast-growing taxa from the Proteobacteria or Bacteroidetes that use available C and N sources (Fierer et al., 2003b). The available nutrients, however, rapidly decrease with Cryosol depth. Oligotrophs have better metabolic capabilities to utilise more complex compounds and can better survive in deeper soil horizons where easily available nutrients are scarce and complex and more recalcitrant compounds dominate. The abundance and diversity of prokaryotes along the soil depth profile thus follows the gradient of C and N availability as well as SOM composition, and together with abiotic conditions (i.e. low to subzero temperatures, anoxia due to water-logging) it results in distinct community patterns and less microbial biomass in deeper soil horizons (Hartmann et al., 2009; Eilers et al., 2012).

However, Gittel et al. (2014a) analysed approximately 100 soil samples and found that C-rich pockets in Turbic Cryosols in eastern Siberia do not follow the trend of decreasing prokaryotic biomass with depth. C-rich pockets harbour a microbial community with significantly higher bacterial and archaeal abundances than found in the surrounding subsoil horizons from similar depths (Figure 10.2; Gittel et al., 2014a). Moreover, prokaryotic abundance and richness were found to be similar to topsoil horizons and reflect the higher amount of C and N in all three different types of tundra investigated (shrubby grass tundra, shrubby tussock tundra and shrubby moss tundra). Multivariate analyses of prokaryotic community revealed that topsoil communities are phylogenetically highly similar and relatively uniform in the composition of major bacterial and archaeal phyla regardless of geographical position and landscape cover (Gittel et al., 2014a). In contrast, communities in C-rich pockets and subsoils are not only distinct from the topsoil communities, but also highly variable as shown by increased distance between sample points (Figure 10.3; Gittel et al., 2014a). This variability in community composition is most likely a result of greater variability in soil properties of C-rich

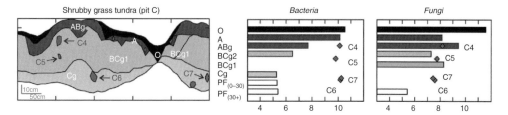

Figure 10.2 Schematic drawings of one representative soil pit profile for shrubby grass tundra: pit C, shrubby tussock tundra: O, organic layer; A, ABg, mineral topsoil; BCg, Cg – mineral subsoil; C4 to C7, C-rich pockets; PF, permafrost (depth in cm below surface indicated in brackets). Bar charts show abundances of Bacteria and Fungi as bacterial and fungal 16S rRNA gene copy numbers g^{-1} dry soil in logarithmic scale (Gittel et al., 2014a).

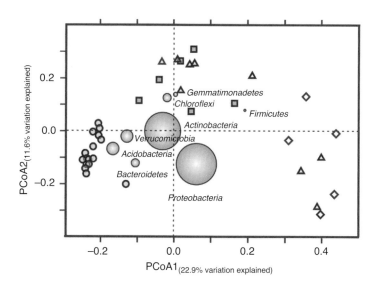

Figure 10.3 Phylogenetic dissimilarity between bacterial communities. Topsoils: circles; subsoils: triangles; buried C-rich organic pockets: squares; permafrost: diamonds. Principal coordinate analysis (PCoA) plot illustrating unweighted UniFrac distances between bacterial communities in individual samples. The coordinates of the eight most abundant phyla are plotted as a weighted average of the coordinates of all samples, where the weights are the relative abundances of the taxon in the samples. The size of the sphere representing a taxon is proportional to the mean relative abundance of the taxon across all samples (Gittel et al., 2014a).

pockets, namely the concentration and quality of the SOM. It may also reflect high differences in C-rich pockets' age and therefore distinct decomposition stages of buried C (Gentsch et al., 2015).

Changes in the relative proportions of prokaryotic functional guilds (e.g. bacteria able to perform anaerobic respiration, fermentation, methano-/methylotrophs) further supports the hypothesis that community structure in Cryosols reflects differences in soil properties. Deeper soil horizons are typically enriched in anaerobic bacteria including fermenting families of Chloroflexi (Anaerolineaceae) and Firmicutes (Clostridia). Members of sulfate-reducing Deltaproteobacteria (Desulfuromonadales) were detected not only at deeper subsoil but in many samples also at topsoil horizons. This demonstrates rapid depletion of oxygen as terminal electron acceptor in topsoil and highly anaerobic nature of almost the entire active layer of Turbic Cryosols. One of the highly enriched anaerobic groups in deeper soil horizons are Fe^{III} reducers and methanogens. Fe^{III} reducers are usually more abundant in mineral Cryosols, while methanogens are more abundant in organic Cryosols. Fe^{III} reducers most probably outcompete methanogens in mineral Cryosols due to the higher Fe^{III} availability, while in organic (peaty) Cryosols' fermentative and methanogenic communities dominate over the anaerobes (Tveit et al., 2013; Mackelprang et al., 2017). Lipson et al. (2013) showed similar findings in their metagenomic study in Alaska. Fe^{III} reducers play an important role in the stability of OM in Turbic Cryosols (Gentsch et al., 2018, 2015); Fe^{III} is reduced to Fe^{II} which can destabilise OM in organo-mineral complexes making it vulnerable to microbial decomposition. While the iron reducers influence the OM decomposition indirectly, the methanogens directly participate on C loss from the soil through the CH_4 production. They are probably metabolically active in permafrost as was demonstrated by the isotopic signatures of CH_4 carbon inside the soil cores from Siberia (Rivkina et al., 2006). Methane production from acetate and bicarbonate was detected at $-16.5°C$ in 3000-year-old loamy peat permafrost soil. In an incubation study of arctic peat soil involving the analyses of metagenomes, metatranscriptomes and metabolomes, methanogenesis was shown to shift from hydrogenotrophic to acetoclastic with increased temperature (Tveit et al., 2015). The hydrogenotrophic methanogens produce only CH_4 and usually cooperate with the syntrophic bacterial partners, which delivers them H_2. Conversely, the acetotrophic methanogens convert acetate (produced by secondary and several primary fermenting bacteria) to CH_4 and CO_2. Higher activity of acetotrophic methanogens can result in higher efflux of GHG from soils with increased temperature but can also increase the rate of metabolism of secondary fermenters (Xiao et al., 2015), which may enhance the OM decomposition in strictly anaerobic conditions.

Anaerobic members from the Firmicutes and Chloroflexi were found to be enriched also in C-rich pockets in Turbic Cryosols; however, Actinobacteria, and in particular the order Actinomycetales, clearly dominate (over 50% of assigned sequences) the prokaryotic community (Gittel et al., 2014a). McMahon et al. (2011) performed a cross-seasonal study on arctic tundra soils and proved that Actinobacteria are the major fraction among the active microbial community. Buried C-rich pockets experience longer freezing and anoxic periods than topsoil, posing a selective pressure on the microbial community. Actinobacteria, which have very efficient DNA repair mechanisms (Johnson et al., 2007), may benefit from these conditions because they can better maintain metabolic activity at low and subzero temperatures. Members of this large and phylogenetically ancient phylum are known for their ability to survive in highly oligotrophic environments (Barka et al., 2016) because of their versatile metabolic capabilities and specific lifestyle. Unlike other bacteria, they can degrade biopolymers, including cellulose, and the most studied model organisms for cellulose utilisation in Actinobacteria are *Thermobifida fusca* and *Cellulomonas fimi* (Lykidis et al., 2007). Actinobacteria can form pseudomycelia and undergo complex morphological differentiation. They produce immense diversity of secondary metabolites including antibiotics or extracellular polymeric substances (EPS). The antibiotics help Actiobacteria outcompete other bacteria when nutrients are scarce (Van Der Heul et al., 2018). The EPS on the other hand serve as the protective cell coat against the low temperatures (Zhou et al., 2016), permit the diffusion of nutrients from cells into the environment (Costa et al., 2018) and help stabilise the OM in soil aggregates (Tisdall & Oades, 1982). In addition, EPS can serve as the C source under C limitation. The production of EPS is an energetically demanding process; however recent studies proved that also fermenting bacteria (i.e. *Xanthomonas campestris*, *Bacillus polymyxa*) produce EPS under anaerobic conditions (Finore et al., 2014), therefore their production in C-rich pockets is highly probable.

Actinobacteria are also able to efficiently survive under different stress conditions (i.e. low temperature, desiccation) and nutrient-limited conditions in the form of semi-dormant spores. The amount of available C and N in C-rich pockets is very low, while lignin and lignin-like compounds dominate (Dao et al., 2018; Gentsch et al., 2018, 2015). As stated above, Actinobacteria are well adapted to low C and N availability (Fierer et al., 2003a), and like fungi they are able to express laccase-like genes and extracellular enzymes which degrade and solubilise lignin and lignocellulose. By this activity they access the associated polysaccharides otherwise protected by the complex lignin matrix (McCarthy, 1987; Le Roes-Hill et al., 2011). The increased evidence that laccase-like genes are present in genomes of diverse bacteria, including Actinobacteria, supports the hypothesis that the capability to modify lignin and decompose lignin derivatives is more widespread among the bacteria than previously thought (Ausec et al., 2011; Bugg et al., 2011).

Because of lower bulk density of cryoturbated C-rich pockets, these undergo periodically increased water-logging after the active layer thaw. This might create a niche for facultative anaerobic bacteria, and because Actinobacteria are able to create hyphal net similar to fungi for better penetration of lignin structure, this can favour presumably actinobacterial anaerobic lignin degraders in cryoturbated C-rich pockets (De Boer et al., 2005; DeAngelis et al., 2011; Schnecker et al., 2014) like the family *Intrasporangiaceae* and class Thermoleophilia (Schellenberger et al., 2010), which was found to be highly abundant in cryoturbated C-rich pockets (Antje Gittel et al., 2014b, 2014a). For example in eastern Greenland it comprised 8% of bacterial community (Gittel et al., 2014b). Because the decomposition of OM in cryoturbated C-rich pockets is slow, it seems that the actinobacterial enzymes are not efficient in lignin degradation or that the bacteria are limited by nutrients like N and P (Wild et al., 2013, 2016, 2018). The recalcitrance, low availability of OM and anoxia therefore results in higher recycling of microbial biomass, which was confirmed by δ^{13} C signatures (Gentsch et al., 2015, 2018) and metagenomic studies where higher abundance and expression of genes responsible for amino-acid and peptide import/degradation and salvage genes were detected (Mackelprang et al., 2017).

10.3.3 Structure and activity of fungal communities

The important role of fungi in ecosystem functions is indisputable. Fungi play a key role in the majority of soil functions and food webs as decomposers, mutualist, plant pathogens, and predators of micro- and mesofauna (Mueller & Schmit, 2007). They represent the most diverse group of Eukaryota, but still only a small fraction of their diversity has been assessed (Hibbett et al., 2011). The structure and shifts in fungal communities are influenced by biotic, abiotic and spatiotemporal factors. For example, the presence of plant communities through the litter and root exudates influence the proportions of distinct fungal guilds including saprotrophs and mycorrhizas. Fungi in the Crysols are diverse and include cold-adapted yeasts, micromycetes, darkly pigmented fungi, lichenised fungi, ectomycorrhizal fungi, pathogens and saprotrophs (Ozerskaya et al., 2009; Kochkina et al., 2012; Bellemain et al., 2013). While all major fungal phyla are present, Ascomycota and Basidiomycota are the most dominant, followed by Chytridiomycota and Mucoromycota (Lydolph et al., 2005; Ozerskaya et al., 2009; Kochkina et al., 2012; A. Gittel et al., 2014a; Hu et al., 2014; Frey et al., 2016). Gilichinsky et al. (2005) isolated 12 distinct taxa of filamentous fungi from permafrost soil. The halophilic species of Gemomyces were the most abundant, and the majority of the isolated yeasts belonged to the Basidiomycota.

The fungi in Cryosols, similarly to bacteria, have to withstand longer periods of anoxia. In the absence of oxygen they are able to use alternative electron acceptors like nitrates maintaining energy generation. Several strains with the ability to perform denitrification were also isolated (Uchimura et al., 2002). Although fungal activity is highest in the top organic layers of the Cryosols, their biomass, unlike bacterial, in cryoturbated C-rich pockets is very low, which was confirmed by phospholipid fatty acid analyses (PLFA), quantitative PCR (qPCR) and DNA gene-targeted sequencing (Gittel et al., 2014a, 2014b). Therefore, fungal influence on the decomposition of recalcitrant organic matter in C-rich pockets is probably very limited. The amount of fungi as determined by qPCR was found to be up to three orders of magnitude lower in cryoturbated C-rich pockets (Figure 10.2; Gittel et al., 2014a) and was at a similar level as in the surrounding mineral subsoil. The low abundance of fungi results in sparse fungal net and together with longer periods of anoxic conditions and low temperature, this may be the other reason for slower OM decomposition in C-rich pockets.

Unlike bacteria, the composition of fungal community in cryoturbated C-rich pockets is more similar to the topsoil and to some extent still reflects the original topsoil community (Figure 10.4; Gittel et al., 2014), mainly the proportion of mycorrhizal fungi (especially the basidiomycetes genera *Russula* and *Thelephora*). This may have consequences for long-term stability of OM in cryoturbated C-rich pockets. Mycorrhizal fungi are known for their ability to preferentially utilise organic N (Hodge et al., 2001; Talbot et al., 2008; Talbot & Treseder, 2010). The presence of mycorrhizal fungi in different soil horizons of Cryosols can therefore influence the amount of available N to the soil microbial community. Mycorrhizal fungi can also to some extent produce enzymes which can break down the C-rich biopolymers like cellulose, pectin or lipids (Talbot & Treseder, 2010). Talbot et al. (2008) suggested several mechanisms by which mycorrhizal fungi contribute to SOM decomposition, one of which was described as the 'Coincidental Decomposer' hypothesis. According to this hypothesis, SOM decomposition by mycorrhizal fungi is a consequence of exploiting soil for nutrients, mainly for N and P (Toljander et al., 2007; Hobbie & Hobbie, 2013). In cryoturbated C-rich pockets their low abundance presumably leads to less mobilisation of N and results in strong N-limitation, which highly retards SOM decomposition (Wild et al., 2014). In addition, the low abundance and proportion of mycorrhizal fungi in cryoturbated C-rich pockets can be the consequence of the lost connection to the host plant, which means that mycorrhizal fungi are no longer 'primed' by plant-derived carbon ('Priming Effect' hypothesis; Talbot et al., 2008). Thus, the synthesis of extracellular fungal enzymes is probably low and depolymerisation of C- and N-rich substrates dramatically decreases.

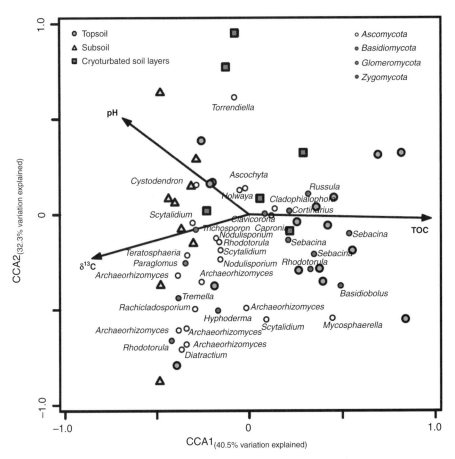

Figure 10.4 CCA ordination plots for the first two dimensions to show the relationship between fungal community structure, fungal taxa and environmental parameters. Organic topsoils: green circles; mineral subsoils: yellow triangles; cryoturbated soil layers: red squares. Correlations between environmental variables and CCA axes are represented by the length and angle of arrows (environmental factor vectors, Gittel et al., 2014a). (A black and white version of this figure will appear in some formats. For the colour version, please refer to the plate section.)

Indeed, Wild et al. (2013) described on average an 84% reduction in protein depolymerisation and 68% reduction in amino acid uptake in C-rich pockets. Thus, the reduced abundance of mycorrhizal fungi, their reduced activity in the depolymerisation of N-rich compounds together with unfavourable abiotic conditions such as subzero temperatures and high moisture presumably lead to a slower decomposition and thus stabilisation of OM in cryoturbated C-rich pockets in northern high-latitude regions.

10.4 Conclusion

Soil microbial communities have a major impact on C vulnerability in Turbic Cryosols. Detailed knowledge about their structure and interactions within and among the domains is therefore crucial. In this chapter we reviewed recent advances in understanding the differences in the composition and activity of bacterial and fungal communities in C-rich pockets in Turbic Cryosols. We identified several factors which are connected to C vulnerability in these specific soil horizons. Extremely low fungal biomass, disconnection with plant root system, N limitation and binding of organic matter to minerals were the most important factors. A change in one or several of these factors may therefore lead to imbalance in C stocks of Turbic Cryosols in future climate change. The future research should focus on understanding: (i) interactions between the different microbial groups including bacteria, fungi and protists; and (ii) spatiotemporal changes of microbial communities and mainly the functional guilds in different soil aggregates (micro- and macroaggregates) and different soil fractions. Combining the data from the meta-omic approaches and soil processes is important for understanding the C and N vulnerability in Cryosols and future predictions of climate change.

References

Ausec, L., van Elsas, J.D., Mandic-Mulec, I. (2011). Two- and three-domain bacterial laccase-like genes are present in drained peat soils. Soil Biology and Biochemistry. *Soil Biology and Biochemistry*, **43**(5), 975–983.

Bailey, V.L., Smith, J.L., Bolton, H. (2002). Fungal-to-bacterial ratios in soils investigated for enhanced C sequestration. *Soil Biology and Biochemistry*, **34**(7), 997–1007.

Bakermans, C., Tsapin, A.I., Souza-Egipsy, V., Gilichinsky, D.A., Nealson, K.H. (2003). Reproduction and metabolism at -10°C of bacteria isolated from Siberian permafrost. *Environmental Microbiology*, **5**(4), 321–326.

Baldrian, P., Valášková, V. (2008). Degradation of cellulose by basidiomycetous fungi. *FEMS Microbiology Reviews*, **32**(3), 501–521.

Barka, E.A., Vatsa, P., Sanchez, L., et al. (2016). Taxonomy, physiology, and natural products of actinobacteria. *Microbiology and Molecular Biology Reviews*, **80**(1), 1–43.

Bellemain, E., Davey, M.L., Kauserud, H., et al. (2013). Fungal palaeodiversity revealed using high-throughput metabarcoding of ancient DNA from arctic permafrost. *Environmental Microbiology*, **15**(4), 1176–1189.

Bugg, T.D.H., Ahmad, M., Hardiman, E.M., Rahmanpour, R. (2011). Pathways for degradation of lignin in bacteria and fungi. *Natural Product Reports*, **28**(12), 1883–1896.

Čapek, P., Diáková, K., Dickopp, J.-E., et al. (2015). The effect of warming on the vulnerability of subducted organic carbon in arctic soils. *Soil Biology and Biochemistry*, **90**, 19–29.

Coolen, M.J.L., Orsi, W.D. (2015). The transcriptional response of microbial communities in thawing Alaskan permafrost soils. *Frontiers in Microbiology*, **6**, 197. doi:10.3389/fmicb.2015.00197

Costa, O.Y.A., Raaijmakers, J.M., Kuramae, E.E. (2018). Microbial extracellular polymeric substances: Ecological function and impact on soil aggregation. *Frontiers in Microbiology*, **9**, 1636. doi:10.3389/fmicb.2018.01636

Czapski, T.R., Trun, N. (2014). Expression of csp genes in E. Coli K-12 in defined rich and defined minimal media during normal growth, and after cold-shock. *Gene*, **547**(1), 91–97. doi:10.1016/j.gene.2014.06.033

Dao, T.T., Gentsch, N., Mikutta, R., et al. (2018). Fate of carbohydrates and lignin in north-east Siberian permafrost soils. *Soil Biology and Biochemistry*, **116**, 311–322.

Davidson, E.A., Janssens, I.A. (2006). Temperature sensitivity of soil carbon decomposition and feedbacks to climate change. *Nature*, **440**(7081), 165–173.

De Boer, W., Folman, L.B., Summerbell, R.C., Boddy, L. (2005). Living in a fungal world: Impact of fungi on soil bacterial niche development. *FEMS Microbiology Reviews*, **29**(4), 795–811.

DeAngelis, K.M., Allgaier, M., Chavarria, Y., et al. (2011). Characterization of trapped lignin-degrading microbes in tropical forest soil. *PLoS ONE*, **6**(4), e19306.

Dungait, J.A.J., Hopkins, D.W., Gregory, A.S., Whitmore, A.P. (2012). Soil organic matter turnover is governed by accessibility not recalcitrance. *Global Change Biology*, **18**(6), 1781–1796.

Eilers, K.G., Debenport, S., Anderson, S., Fierer, N. (2012). Digging deeper to find unique microbial communities: The strong effect of depth on the structure of bacterial and archaeal communities in soil. *Soil Biology and Biochemistry*, **50**, 58–65.

Fierer, N., Allen, A.S., Schimel, J.P., Holden, P.A. (2003a). Controls on microbial CO2 production: A comparison of surface and subsurface soil horizons. *Global Change Biology*, **9**(9), 1322–1332.

Fierer, N., Schimel, J.P., Holden, P.A. (2003b). Variations in microbial community composition through two soil depth profiles. *Soil Biology and Biochemistry*, **35**(1), 167–176.

Finore, I., Di Donato, P., Mastascusa, V., Nicolaus, B., Poli, A. (2014). Fermentation technologies for the optimization of marine microbial exopolysaccharide production. *Marine Drugs*, **12**(5), 3005–3024.

Frey, B., Rime, T., Phillips, M., et al. (2016). Microbial diversity in European alpine permafrost and active layers. *FEMS Microbiology Ecology*, **92**(3).

Gentsch, N., Mikutta, R., Alves, R.J.E., et al. (2015). Storage and transformation of organic matter fractions in cryoturbated permafrost soils across the Siberian Arctic. *Biogeosciences*, **12**(14), 4525–4542.

Gentsch, N., Wild, B., Mikutta, R., et al. (2018). Temperature response of permafrost soil carbon is attenuated by mineral protection. *Global Change Biology*, **24**, 3401–3415. doi:10.1111/gcb.14316

Gilichinsky, D., Rivkina, E., Bakermans, C., et al. (2005). Biodiversity of cryopegs in permafrost. *FEMS Microbiology Ecology*, **53**, 117–128. doi:10.1016/j.femsec.2005.02.003

Gittel, A., Bárta, J., Kohoutováj, I., et al. (2014a). Site- and horizon-specific patterns of microbial community structure and enzyme activities in permafrost-affected soils of Greenland. *Frontiers in Microbiology*, **5**. doi:10.3389/fmicb.2014.00541

Gittel, A., Bárta, J., Kohoutová, I., et al. (2014b). Distinct microbial communities associated with buried soils in the Siberian tundra. *ISME Journal*, **8**(4), 841–853. doi:10.1038/ismej.2013.219

Orwin, K.H., Kirschbaum, M.U.F., St John, M.G., Dickie, I.A. (2011). Organic nutrient uptake by mycorrhizal fungi enhances ecosystem carbon storage: a model-based assessment. *Ecology Letters*, **14**, 493–502. doi:10.1111/j.1461-0248.2011.01611.x

Harden, J.W., Koven, C.D., Ping, C.L., et al. (2012). Field information links permafrost carbon to physical vulnerabilities of thawing. *Geophysical Research Letters*, **39**(15). doi:10.1029/2012GL051958

Hartmann, M., Lee, S., Hallam, S.J., Mohn, W.W. (2009). Bacterial, archaeal and eukaryal community structures throughout soil horizons of harvested and naturally disturbed forest stands. *Environmental Microbiology*, **11**(12), 3045–3062. doi:10.1111/j.1462-2920.2009.02008.x

Hibbett, D.S., Ohman, A., Glotzer, D., et al. (2011). Progress in molecular and morphological taxon discovery in Fungi and options for formal classification of environmental sequences. Fungal Biology Reviews, **25**(1), 38–47. doi:10.1016/j.fbr.2011.01.001

Hobbie, J.E., Hobbie, E.A. (2013). Microbes in nature are limited by carbon and energy: The starving-survival lifestyle in soil and consequences for estimating microbial rates. *Frontiers in Microbiology*, **4**. doi:10.3389/fmicb.2013.00324

Hodge, A., Campbell, C.D., Fitter, A.H. (2001). An arbuscular mycorrhizal fungus accelerates decomposition and acquires nitrogen directly from organic material. *Nature*, **413**(6853), 297–299.

Hoshino, T., Xiao, N., Tkachenko, O.B. (2009). Cold adaptation in the phytopathogenic fungi causing snow molds. *Mycoscience*, **50**(1), 26–38.

Hu, W., Zhang, Q., Li, D., et al. (2014). Diversity and community structure of fungi through a permafrost core profile from the Qinghai-Tibet Plateau of China. *Journal of Basic Microbiology*, **54**(12), 1331–1341.

Hugelius, G., Kuhry, P., Tarnocai, C. (2016). Ideas and perspectives: Holocene thermokarst sediments of the Yedoma permafrost region do not increase the northern peatland carbon pool. *Biogeosciences*, **13**(7), 2003–2010.

IPCC (2014). AR5 Climate Change 2014: Impacts, Adaptation, and Vulnerability. www.ipcc.ch/report/ar5/wg2/

Iversen, C.M., Sloan, V.L., Sullivan, P.F., et al. (2015). The unseen iceberg: Plant roots in arctic tundra. *New Phytologist*, **205**(1), 34–58.

Jansson, J.K., Taş, N. (2014). The microbial ecology of permafrost. *Nature Reviews Microbiology*, **12**(6), 414–425.

Jobbágy, E.G., Jackson, R.B. (2000). The vertical distribution of soil organic carbon and its relation to climate and vegetation. *Ecological Applications*, **10**(2), 423–436.

Johnson, S.S., Hebsgaard, M.B., Christensen, T.R., et al. (2007). Ancient bacteria show evidence of DNA repair. *Proceedings of the National Academy of Sciences of the United States of America*, **104**(36), 14401–14405.

Kaiser, C., Meyer, H., Biasi, C., et al. (2007). Conservation of soil organic matter through cryoturbation in arctic soils in Siberia. *Journal of Geophysical Research: Biogeosciences*, **112**(2).

Kochkina, G., Ivanushkina, N., Ozerskaya, S., et al. (2012). Ancient fungi in Antarctic permafrost environments. *FEMS Microbiology Ecology*, **82**(2), 501–509.

Koven, C.D., Ringeval, B., Friedlingstein, P., et al. (2011). Permafrost carbon-climate feedbacks accelerate global warming. *Proceedings* of the National Academy of Sciences of the United States of America, **108**(36), 14769–14774.

Lawrence, D.M., Koven, C.D., Swenson, S.C., Riley, W.J., Slater, A.G. (2015). Permafrost thaw and resulting soil moisture changes regulate projected high-latitude CO2 and CH4 emissions. *Environmental Research Letters*, **10**(9). doi:10.1088/1748-9326/10/9/094011

Le Roes-Hill, M., Khan, N., Burton, S.G. (2011). Actinobacterial peroxidases: An unexplored resource for biocatalysis. *Applied Biochemistry and Biotechnology*, **164**(5), 681–713.

Lipson, D.A., Haggerty, J.M., Srinivas, A., et al. (2013). metagenomic insights into anaerobic metabolism along an arctic peat soil profile. *PLoS ONE*, **8**(5). doi:10.1371/journal.pone.0064659

Lydolph, M.C., Jacobsen, J., Arctander, P., et al. (2005). Beringian paleoecology inferred from permafrost-preserved fungal DNA. *Applied and Environmental Microbiology*, **71**(2), 1012–1017.

Lykidis, A., Mavromatis, K., Ivanova, N., et al. (2007). Genome sequence and analysis of the soil cellulolytic actinomycete Thermobifida fusca YX. *Journal of Bacteriology*, **189**(6), 2477–2486.

MacKelprang, R., Waldrop, M.P., Deangelis, K.M., et al. (2011). Metagenomic analysis of a permafrost microbial community reveals a rapid response to thaw. *Nature*, **480**(7377), 368–371. doi:10.1038/nature10576

MacKelprang, R., Burkert, A., Haw, M., et al. (2017). Microbial survival strategies in ancient permafrost: Insights from metagenomics. *ISME Journal*, **11**(10), 2305–2318.

McCarthy, A.J. (1987). Lignocellulose-degrading actinomycetes. *FEMS Microbiology Letters*, **46**(2), 145–163.

McMahon, S.K., Wallenstein, M.D., Schimel, J.P. (2011). A cross-seasonal comparison of active and total bacterial community composition in Arctic tundra soil using bromodeoxyuridine labeling. *Soil Biology and Biochemistry*, **43**(2), 287–295.

Mondav, R., McCalley, C.K., Hodgkins, S.B., et al. (2017). Microbial network, phylogenetic diversity and community membership in the active layer across a permafrost thaw gradient. *Environmental Microbiology*, **19**(8), 3201–3218.

Mueller, G.M., Schmit, J.P. (2007). Fungal biodiversity: What do we know? What can we predict? *Biodiversity and Conservation*, **16**(1), 1–5.

Ozerskaya, S., Kochkina, G., Ivanushkina, N., Gilichinsky, D.A. (2009). Fungi in permafrost. In: Margesin, R. (ed.) Permafrost Soils. Soil Biology, vol 16. Springer, Berlin, Heidelberg.

Phadtare, S. (2004). Recent developments in bacterial cold-shock response. *Current Issues in Molecular Biology*, **6**(2), 125–136.

Ping, C.L., Jastrow, J.D., Jorgenson, M.T., Michaelson, G.J., Shur, Y.L. (2015). Permafrost soils and carbon cycling. *Soil*, **1**(1), 147–171.

Rivkina, E.M., Kraev, G.N., Krivushin, K.V., et al. (2006). Methane in permafrost of Northeastern Arctic. *Earth's Cryosphere*, **10**(3), 23–41.

Robinson, C.H. (2001). Cold adaptation in Arctic and Antarctic fungi. *New Phytologist*, **151**(2), 341–353.

Šantrůčková, H., Kotas, P., Bárta, J., et al. (2018). Significance of dark CO2 fixation in arctic soils. *Soil Biology and Biochemistry*, **119**. doi:10.1016/j.soilbio.2017.12.021

Schellenberger, S., Kolb, S., Drake, H.L. (2010). Metabolic responses of novel cellulolytic and saccharolytic agricultural soil Bacteria to oxygen. *Environmental Microbiology*, **12**(4), 845–861.

Schmidt, G. (2011). Climate change and climate modeling. *Eos, Transactions American Geophysical Union*, **92**(23), 198–199.

Schnecker, J., Wild, B., Hofhansl, F., et al. (2014). Effects of soil organic matter properties and microbial community composition on enzyme activities in cryoturbated arctic soils. *PLoS ONE*, **9**(4).

Schuur, E.A.G., Abbott, B. (2011). Climate change: High risk of permafrost thaw. *Nature*, **480**(7375), 32–33.

Schuur, E.A.G., Bockheim, J., Canadell, J. G., et al. (2008). Vulnerability of permafrost carbon to climate change: Implications for the global carbon cycle. *BioScience*, **58**(8), 701–714.

Soil Survey Staff (2010). Keys to Soil Taxonomy, 11th ed. USDA-NRCS, Washington DC.

Soina, V.S., Vorobiova, E.A., Zvyagintsev, D.G., Gilichinsky, D.A. (1995). Preservation of cell structures in permafrost: A model for exobiology. *Advances in Space Research*, **15**, 237–242. doi:10.1016/S0273-1177(99)80090-8

Talbot, J.M., Treseder, K.K. (2010). Controls over mycorrhizal uptake of organic nitrogen. *Pedobiologia*, **53**(3), 169–179. doi:10.1016/j.pedobi.2009.12.001

Talbot, J.M., Allison, S.D., Treseder, K.K. (2008). Decomposers in disguise: Mycorrhizal fungi as regulators of soil C dynamics in ecosystems under global change. *Functional Ecology*, **22**(6), 955–963.

Tarnocai, C. (2018). The amount of organic carbon in various soil orders and Ecological Provinces in Canada. In: Harms, D., Korschens, M., Olson, G.,

et al. (eds) Soil Processes and the Carbon Cycle. CRC Press, Boca Raton, FL. doi:10.1201/9780203739273

Tarnocai, C. (2006). The effect of climate change on carbon in Canadian peatlands. *Global and Planetary Change* **53**(4), 222–232.

Tarnocai, C., Bockheim, J. (2011). Cryosolic soils of Canada: Genesis, distribution, and classification. *Canadian Journal of Soil Science*, **91**(5), 749–762.

Tibbett, M., Sanders, F.E., Cairney, J.W.G. (2002). Low-temperature-induced changes in trehalose, mannitol and arabitol associated with enhanced tolerance to freezing in ectomycorrhizal basidiomycetes (Hebeloma spp.). *Mycorrhiza*, **12**(5), 249–255.

Tisdall, J.M., Oades, J.M. (1982). Organic matter and water-stable aggregates in soils. *Journal of Soil Science*, **33**(2), 141–163.

Toljander, J.F., Lindahl, D., Paul, L.R., Elfstrand, M., Finlay, R.D. (2007). Influence of arbuscular mycorrhizal mycelial exudates on soil bacterial growth and community structure. *FEMS Microbiology Ecology*, **61**, 295–304. doi:10.1111/j.1574-6941.2007.00337.x

Tveit, A., Schwacke, R., Svenning, M.M., Urich, T. (2013). Organic carbon transformations in high-Arctic peat soils: Key functions and microorganisms. *ISME Journal*, **7**(2), 299–311.

Tveit, A.T., Urich, T., Frenzel, P., Svenning, M.M. (2015). Metabolic and trophic interactions modulate methane production by Arctic peat microbiota in response to warming. *Proceedings of the National Academy of Sciences of the United States of America*, **112**(19), E2507–2516.

Uchimura, H., Enjoji, H., Seki, T., et al. (2002). Nitrate reductase-formate dehydrogenase couple involved in the fungal denitrification by Fusarium oxysporum. *Journal of Biochemistry*, **131**(4), 579–586.

Van Der Heul, H.U., Bilyk, B.L., McDowall, K.J., Seipke, R.F., Van Wezel, G.P. (2018). Regulation of antibiotic production in Actinobacteria: New perspectives from the post-genomic era. *Natural Product Reports*, **35**(6), 575–604.

Weinstein, R.N., Montiel, P.O., Johnstone, K. (2007). Influence of growth temperature on lipid and soluble carbohydrate synthesis by fungi isolated from Fellfield Soil in the Maritime Antarctic. *Mycologia*, **92**(2), 222–229.

Wild, B., Schnecker, J., Bárta, J., et al. (2013). Nitrogen dynamics in Turbic Cryosols from Siberia and Greenland. *Soil Biology and Biochemistry*, **67**. doi:10.1016/j.soilbio.2013.08.004

Wild, B., Schnecker, J., Alves, R.J.E., et al. (2014). Input of easily available organic C and N stimulates microbial decomposition of soil organic matter in arctic permafrost soil. *Soil Biology and Biochemistry*, **75**, 143–151. doi:10.1016/j.soilbio.2014.04.014

Wild, B., Schnecker, J., Knoltsch, A., et al. (2015). Microbial nitrogen dynamics in organic and mineral soil horizons along a latitudinal transect in western Siberia. *Global Biogeochemical Cycles*, **29**(5), 567–582. doi:10.1002/2015GB005084

Wild, B., Gentsch, N., Capek, P., et al. (2016). Plant-derived compounds stimulate the decomposition of organic matter in arctic permafrost soils. *Scientific Reports*, **6**. doi:10.1038/srep25607

Wild, B., Alves, R.J.E., Bárta, J., et al. (2018). Amino acid production exceeds plant nitrogen demand in Siberian tundra. *Environmental Research Letters*, **13**. doi:10.1088/1748-9326/aaa4fa

Wilhelm, R.C., Niederberger, T.D., Greer, C., Whyte, L.G. (2011). Microbial diversity of active layer and permafrost in an acidic wetland from the Canadian High Arctic. *Canadian Journal of Microbiology*, **57**(4), 303–315.

Xiao, D., Peng, S.P., Wang, E.Y. (2015). Fermentation enhancement of methanogenic archaea consortia from an Illinois basin coalbed via DOL emulsion nutrition. *PLoS ONE*, **10**(4).

Yergeau, E., Hogues, H., Whyte, L.G., Greer, C.W. (2010). The functional potential of high Arctic permafrost revealed by metagenomic sequencing, qPCR and microarray analyses. *ISME Journal*, **4**(9), 1206–1214.

Yong-Liang, C., Ye, D., Jin-Zhi, D., et al. (2017). Distinct microbial communities in the active and permafrost layers on the Tibetan Plateau. *Molecular Ecology*, **26**, 6608–6620. doi:10.1111/mec.14396

Zhou, L., Xing, X., Peng, B., Fang, G. (2016). Extracellular polymeric substance (EPS) characteristics and comparison of suspended and attached activated sludge at low temperatures. *Qinghua Daxue Xuebao/Journal of Tsinghua University*, **56**(9), 1009–1015.

CHAPTER ELEVEN

Chemical ecology in the Southern Ocean

CARLOS ANGULO-PRECKLER
University of Barcelona
PAULA DE CASTRO-FERNANDEZ
University of Barcelona
RAFAEL MARTÍN-MARTÍN
University of Barcelona
BLANCA FIGUEROLA
Smithsonian Tropical Research Institute

and

CONXITA AVILA
University of Barcelona

11.1 Introduction

This chapter aims to review the most recent findings regarding chemical ecology in Antarctic marine macroorganisms, provide some insights into how environmental changes may affect the production of natural compounds, and how species may adapt (or not) to new scenarios related to climate change. The ecological significance of bioactive compounds in the marine environment remains as one of the most understudied topics of recent years. Even if many compounds have been described from marine organisms (Blunt et al., 2018, and previous reports), only a few have been investigated for their role in the environment where the organisms actually live (Puglisi et al., 2019 and previous reports, Puglisi & Becerro, 2018), and this is especially significant for antarctic areas (Avila et al., 2008; Núñez-Pons & Avila, 2015; Principe & Fisher, 2018). In fact, many organisms' interactions in the marine benthos may be mediated by chemicals that are currently unknown or undescribed. Recent reviews have covered the ca. 600 natural compounds described from antarctic marine benthic organisms (Lebar et al., 2007; Soldatou & Baker, 2017; Tian et al., 2017), but only a few have reported the ecological role of their compounds and/or extracts (Avila et al., 2008; McClintock et al., 2010; Núñez-Pons & Avila, 2015; Avila 2016a; Angulo-Preckler et al., 2018; Núñez-Pons et al., 2018; von Salm et al., 2018). Thus, antarctic marine benthos is still an

(a)

Figure 11.1 Antarctic benthic communities photographed at shallow waters in Deception Is. (Antarctica). (a) Typical invertebrate associations in a hard-bottom substrate (18-m depth). (b) Common algal communities on a rocky wall (15-m depth). (A black and white version of this figure will appear in some formats. For the colour version, please refer to the plate section.)

(b)

unexplored source of interesting natural products yet to be discovered. Southern Ocean ecosystems hold a huge amount of biodiversity, much higher than ever thought before, with many cryptic species being recently described, and thus chemical diversity is expected to be very high too (Downey et al., 2012; Wilson et al., 2013; De Broyer et al., 2014; Avila 2016a, 2016b). How antarctic organisms use these compounds is the subject of this review, along with the possible changes that we could expect in relation to environmental changes, particularly climate change (Figure 11.1).

Anthropogenic environmental change is a global phenomenon, with strong impact on biodiversity all around the planet (IPCC 2014). Polar regions are suffering the fastest rates of warming, with a loss of sea ice and the retreat of coastal glaciers and ice shelves too (IPCC 2014). The Antarctic Peninsula, in particular, is one of the areas with the fastest change over the last 50 years (Turner et al., 2009; Ducklow et al., 2013). Polar benthic marine species are exposed to major challenges due to environmental changes, mainly consisting in higher temperatures, ocean acidification, increasing UV radiation, and altered levels of sea ice and iceberg scouring (Núñez-Pons et al., 2018; Peck, 2018), although other potentially important stressors have also been reported, such as salinity and hypoxia (Clark & Peck, 2009a, 2009b; Tremblay and Abele 2016). Antarctic organisms are thus very vulnerable to environmental changes (Peck, 2018). Their unique characteristics and evolution under extreme physical and biological conditions make them very interesting examples to test possible changes in our planet. In fact, Peck (2018) pointed out recently that there is a very urgent need to do more research in polar areas because they are the fastest changing regions in the planet due to climate change impacts, and contain faunas that are possibly the least capable of resisting change globally.

The effects of climate change on population and community ecology currently receives a lot of attention in research, and many articles describe either observed or potential changes in marine invertebrate distribution and population dynamics (e.g. Fabry et al., 2008; Wang, 2014; Griffiths et al., 2017). However, very few studies deal with possible changes in the chemical ecology of the organisms (Campbell et al., 2011). Thus, the assessment of how climate change may affect the synthesis of natural products in marine organisms appears to be a relatively uncharted field of research. Some studies, however, suggest the possible influence of environmental factors, such as seasonal changes, depth and light on the biosynthesis of natural compounds in non-polar species (Turon et al., 1996; Swearingen & Pawlik, 1998; Duckworth & Battershill, 2001; Peters et al., 2004; Ferretti et al., 2009), most of them with the aim of finding suitable conditions for culturing marine invertebrates, e.g. sponges, to obtain compounds with pharmaceutical or biotechnological properties (Ferretti et al., 2009).

Marine natural products comprise mainly secondary metabolites that regulate the biology, coexistence and coevolution of the species, without directly participating in their primary metabolism (i.e. growth, development and reproduction; Torssel, 1983). Natural products often play important roles in predator–prey interactions, but also in symbiosis, competition, antifouling, reproduction, larval settlement and other relationships (Amsler et al., 2001; Figuerola et al., 2012b; Puglisi et al., 2019). One of the most studied roles of natural compounds in antarctic communities is the antipredatory activity, with many protected species already described in areas such as the Ross Sea, the Western Antarctic Peninsula, the Eastern Weddell Sea and Bouvet Island (Amsler et al., 2001, 2014; Avila et al., 2008; McClintock et al., 2010; Figuerola et al., 2013a; Taboada et al., 2013; Núñez-Pons & Avila, 2014a, 2014b). Not surprisingly, these areas are the closest to scientific research stations, while vast unstudied areas remain to be investigated. Other ecological activities have been less studied so far. We review here the data on chemical ecology of selected antarctic organisms from 2000 to 2018.

11.2 Types of molecules

A characteristic of secondary metabolites is their limited phylogenetic distribution; while primary metabolites, such as the common amino acids, carbohydrates and nucleosides, are chemically identical in virtually all organisms, both simple and complex secondary metabolites are generally limited to a given species, genus or family, or even a species chemotype (Torssel, 1983; Blunt et al., 2018). There are a number of classes of natural products, recognised on the basis of their biosynthetic origin, such as polyketides, terpenes and alkaloids (Torssel, 1983). Chemical studies on selected antarctic phyla are briefly discussed here.

Even though the number of described metabolites from macroalgae has increased since the mid-twentieth century, the proportion of antarctic seaweed species studied in this field is still smaller than in other geographical areas (Wiencke & Clayton, 2002; Wiencke et al., 2014; von Salm et al., 2018). Antarctic macroalgae possess a relatively rich diversity of molecules with different roles, which are ecologically important because they are key players in antarctic shorelines, structuring their communities (Wiencke & Clayton, 2002; Wiencke et al., 2014). An important part of the metabolites described from algae are halogenated, but each group (Rhodophyta, Chlorophyta and Ochrophyta) tends to produce its own unique metabolites. Most compounds are isoprenoids (terpenes, carotenoids, cystones, meroterpenoids, cystodiones and steroids), but polyketides and shikimates (mostly aromatic products such as quinones, prenylated hydroquinones and tannins) are also abundant (Young et al., 2007; Blunt et al., 2018). Only 17 antarctic green seaweed taxa have been described so far (Wiencke & Clayton, 2002; Amsler & Fairhead, 2006; Wiencke et al., 2014). Most molecules described from chlorophytes are terpenoids, but contrary to other seaweed groups, such as red algae, they do not have a high level of halogenation (Blunt et al., 2007; Amsler et al., 2008; Young et al., 2015; von Salm et al., 2018; Blunt et al., 2018). Most of them produce volatile halogenated organic (VHO) compounds (Laturnus et al., 1996), and also some UVR-absorbing pigments (Núñez-Pons et al., 2018). Compounds produced by antarctic Ochrophyta include mostly phlorotannins (polyphenols), and also diterpenes and acetogenins (Blunt et al., 2007; Amsler et al., 2008, 2009; von Salm et al., 2018). Red algae (ca. 80 spp.) are the most diverse group of macroalgae both in species number and metabolites described, possessing a wide variety of chemical structures (Wiencke & Clayton, 2002; Amsler & Fairhead, 2006; Amsler et al., 2008; Wiencke et al., 2014). Remarkably, in contrast with brown algae, they lack phlorotannins (Blunt et al., 2018; Núñez-Pons et al., 2018).

Sponges are dominant components of the antarctic benthos and they play an important role in the structure and dynamics of benthic communities (Dayton et al., 1974; McClintock, 1987; Dayton, 1989). In the past century, 15 species of antarctic sponges, belonging to 14 genera, were chemically investigated (reviewed in Avila et al., 2008). A variety of structural types and new metabolites were isolated from *Latrunculia* (Ford & Capon, 2000; Furrow et al., 2003; Li et al., 2018), *Isodictya* (Moon et al., 2000; Vankayala et al., 2017), *Crella* (Ma et al., 2009) and hexactinellid sponges (Núñez-Pons et al., 2012a; Carbone et al., 2014), among others (Table 11.1). Suberitane derivatives from the sponge genus *Suberites* have been proposed as compounds with high taxonomic relevance (Díaz-Marrero et al., 2003, 2004). However, in a recent study, we reported very close analogues from another antarctic sponge, *Phorbas areolatus*, thus adding *Phorbas* as another genus of interest for the discovery of novel sesterterpenoids (Solanki et al., 2018). A literature review pointed out many other interesting sesterterpenoids from other species of the genus *Phorbas* in other geographical areas (Daoust et al., 2013; Wang et al., 2016). In light of these findings, we suggested a taxonomical reinvestigation of the species of the genera *Suberites* and *Phorbas* (Solanki et al., 2018).

Soft corals are sessile organisms often without physical and/or behavioural defences. This has led to a great development of chemical defences. Thus, the vast majority of natural products described from cnidarians are from Anthozoa, the largest of the four cnidarian classes, and particularly from Octocorallia, with >80% of the compounds identified (Harper et al., 2001; Blunt et al., 2018). Here, the typical defensive chemicals are terpenoids and steroids (Paul, 1992; von Salm et al., 2014, and reviewed in Núñez-Pons & Avila, 2015), although they may also include potent toxins (Slattery & McClintock, 1995; Jouiaei et al., 2015). In recent years, some new natural products, mainly terpenoids, have been isolated from antarctic octocorals. *Alcyonium antarcticum* (*A. paessleri*) has been found to possess several terpenoids, including paesslerins A and B (Rodríguez Brasco et al., 2001), alcyopterosins (Palermo et al., 2000; Carbone et al., 2009), alcyonicene and deacetoxy-alcyonicene, and some other sesquiterpenes (Manzo et al., 2009). Alcyopterosins are illudalane sesquiterpenoids also described for other species of the same genus, namely *A. grandis*, *A. haddoni*, *A. paucilobulatum* and *A. roseum* (Núñez-Pons & Avila, 2015). The gorgonian *Dasystenella acanthina* also produces sesquiterpenes (Gavagnin et al., 2003). In addition to terpenoids, new steroids have also been described in *Anthomastus bathyproctus* (Mellado et al., 2005).

Bryozoans are particularly well represented sessile suspension feeders in the antarctic benthic communities (ca. 390 spp.) (De Broyer & Danis, 2011)

Table 11.1 *Chemicals from marine benthic macroorganisms from Antarctica reported from 2000 to 2018*

Natural products	Phylum	Taxa	Location	References
Cystos phaerol	Ochrophyta	*Cystosphaera jacquinotii*	WAP, South Shetland Islands	Ankisetty et al., 2004
Bromoform		*Desmarestia anceps*	WAP, South Shetland Islands	Ankisetty et al., 2004
Plastoquinones		*Desmarestia menziesii*	WAP, South Shetland Islands	Ankisetty et al., 2004
Phlorotanins, Acetogenins, Diterpenes		*Desmarestia menziesii, D. anceps, D. antarctica, Himantothallus grandifolius, Ascoseira mirabilis, Cystosphaera jacquinotii*	WAP, South Shetland Islands	Ankisetty et al., 2004
Halogenated furanones, Pulchralides, Fimbrolides, Acetoxyfimbrolide, Hydroxyfimbrolide	Rhodophyta	*Delisea pulchra*	WAP, South Shetland Islands	Ankisetty et al., 2004
Halogenated organic compounds		*Delisea pulchra, Plocamium cartilagineum*	WAP, South Shetland Islands	Maschek & Baker, 2008
Halogenated terpenes		*Delisea pulchra, Plocamium cartilagineum, Pantoneura plocamioides*	WAP, South Shetland Islands	Argandona et al., 2002
Sulfated polysaccharides		*Iridaea cordata*	WAP, South Shetland Islands	Kim et al., 2017

Compound	Group	Species	Location	Reference
P-hydroxybenzaldehyde, P-methoxyphenol		*Myriogramme smithii*	WAP, South Shetland Islands	Ankisetty et al., 2004
Epi-plocameneD		*Plocamium cartilagineum*	WAP, South Shetland Islands	Ankisetty et al., 2004
Keto-steroid	Porifera	*Anoxycalyx ijimai, A. joubini, Rossella antarctica, R. fibulata, R. nuda, R. racovitzae, R. villosa*	Weddell Sea	Núñez-Pons et al., 2012a; Núñez-Pons & Avila, 2014a
Glassponsine		*Anoxycalyx joubini*	Weddell Sea	Carbone et al., 2014
Norselic acids A-E		*Crella* sp.	Anvers island	Ma et al., 2009
Darwinolide, Membranolide B, C, and D, Dihydrogracilin A		*Dendrilla membranosa*	Anvers Island	Ankisetty et al., 2004; Witowski, 2015; von Salm et al., 2016; Ciaglia et al., 2017
Antifreeze peptide		*Homaxinella balfourensis*	McMurdo Sound	Wilkins et al., 2002
Erebusinone, 3-Hydroxykyrunenine, Methyl 3-Hydroxyanthranilate		*Isodyctia erinacea*	Ross Island	Moon et al., 2000; Vankayala et al., 2017
Organohalogens		*Kirkpatrickia variolosa, Artemisina apollinis, Phorbas glaberrima, Halichondria* sp., *Leucetta antarctica*	King George Island	Vetter & Janussen, 2005
Discorhabdin G		*Latrunculia apicalis*	McMurdo Sound	Furrow et al., 2003
Discorhabdins, Tsitsikamma mines		*Latrunculia biformis*	Weddell Sea	Li et al, 2018
Discorhabdin R		*Latrunculia* sp.	Prydz Bay	Ford & Capon, 2000
Flabellone		*Lissodendoryx (Lissodendoryx) flabellata*	Terranova Bay	Cutignano et al., 2012

Table 11.1 (*cont.*)

Natural products	Phylum	Taxa	Location	References
Suberitenones A and B, Oxapyrosuberiterone, Isosuberiterone B, 19-episuberiterone B, Isoxaspirosuberiterone		*Phorbas areolatus*	Deception Island	Solanki et al., 2018
Caminatal, Oxaspirosuberitenone, 19-episuberiterone B, Suberiterone B		*Suberites caminatus*	King George Island	Díaz-Marrero et al., 2003, 2004
Suberitenones C and D, Suberiphenol		*Suberites* sp.	King George Island	Lee et al., 2004
Linderazulene, Ketolactone, C-16-Azulenoid	Cnidaria	*Acanthogorgia laxa*	South Shetland Islands	Patiño Cano et al., 2018
Ainigmaptilone A and B		*Ainigmaptilon antarcticus*	Weddell Sea	Iken & Baker, 2003
Illudalane sesquiterpenes		*Alcyonium antarcticum, A. grandis, A. haddoni, A. paucilobulatum, A. roseum*	Weddell Sea, Deception Is.	Carbone et al., 2009; Nunez-Pons et al., 2013
Alcyonicene, deacetoxy-alcyonicene, and other sesqui terpenes		*Alcyonium antarcticum*	Terra Nova Bay	Manzo et al., 2009
Paesslerins A and B		*Alcyonium antarcticum* (*A. paessleri*)	South Georgia Islands	Rodriguez Brasco et al., 2001
Alcyopterosins A-O		*Alcyonium antarcticum* (*A. paessleri*)	South Georgia Islands	Palermo et al., 2000
Steroids		*Anthomastus bathyproctus*	South Shetland Island	Mellado et al., 2005

Metabolite	Phylum	Species	Location	References
Trans-beta-farnesene, Isofuranodiene, Furanoeudesmane		Dasystenella acanthina	Terra Nova Bay	Gavagnin et al., 2003
Hodgsonal	Mollusca	Bathydoris hodgsoni	Weddell Sea	Iken et al., 1998; Avila et al., 2000
Granuloside		Charcotia granulosa	Deception Is., Livingston Is.	Cutignano et al., 2015; Moles et al., 2016
Diterpene glycerides		Doris kerguelenensis	Weddell Sea, Ross Sea and Antarctic Peninsula	Iken et al., 2002; Cutignano et al., 2011; Maschek et al., 2012; Wilson et al., 2013
Tambjamine A	Bryozoa	Bugula longissima	Antarctica (unknown locality)	Lebar et al., 2007
Asterosaponins and steroids	Echinodermata	Diplasterias brucei	Terra Nova Bay	Ivanchina et al., 2006, 2011
Disulphated polyhydroxysteroids		Gorgonocephalus chilensis	Antarctica (unknown locality)	Maier et al., 2000
Liouvilloside A and B		Staurocucumis liouvillei	Antarctica (unknown locality)	Maier et al., 2001
Triterpene glycosides		Staurocucumis liouvillei, S. turqueti, Achlionice violaecuspidata	Weddell Sea	Antonov et al., 2008, 2009, 2011; Silchenko et al., 2013
Rossinones	Tunicata	A. falklandicum, A. fuegiense, A. meridianum, A. millari, Synoicum adareanum	Weddell Sea	Nunez-Pons et al., 2010; 2012b
Aplicyanins A-F		Aplidium cyaneum	Weddell Sea	Reyes et al., 2008
Meridianins		Aplidium falklandicum, A. meridianum	Weddell Sea	Nunez-Pons et al., 2015
Rossinone B and others		Aplidium fuegiense	Weddell Sea	Carbone et al., 2012; Nunez-Pons et al., 2012b

Table 11.1 *(cont.)*

Natural products	Phylum	Taxa	Location	References
Meridianins		*Aplidium meridianum, Synoicum* sp.	South Georgia Is.., Anvers Island	Lebar & Baker, 2010
Rossinone A and B		*Aplidium* sp.	Ross Sea	Appleton et al., 2009
Palmerolide A, Hyousterones A–D and Abeohyousterone, Palmerolides D-G		*Synoicum adareanum*	Anvers Island	Diyabalanage, 2006; Miyata et al., 2007 Noguez et al., 2011

WAP: Western Antarctic Peninsula.

and some species show circumpolar distributions and broad bathymetrical ranges (Figuerola et al., 2012a). This relevant worldwide benthic group is a source of pharmacologically interesting substances such as alkaloids and terpenoids with various ecological defensive activities from antifouling to antipredation (Lebar et al., 2007; Sharp et al., 2007). A remarkable number of antarctic species has been proven to be bioactive (Angulo-Preckler et al., 2015; Figuerola et al., 2012b, 2013a, 2013b, 2014a, 2014b, 2017; Taboada et al., 2013), although only one metabolite, the alkaloid tambjamine A from *Bugula longissima,* has been identified so far (Lebar et al., 2007).

Tunicata (Chordata) are exclusively marine animals, sessile in adult stages, and protected by a more or less tough tunic. They have developed a great variety of defensive mechanisms to avoid predation and overgrowth, including physical protection, but mostly chemistry-based defences. These include accumulation of heavy metals or acids within their tissues and the use of bioactive compounds (Núñez-Pons et al., 2012b). Different strategies do exist, while colonial ascidians tend to produce antifouling or repellent chemicals, some solitary ascidians instead tend to be overgrown by epibionts to hide from possible predators (Stoecker, 1980; Bryan et al., 2003; Lambert, 2005). Ascidians mostly possess nitrogenated compounds, particularly aromatic heterocycles, like peptides, alkaloids and amino acid derived metabolites (Blunt et al., 2018). Also, in smaller amounts, they present some non-nitrogenous compounds, such as lactones, terpenoids or quinones (Blunt et al., 2018). Antarctic ascidians coming from shallow and deep bottoms present bioactive natural products such as palmerolide A, a group of ecdysteroids, meridianins, aplicyanins and rossinones (Diyabalanage et al., 2006; Miyata et al., 2007; Seldes et al., 2007; Appleton et al., 2009). It is often unclear whether the tunicates are the true producers of the molecules or if associated microbes may play a role in their chemical ecology (Núñez-Pons et al., 2012b).

Natural products from antarctic molluscs have very recently been reviewed in Avila et al. (2018). Very few new studies have dealt with other groups, such as echinoderms or other minor groups (Table 11.1). Overall, secondary metabolites in antarctic marine organisms are critical for structuring marine benthic communities (Avila et al., 2008; Figuerola et al., 2012b; von Salm et al., 2018). Chemical marine natural products display unique carbon skeletons and functional groups (Table 11.1), terpenoids, acetogenins and compounds of mixed biosynthesis being the major classes of compounds found. The total number of antarctic benthic macroorganisms chemically studied from 2000 to 2018 was 45, being 12 macroalgae (6 Ochrophyta and 6 Rhodophyta), 14 Porifera, 9 Cnidaria, 2 Mollusca, 3 Echinodermata and 5 Tunicata. The number and diversity of natural products being found is quickly increasing, and the next question is how these compounds function in nature.

11.3 Ecological activity

Polar regions are more difficult to access than other areas in the planet, and thus scientific progress has been slower there. Nevertheless, the extreme and often unique marine environments surrounding Antarctica, as well as the many unusual trophic interactions in antarctic marine communities, may be expected to favour the development of new natural products and/or for finding novel biological roles for them (Amsler et al., 2001; Avila et al., 2008). A number of new secondary metabolites with various activities, such as unpalatability, antibacterial, antitumor, cytotoxicity and others, has been reported from antarctic organisms, including sponges, cnidarians, bryozoans, molluscs, echinoderms, tunicates, microorganisms and symbiotic microorganisms, such as sponge-associated microbes (Avila et al., 2008; Papaleo et al., 2012; Núñez-Pons et al., 2012c; Núñez-Pons & Avila, 2015; von Salm et al., 2018). A complete list of the ecological activities of identified molecules from antarctic marine benthic macroorganisms recently described (from 2000 to 2018) is shown in Table 11.2. In many cases it is not clear whether compounds are produced through *de novo* biosynthesis by the organism itself or whether they are acquired through diet or have microbial symbiont origin (Avila et al., 2008; von Salm et al., 2018). Ecological studies on selected antarctic natural products are briefly discussed here.

Isolated compounds from macroalgae are few, but chemical extracts of macroalgae showed several remarkable activities, including predator deterrence, antibiotic and UVR protection (McClintock & Karentz, 1997; Schnitzler et al., 2001; Amsler et al., 2005, 2009; Fairhead et al., 2005; Erickson et al., 2006; Rhimou et al., 2010; Figuerola et al., 2012b). A specific review on UV-protecting compounds from antarctic organisms has been published recently (Núñez-Pons et al., 2018), and therefore we will not discuss these here. Deterrency has been described for extracts and tissue of brown algae species against sympatric herbivores (Ankisetty et al., 2004; Amsler et al., 2005, 2008, 2009, 2014; Huang et al., 2006), such as in the genus *Desmarestia* (*D. antarctica*, *D. anceps* and *D. menziesii*), as well as in *Himantothallus grandifolius*, *Cystosphaera jacquinotii* and *Ascoseira mirabilis*. Brown seaweeds are very important in terms of biomass in Antarctica, as well as in species diversity (ca. 27 spp.) and grade of endemism (12 spp.) (Wiencke & Clayton, 2002; Amsler & Fairhead, 2006; Wiencke et al., 2014). Therefore, this group and their chemical interactions are a very relevant component of antarctic benthic communities (Amsler et al., 2009). Furthermore, an interesting example of allopathic activity has been described in *D. menziesii* in which plastoquinones act against herbivores, affect fertility in sea urchins, deter seastars, and prevent proliferation of some bacteria (Rivera, 1996; Ankisetty et al., 2004).

Some examples of highly active molecules from red algae are halogenated furanones, such as pulchralide, fimbrolide, acetoxyfimbrolide and hydroxyfimbrolide, the last two with a strong antimicrobial activity (Ankisetty et al., 2004). One of the most well-known red algae from Antarctica is *Plocamium cartilagineum*, which shows a great profusion of compounds both to deter predators and to control microbial proliferation, ranging from epiplocamene, pyranoids, to cyclic and acyclic halogenated monoterpenes (Fries, 2016; von Salm et al., 2018).

As mentioned above, examples of ecologically relevant chemically mediated relationships include repellent substances from a range of antarctic macroalgae (Amsler et al., 2005; Aumack et al., 2010; Bucolo et al., 2011), and defensive molecules from diverse invertebrates, such as the sponges *Latrunculia apicalis* (Furrow et al., 2003), *Rossella* spp. (Núñez-Pons et al., 2012a), *Phorbas areolatus* (Solanki et al., 2018) and several other antarctic sponges (Peters et al., 2009; Núñez-Pons et al., 2012a; Angulo-Preckler et al., 2018); the cnidaria *Alcyonium* spp. (Carbone et al., 2009; Núñez-Pons et al., 2013) and three other soft corals (Slattery & McClintock, 1995); the brachiopod *Liothyrella uva* (McClintock et al., 1993; Mahon et al., 2003); the nudibranch molluscs *Bathydoris hodgsoni* (Avila et al., 2000) and *Doris kerguelenensis* (Iken et al., 2002); the ascidians *Distaplia cylindrica* (McClintock et al., 2004), *Cnemidocarpa verrucosa* (McClintock et al., 1991), *Aplidium* spp., *Synoicum* spp. (Núñez-Pons et al., 2010, 2012b) and several other ascidians (Koplovitz et al., 2009); as well as eggs, embryos and larvae of a range of invertebrate species (McClintock & Baker, 1997; Moles et al., 2017). Activity in molluscs is reviewed in Avila et al. (2018). Furthermore, in several multispecies studies, many antarctic marine species were found to contain lipophilic fractions that repelled the starfish *Odontaster validus* and the amphipod *Cheirimedon femoratus* (Taboada et al., 2013; Núñez-Pons & Avila, 2014b; Moles et al., 2015). A more recent study with 20 antarctic sponges evaluated repellence against seastar and antimicrobial activity against sympatric bacteria revealing a striking antimicrobial activity (100%) and 22% repellency (Angulo-Preckler et al., 2018).

Some alcyopterosins are also feeding deterrents against the omnivorous antarctic seastar *Odontaster validus* (Carbone et al., 2009). No ecological activity has been described for the rest of *Alcyonium* spp. natural products, except for moderate cytotoxicity of paesslerins A and B (Rodríguez Brasco et al., 2001). Ainigmaptilone A, isolated from the gorgonian coral *Ainigmaptilon antarcticus*, also showed repellence towards *O. validus*, along with antifouling properties, while ainigmaptilone B did not show any of these activities (Iken & Baker, 2003). Pigments and pigment derivatives may also be used for defensive purposes. Two sesquiterpenoids from *Acanthogorgia laxa* present antifouling activity against a wide array of microorganisms (Patiño Cano et al., 2018). Moreover, the seven steroids from the octocoral *Anthomastus bathyproctus* displayed weak cytotoxicity (Mellado et al., 2005). Even if compounds have not been identified yet, extracts from many other cnidarian species have proven to be repellent to *O. validus* too (Avila, 2016a). Further studies should be directed to identify the chemicals behind these ecological activities.

Interestingly, a variety of chemical defensive strategies are common in several antarctic bryozoan species against microorganisms and abundant and ubiquitous sympatric predators with circumpolar and eurybathic distributions. This evidences that their natural products are used for a wide array of ecological roles. In particular, several experiments showed antibacterial activity against a range of antarctic bacteria (Figuerola et al., 2014b, 2017), while the same and/or other species displayed repellent activities against the seastar *Odontaster validus* and the amphipod *Cheirimedon femoratus* (Taboada et al., 2013; Figuerola et al., 2017), and cytotoxic activities against the sea urchin *Sterechinus neumayeri*, reducing its reproductive success (Figuerola et al., 2013b, 2014a). These studies also demonstrated the presence of lipophilic and/or hydrophilic bioactive compounds in different antarctic bryozoan species, and these activities sometimes differed within populations of the same species (Figuerola et al., 2013a, 2017). In particular, some species exhibited antibacterial, cytotoxic and/or repellent activities of different nature (lipophilic or hydrophilic) at different sites (e.g. *Notoplites drygalskii*) or in the same area at similar (e.g. *Cornucopina pectogemma*) or different depths (e.g. *Camptoplites angustus* and *C. tricornis*; Figuerola et al., 2013a, 2014b, 2017). These inter- and intraspecific variabilities found in bryozoan species could be attributed to environmental-induced responses, genetic variability among populations and/or bacterial symbiotic associations (Davidson & Haygood, 1999; Morris & Prinsep, 1999; McGovern & Hellberg, 2003; Pawlik, 2012). However, no clear prove of these has been provided so far for these species.

Chemical defences might be more prevalent in particular species with the lack of apparent physical defensive structures. The best group to test this hypothesis is the ctenostome bryozoans, which are not protected by calcarious skeletons. In fact, the antarctic ctenostome *Alcyonidium flabeliforme* showed cytotoxic activity against the sperm of the sea urchin and significant repellency towards *O. validus* (Figuerola et al., 2013b; Taboada et al., 2013). In agreement, other species of the same genus from diverse regions produce bioactive compounds (Sharp et al., 2007). Several species with lightly calcified frontal walls like the antarctic *Melicerita obliqua* seem to compensate their vulnerability to predation with chemical defences (Figuerola et al., 2013a). Other species did not show chemical defences, suggesting the presence of alternative chemical and/or physical defensive mechanisms. Indeed, it is well known that cheilostome bryozoans exhibit rigid exoskeletons and/or a wide variety of skeletal structures, including spines, avicularia and vibracula, with protective function (Winston 2010). For example, defensive chemicals against the seastar *O. validus* were not detected in *Dakariella dabrowni*, which possess a rigid well-calcified skeleton (Figuerola et al., 2013a). Also, some authors suggest encrusting species are more capable of resisting damage by predators and effectively repair from grazing injuries, compared to erect colonies (Best & Winston, 1984). Therefore, chemical defences might be more common in erect and flexible forms as suggested for the antarctic species *Klugella echinata* and *N. drygalskii* (Figuerola et al., 2013a). Contrary to this hypothesis, the erect and flexible *C. pectogemma*, *C. polymorpha* and *Nematoflustra flagellata* did not show repellence to the seastar *O. validus* (Figuerola et al., 2013a, 2017) although both species appeared to be defended against the amphipod *C. femoratus*. Feeding repellent responses of a wide range of flexible and non-flexible bryozoans were also more frequent in the assays with *C. femoratus* than towards the seastars (Figuerola et al., 2013a). These suggest that amphipods might exert a higher localised pressure in bryozoans than seastars. In addition, these species possess defensive physical devices such as avicularia and vibracula. Although there is evidence of trade-offs between chemical and physical defensive mechanisms in antarctic bryozoans, previous studies showed a general trend in combining both defensive strategies, suggesting complementary traits. Further isolation and characterisation of the metabolites involved in these chemoecological interactions should be conducted.

Remarkably, a strong antifouling activity was reported *in situ* for different antarctic species, including the hydroid *Eudendrium* sp., the sponges *Phorbas glaberrima* and an *Hadromerida* sp., the tunicate *Synoicum adareanum* (Angulo-Preckler et al., 2015). Recently, the crude extracts of *Mycale tylotornota* (sponge) and *Cornucopina pectogemma* (bryozoan) avoided *in situ* fouling by eukaryotic organisms in field experiments, showing that invertebrates may also modulate the attachment of different microbial communities, either by natural products of the invertebrate itself or by natural products produced by the microbial community, resulting in different levels of fouling.

Although the ecological function of many metabolites from most tunicates remains undetermined, it is known that at least some of them are used as predator deterrents (Núñez-Pons et al., 2010) as well as antifoulants (Davis & Bremner, 1999). Antarctic examples are reported in Table 11.2. Most compounds come from the genera *Aplidium* and *Synoicum*. Cytotoxicity has been reported for aplycianins, rossinones and palmerolide A, while feeding repellence has been described for meridianins, rossinones and ecdysteroids.

In general, most authors keep looking at predator deterrence as the main ecological assay, although in recent years we observe an increasing number of studies widening their scope to different ecological roles, such as antimicrobial inhibition and cytotoxic effects (Table 11.2).

11.4 Adaptation to climate change?

According to the available information, climate change may already be affecting many antarctic organisms in different ways (Constable et al., 2014; Turner

Table 11.2 *Ecological activity of identified molecules from Antarctic marine benthic macroorganisms from 2000 to 2018*

Phylum	Taxa	Feeding deterrence		Antimicrobial activity	Bioactivity				
		Macropredator	Mesopredator		Cytotoxicity	Antifouling	Antifreezing	Growth inhibition	Allelopathy
Ochrophyta	*Ascoseira mirabilis*	*	*	*					
	Cystosphaera jacquinotii	*	*	*					
	Desmarestia anceps	*	*	*					
	Desmarestia antarctica	*	*	*					
	Desmarestia menziesii	*	*	*					
	Himantothallus grandifolius	*	*	*	*				
Rhodophyta	*Delisea pulchra*	*	*	*					
	Iridaea cordata	*							
	Myriogramme smithii		*						
	Palmaria decipiens	*							
	Pantoneura plocamioides		*						
	Plocamium cartilagineum		*	*					
Porifera	*Artemisina apollinis*	*							
	Crella sp.		*	*					*
	Dendrilla membranosa			*	*	*			
	Halichondria sp.								*
	Homaxinella balfourensis						*		
	Isodictya erinacea	*						*	
	Kirkpatrick ia variolosa			*					*
	Latrunculia apicalis	*			*				
	Latrunculia biformis				*				

Group	Species	1	2	3	4	5	6	7
	Latrunculia sp.			*				
	Lissodendoryx flabellata				*			
	Phorbas areolatus	*		*	*			
	Phorbas glaberrima							*
	Suberites caminatus				*			
Cnidaria	Acanthogorgia laxa	*						
	Ainigmaptilon antarcticus	*						
	Alcyonium antarcticum (A. paessleri)	*			*			
	Alcyonium grandis	*						
	Alcyonium haddoni	*						
	Alcyonium paucilobulatum	*						
	Alcyonium roseum	*						
	Anthomastus bathyproctus	*			*			
	Dasystenella acanthina	*				*		
Mollusca	Bathydoris hodgsoni	*				*		
	Doris k erguelenensis	*			*			
Echinodermata	Diplasterias brucei				*			
	Gorgonocephalus chilensis				*			
	Staurocucumis liouvillei			*	*			
Tunicata	Aplidium cyaneum	*			*			
	Aplidium falklandicum	*	*		*			
	Aplidium fuegiense	*	*		*			
	Aplidium meridianum	*	*		*			
	Synoicum adareanum	*	*		*		*	

et al., 2014; Poloczanska et al., 2016; Ashton et al., 2017; Griffiths et al., 2017). For example, penguins are affected by many more parasites (Díaz et al., 2017), while some benthic species may experience changes in their distribution ranges (Barnes et al., 2009; Gutt et al., 2011; Fillinger et al., 2013; Pasotti et al., 2015), or even reduce some of their interspecific relationships (Barnes et al., 2014). Since Antarctica plays an important role in the Earth's climate regulation system, knowledge of climate-change related processes is vital to understand and predict future scenarios. From this perspective, information on how Antarctic organisms relate with one another and with other components of the communities via natural products is key to advance in our understanding of a climate-changing world.

The ability to physiologically respond to temperature stress has been studied in many different antarctic taxa over many years (reviewed by Peck, 2018). Results showed poor survival capacities, but there seems to be some variation at species level (Ashton et al., 2017). Regarding natural products, we do not know yet whether changes in temperature, pH, calcification and others, may affect the production and/or use of chemicals by marine benthic organisms in Antarctica. The potential calcification problems in groups like bryozoans (see above) could dramatically affect the trade-offs between chemical and physical defences in these species, and thus their survival in the years to come. Metabolites related to unpalatability, as halogenated monoterpenes (like anverene and epi-plocamene) which define relations between macroalgae and sympatric herbivores, may vary upon environmental conditions, and therefore, trophic relationships in the antarctic ecosystems could be strongly affected by climate change. The macroalgae *Desmarestia menziesii*, for instance, increases phlorotannin production when exposed to acidification (Schoenrock et al., 2015). Another example could perhaps be the potential stress-induced compounds found in the nudibranch *D. kerguelenensis*, austrodoral and austrodoric acid (Gavagnin et al., 2003), among the highly diverse chemical arsenal of this mollusc (Avila et al., 2018), although this requires further investigation. Some variation in chemicals has also been cited for the macroalgae *Plocamium cartilagineum* from different localities (Young et al., 2013, 2015), the sponge *Dendrilla membranosa* (Witowski, 2015) and some bryozoan species (see above), which could be related to changes related to habitat or geographical specificity. Whether this could be related to adaptation to environmental change remains to be further investigated.

As an *a priori* assumption, one could think that, for instance, increasing water temperature would be a stressful factor that could induce organisms to stop producing secondary metabolites, since this is an expensive strategy. On the contrary, for some species, a higher temperature could lead to an increase in chemical defences as a reaction to gather protection against the same stress. Our preliminary experiments suggest that this could be a species-specific trend (unpublished data from the authors) and that it deserves further experimentation and analysis. For example, it is not the same to study a nudibranch mollusc that biosynthesises its own chemical defences than a sponge that contains a very rich microbiome, which may in fact be the producer of the bioactive compounds. How to test the effects of climate change in natural compounds production may therefore be a challenging task that should consider not only the different variables associated to environmental change but also the different strategies of the benthic organisms studied.

To conclude, we believe that even if advances have been made recently in understanding the ecological role of natural products in antarctic marine benthos, there is still a lot to be done to clarify the potential use of chemicals by cold-water organisms. Among other priorities, it is essential that further studies address the topic of compound production related to climate change and how this may affect species survival, before it is too late.

References

Amsler, C.D., Fairhead V.A. (2006). Defensive and sensory chemical ecology of brown algae. *Advances in Botanical Research*, **43**, 1–91.

Amsler C.D., Iken K.B., McClintock J.B., Baker B.J. (2001). Secondary metabolites from Antarctic marine organisms and their ecological implications. In: J.B. McClintock, B.J. Baker (eds) *Marine Chemical Ecology*. CRC Press, Boca Raton, FL, pp. 267–300.

Amsler, C.D., Iken, K.B., McClintock, J.B., et al. (2005). Comprehensive evaluation of the palatability and chemical defences of subtidal macroalgae from the Antarctic Peninsula. *Marine Ecology Progress Series*, **294**, 141–159.

Amsler C.D., McClintock J.B., Baker B.J. (2008). Macroalgal chemical defenses in polar marine communities. In: C. D. Amsler (ed.) *Algal Chemical Ecology*. Springer-Verlag, Berlin, pp. 91–103.

Amsler, C.D., Iken, K., McClintock, J.B., Baker, B.J. (2009). Defenses of polar macroalgae against herbivores and biofoulers. *Botanica Marina*, **52**, 535–545.

Amsler, C.D., McClintock, J.B., Baker, B.J. (2014). Chemical mediation of mutualistic interactions between macroalgae and mesograzers structure unique coastal communities along the western Antarctic Peninsula. *Journal of Phycology*, **50**, 1–10.

Angulo-Preckler, C., Cid, C., Oliva, F., Avila, C. (2015). Antifouling activity in some benthic Antarctic invertebrates by 'in situ' experiments at Deception Island, Antarctica. *Marine Environmental Research*, **105**, 30–38.

Angulo-Preckler, C., San Miguel, O., Garcia-Aljaro, C., Avila, C. (2018). Antibacterial defenses and palatability of shallow-water Antarctic sponges. *Hydrobiologia*, **806**, 123–128.

Ankisetty, S., Nandiraju, S., Win, H., et al. (2004). Chemical investigation of predator-deterred macroalgae from the Antarctic Peninsula. *Journal of Natural Products*, **67**, 1295–1302.

Antonov, A.S., Avilov, S.A., Kalinovsky, A.I., et al. (2008). Triterpene glycosides from Antarctic sea cucumbers. 1. Structure of Liouvillosides A1, A2, A3, B1, and B2 from the sea cucumber Staurocucumis liouvillei: new procedure for separation of highly polar glycoside fractions and taxonomic revision. *Journal of Natural Products*, **71**, 1677–1685.

Antonov, A.S., Avilov, S.A., Kalinovsky, A.I., et al. (2009). Triterpene glycosides from Antarctic sea cucumbers. 2. Structure of Achlioniceosides A (1), A (2), and A (3) from the sea cucumber Achlionice violaecuspidata (=Rhipidothuria racowitzai). *Journal of Natural Products*, **72**, 33–38.

Antonov, A.S., Avilov, S.A., Kalinovsky, A.I., et al. (2011). Triterpene glycosides from Antarctic sea cucumbers III. Structures of liouvillosides A (4) and A (5), two minor disulphated tetraosides containing 3-O-methylquinovose as terminal monosaccharide units from the sea cucumber Staurocumis liouvillei (Vaney). *Natural Product Research*, **25**, 1324–1333.

Appleton, D.R., Chuen, C.S., Berridge, M.V., Webb, V.L., Copp, B.R. (2009). Rossinones, A., B, biologically active meroterpenoids from the Antarctic Ascidian, Aplidium species. *Journal of Organic Chemistry*, **74**, 9195–9198.

Argandona, V.H., Rovirosa, J., San-Martin, A., et al. (2002). Antifeedant effect of marine halogenated monoterpenes. *Journal of Agricultural and Food Chemistry*, **50**, 7029–7033.

Ashton, G., Morley, S.A., Barnes, D.K.A., Clark, M.S., Peck, L.S. (2017). Warming by 1°C drives species and assemblage level responses in Antarctica's marine shallows. *Current Biology*, **27**, 2698–2705.

Aumack, C.F., Amsler, C.D., McClintock, J.B., Baker, B.J. (2010). Chemically mediated resistance to meso-herbivory in finely branched macroalgae along the western Antarctic Peninsula. *European Journal of Phycology*, **45**, 19–26.

Avila, C. (2016a). Ecological and pharmacological activities of Antarctic marine natural products. *Planta Medica*, **82**, 767–774.

Avila, C. (2016b). Biological and chemical diversity in Antarctica: from new species to new natural products. *Biodivers*, **17**, 5–11.

Avila, C., Iken, K.B., Fontana, A., Gimino, G. (2000). Chemical ecology of the Antarctic nudibranch Bathydoris hodgsoni Eliot, 1907: defensive role and origin of its natural products. *Journal of Experimental Marine Biology and Ecology*, **252**, 27–44.

Avila, C., Taboada, S., Núñez-Pons, L. (2008). Antarctic marine chemical ecology: what is next? *Marine Ecology*, **29**, 1–71.

Avila, C., Núñez-Pons, L., Moles, J. (2018). From the tropics to the poles: chemical defensive strategies in sea slugs (Mollusca: Heterobranchia). In: M. P. Puglisi, M. A. Becerro (eds) *Chemical Ecology: The Ecological Impacts of Marine Natural Products*. CRC Press, Boca Raton, FL, pp. 71–163.

Barnes, D.K.A., Griffiths, H.J., Kaiser, S. (2009). Geographic range shift responses to climate change by Antarctic benthos: where we should look. *Marine Ecology Progress Series*, **393**, 13–26.

Barnes, D.K.A., Fenton, M., Cordingley, A. (2014). Climate-linked iceberg activity massively reduces spatial competition in Antarctic shallow waters. *Current Biology*, **24**, 553–554.

Best, B.A., Winston, J.E. (1984). Skeletal strength of encrusting cheilostome bryozoans. *Biological Bulletin*, **167**, 390–409.

Blunt, J.W., Copp, B.R., Hu, W.P., et al. (2007). Marine natural products. *Natural Product Report*, **24**, 31–86.

Blunt, J.W., Carroll, A.R., Copp, B.R., Keyzers, R. A., Davis, R.A. (2018). Marine natural products. *Natural Product Report*, **35**, 8–53.

Bryan, P., McClintock, J., Slattery, M., Rittschof, D. (2003). A comparative study of the non-acidic chemically mediated antifoulant properties of three sympatric species of ascidians associated with seagrass habitats. *Biofouling*, **19**, 235–245.

Bucolo, P., Amsler, C.D., McClintock, J.B., Baker, B.J. (2011). Palatability of the Antarctic rhodophyte Palmaria decipiens (Reinsch) RW Ricker and its endo/epiphyte Elachista antarctica Skottsberg to sympatric amphipods. *Journal of*

Experimental Marine Biology and Ecology, **396**, 202–206.

Campbell, A.H., Harder, T., Nielsen, S., Kjelleberg, S., Steinberg, P.D. (2011). Climate change and disease: bleaching of a chemically defended seaweed. *Global Change Biology*, **17**, 2958–2970.

Carbone, M., Nuñez-Pons, L., Castelluccio, F., Avila, C., Gavagnin, M. (2009). Illudalene sesquiterpenoids of the alcyopterosin series from the Antarctic marine soft-coral Alcyonium grandis. *Journal of Natural Products*, **72**, 1357–1360.

Carbone, M., Núñez-Pons, L., Paone, M., et al. (2012). Rossinone-related meroterpenes from the Antarctic ascidian Aplidium fuegiense. *Tetrahedron*, **68**, 3541–3544.

Carbone, M., Nunez-Pons, L., Ciavatta, M.L., et al. (2014). Occurrence of a taurine derivative in an Antarctic glass sponge. *Natural Product Communications*, **9**, 469–470.

Ciaglia, E., Malfitano, A.M., Laezza, C., et al. (2017). Immuno-modulatory and anti-inflammatory effects of dihydrogracilin A, a terpene derived from the marine sponge Dendrilla membranosa. *International Journal of Molecular Sciences*, **18**, 1643.

Clark, M.S., Peck, L.S. (2009a). HSP70 Heat shock proteins and environmental stress in Antarctic marine organisms: a mini-review. *Marine Genomics*, **2**, 11–18.

Clark, M.S., Peck, L.S. (2009b). Triggers of the HSP70 stress response: environmental responses and laboratory manipulation in an Antarctic marine invertebrate (Nacella concinna). *Cell Stress Chaperones*, **14**, 649–660.

Constable, A.J., Melbourne-Thomas, J., Corney, S.P., et al. (2014). Climate change and Southern Ocean ecosystems I: how changes in physical habitats directly affect marine biota. *Global Change Biology*, **20**, 3004–3025.

Cutignano, A., Moles, J., Avila, C., Fontana, A. (2015). Granuloside, a unic linear homosesterterpene from the Antarctic nudibranch Charcotia granulosa. *Journal of Natural Products*, **78**, 1761–1764.

Cutignano, A., Zhang, W., Avila, C., Cimino, G., Fontana, A. (2011). Intrapopulation variability in the terpene metabolism of the Antarctic opisthobranch mollusc Austrodoris kerguelenensis. *European Journal of Organic Chemistry*, **27**, 5383–5389.

Cutignano, A., De Palma, R., Fontana, A. (2012). A chemical investigation of the Antarctic sponge Lyssodendoryx flabellata. *Natural Product Research*, **26**, 1240–1248.

Daoust, J., Chen, M., Wang, M., et al. (2013). Sesterterpenoids isolated from a northeastern Pacific Phorbas sp. *Journal of Organic Chemistry*, **78**, 8267–8273.

Davidson, S.K., Haygood, M.G. (1999). Identification of sibling species of the bryozoan Bugula neritina that produce different anticancer bryostatins and harbor distinct strains of the bacterial symbiont Candidatus endobugula sertula. *Biological Bulletin*, **196**, 273–280.

Davis, A.R., Bremner, J.B. (1999). Potential antifouling natural products from ascidians: a review. In: M. Fingerman, R. Nagabhushanam, M.F. Thompson (eds) *Recent Advances in Marine Biotechnology*, Vol. **III**. Science Publishers, New Hampshire, pp. 259–308.

Dayton, P.K. (1989). Interdecadal variation in an Antarctic sponge and its predators from oceanographic climate shifts. *Science*, **245**, 1484–1486.

Dayton, P.K., Robilliard, G.A., Paine, R.T., Dayton, L. B. (1974). Biological accommodation in the benthic community at McMurdo Sound, Antarctica. *Ecological Monographs*, **44**, 105–128.

De Broyer, C., Danis, B. (2011). How many species in the Southern Ocean? Towards a dynamic inventory of the Antarctic marine species. *Deep-Sea Research Pt II*, **58**, 5–17.

De Broyer, C., Koubbi, P., Griffiths, H.J., et al. (2014). *Biogeographic Atlas of the Southern*

Ocean. Scientific Committee on Antarctic Research, Cambridge, UK.

Díaz, J.I., Fusaro, B., Vidal, V., et al. (2017). Macroparasites in Antarctic penguins. In: S. Klimpel, T. Kuhn, H. Mehlhorn (eds) *Biodiversity and Evolution of Parasitic Life in the Southern Ocean.* Springer International Publishing, Cham, Switzerland, pp. 183–204.

Díaz-Marrero, A.R., Brito, I., Dorta, E., et al. (2003). Caminatal, an aldehyde sesterterpene with a novel carbon skeleton from the Antarctic sponge Suberites caminatus. *Tetrahedron Letters,* **44,** 5939–5942.

Díaz-Marrero, A.R., Brito, I., Cueto, M., San-Martin, A., Darias, J. (2004). Suberitane network, a taxonomical marker for Antarctic sponges of the genus Suberites? Novel sesterterpenes from Suberites caminatus. *Tetrahedron Letters,* **45,** 4707–4710.

Diyabalanage, T., Amsler, C.D., McClintock, J.B., Baker, B.J. (2006). Palmerolide A, a cytotoxic Macrolide from the Antarctic tunicate Synoicum adareanum. *Journal of the American Chemical Society,* **128,** 5630–5631.

Downey, R.V., Griffiths, H.J., Linse, K., Janussen, D. (2012). Diversity and distribution patterns in high southern latitude sponges. *PLoS One,* **7,** e41672.

Ducklow, H.W., Fraser, W.R., Meredith, M.P., et al. (2013). West Antarctic Peninsula: an ice-dependent coastal marine ecosystem in transition. *Oceanography,* **26,** 190–203.

Duckworth, A.R., Battershill, C.N. (2001). Population dynamics and chemical ecology of New Zealand Demospongiae Latrunculia sp. nov., Polymastia croceus (Poecilosclerida: Latrunculiidae: Polymastiidae). *New Zealand Journal of Marine and Freshwater Research,* **35,** 935–949.

Erickson, A.A., Paul, V.J., Alstyne, K.L., Van Kwiatkowski, L.M. (2006). Palatability of macroalgae that use different types of chemical defenses. *Journal of Chemical Ecology,* **32,** 1883–1895.

Fabry, V.J., Seibel, B.A., Feely, R.A., Orr, J.C. (2008). Impacts of ocean acidification on marine fauna and ecosystem processes. *ICES Journal of Marine Sciences,* **65,** 414–432.

Fairhead, V.A., Amsler, C.D., Mcclintock, J.B., Baker, B.J. (2005). Within-thallus variation in chemical and physical defenses in two species of ecologically dominant brown macroalgae from the Antarctic Peninsula. *Journal of Experimental Marine Biology and Ecology,* **322,** 1–12.

Ferretti, C., Vacca, S., De Ciucis, C., et al. (2009). Growth dynamics and bioactivity variation of the Mediterranean demosponges Agelas oroides (Agelasida, Agelasidae) and Petrosia ficiformis (Haplosclerida, Petrosiidae). *Marine Ecology,* **30,** 1–10.

Figuerola, B., Monleón-Getino, T., Ballesteros, M., Avila, C. (2012a). Spatial patterns and diversity of bryozoan communities from the Southern Ocean: South Shetland Islands, Bouvet Island and Eastern Weddell Sea. *Systematics and Biodiversity,* **10,** 109–123.

Figuerola, B., Núñez-Pons, L., Vázquez, J., et al. (2012b). Chemical interactions in Antarctic marine benthic ecosystems. In: A. Cruzado (ed.) *Marine Ecosystems.* InTech, Rijeka, pp. 105–126.

Figuerola, B., Núñez-Pons, L., Moles, J., Avila, C. (2013a). Feeding repellence in Antarctic bryozoans. *Naturwissenschaften,* **100,** 1069–1081.

Figuerola, B., Taboada, S., Monleón-Getino, T., Vázquez, J., Avila, C. (2013b). Cytotoxic activity of Antarctic benthic organisms against the common sea urchin Sterechinus neumayeri. *Oceanography,* **1,** 2.

Figuerola, B., Núñez-Pons, L., Monleón-Getino, T., Avila, C. (2014a). Chemo-ecological interactions in Antarctic bryozoans. *Polar Biology,* **37,** 1017–1030.

Figuerola, B., Sala-Comorera, L., Angulo-Preckler, C., et al. (2014b). Antimicrobial activity of Antarctic bryozoans: an ecological perspective with potential for clinical applications. *Marine Environmental Research*, **101**, 52–59.

Figuerola, B., Angulo-Preckler, C., Núñez-Pons, L., et al. (2017). Experimental evidence of chemical defence mechanisms in Antarctic bryozoans. *Marine Environmental Research*, **129**, 68–75.

Fillinger, L., Janussen, D., Lundälv, T., Richter, C. (2013). Rapid glass sponge expansion after climate-induced Antarctic ice shelf collapse. *Current Biology*, **23**, 1330–1334.

Ford, J., Capon, R.J. (2000). Discorhabdin R: a new antibacterial pyrroloiminoquinone from two latrunculiid marine sponges, Latrunculia sp., Negombata sp. *Journal of Natural Products*, **63**, 1527–1528.

Fries, J.L. (2016). Chemical investigation of Antarctic marine organisms and their role in modern drug discovery. University of South Florida.

Furrow, F.B., Amsler, C.D., McClintock, J.B., Baker, B.J. (2003). Surface sequestration of chemical feeding deterrents in the Antarctic sponge Latrunculia apicalis as an optimal defence against sea star spongivory. *Marine Biology*, **143**, 443–449.

Gavagnin, M., Carbone, M., Mollo, E., Cimino, G. (2003). Austrodoral and austrodoric acid: nor-sesquiterpenes with a new carbon skeleton from the Antarctic nudibranch *Austrodoris kerguelenensis. Tetrahedron Letters*, **44**, 1495–1498.

Griffiths, H.J., Meijers, A., Bracegirdle, T. (2017). More losers than winners in a century of future Southern Ocean seafloor warming. *Nature Climate Change*, **7**, 749–754.

Gutt, J., Barratt, I., Domack, E., et al. (2011). Biodiversity change after climate-induced iceshelf collapse in the Antarctic. *Deep-Sea Research Pt. II*, **58**, 74–83.

Harper, M.K., Bugni, T.S., Copp, B.R., et al. (2001). Introduction to the chemical ecology of marine natural products. In: J. B. McClintock and B.J. Baker (eds) *Marine Chemical Ecology*. CRC Press, Boca Raton, FL, pp. 3–69.

Huang, Y.M., McClintock, J.B., Amsler, C.D., Peters, K.J., Baker, B.J. (2006). Feeding rates of common Antarctic gammarid amphipod on ecologically important sympatric macroalgae. *Journal of Experimental Marine Biology and Ecology*, **329**, 55–65.

Iken, K.B., Baker, B.J. (2003). Ainigmaptilones, sesquiterpenes from the Antarctic gorgonian coral *Ainigmaptilon antarcticus. Journal of Natural Products*, **66**, 888–890.

Iken, K., Avila, C., Ciavatta, M.L., Fontana, A., Cimino, G. (1998). Hodgsonal, a new drimane sesquiterpene from the mantle of the Antarctic nudibranch *Bathydoris hodgsoni. Tetrahedron Letters*, **39**, 5635–5638.

Iken, K., Avila, C., Fontana, A., Gavagnin, M. (2002). Chemical ecology and origin of defensive compounds in the Antarctic nudibranch Austrodoris kerguelenensis (Opisthobranchia: Gastropoda). *Marine Biology*, **141**, 101–109.

IPCC Core Writing Team (2014). In: R. K. Pachauri, L.A. Meyer (eds) *Climate Change 2014: Synthesis Report. Contribution of Working Groups I, II and III to the Fifth Assessment Report of the Intergovernmental Panel on Climate Change*. IPCC, Geneva.

Ivanchina, N.V., Kicha, A.A., Kalinovsky, A.I., et al. (2006). Polar steroidal compounds from the Far Eastern starfish *Henricia leviuscula. Journal of Natural Products*, **69**, 224–228.

Ivanchina, N.V., Kicha, A.A., Stonik, V.A. (2011). Steroid glycosides from marine organisms. *Steroids*, **76**, 425–454.

Jacob, U., Terpstra, S., Brey, T. (2003). High-Antarctic regular sea urchins – the role of depth and feeding in niche separation. *Polar Biology*, **26**, 99–104.

Jouiaei, M., Yanagihara, A., Madio, B., et al. (2015). Ancient venom systems: a review on cnidaria toxins. *Toxins*, **7**, 2251–2271.

Kim, H.J., Kim, W.J., Koo, B-W., et al. (2017). Anticancer activity of sulfated polysaccharides isolated from the Antarctic red seaweed *Iridaea cordata. Ocean and Polar Research*, **38**, 129–137.

Koplovitz, G., McClintock, J.B., Amsler, C.D., Baker, B.J. (2009). Palatability and chemical anti-predatory defenses in common ascidians from the Antarctic Peninsula. *Aquatic Biology*, **7**, 81–92.

Lambert, G. (2005). Ecology and natural history of the protochordates. *Canadian Journal of Zoology*, **83**, 34–50.

Laturnus, F., Wiencke, C., Klöser, H. (1996). Antarctic macroalgae – sources of volatile halogenated organic compounds. *Marine Environmental Research*, **41**, 169–181.

Lebar, M.D., Baker, B.J. (2010). Synthesis and structure reassessment of Psammopemmin A. *Australian Journal of Chemistry*, **63**, 862–866.

Lebar, M.D., Heimbegner, J.L., Baker, B.J. (2007). Cold-water marine natural products. *Natural Product Report*, **24**, 774–797.

Lee, H., Ahn, J., Lee, Y., Rho, J., Shin, J. (2004). New sesterterpenes from the Antarctic sponge Suberites sp. *Journal of Natural Products*, **67**, 672–674.

Li, F., Janussen, D., Peifer, C., Pérez-Victoria, I., Tasdemir, D. (2018). Targeted isolation of Tsitsikammamines from the Antarctic deep-sea sponge Latrunculia biformis by molecular networking and anticancer activity. *Marine Drugs*, **16**, 268.

Ma, W.S., Mutka, T., Vesley, B., et al. (2009). Norselic acids A–E, highly oxidized anti-infective steroids that deter mesograzer predation, from the Antarctic sponge Crella sp. *Journal of Natural Products*, **72**, 1842–1846.

Mahon, A.R., Amsler, C.D., McClintock, J.B., Amsler, M.O., Baker, B.J. (2003). Tissue-specific palatability and chemical defences against macropredators and pathogens in the common articulate brachiopod Liothyrella uva from the Antarctic Peninsula. *Journal of Experimental Marine Biology and Ecology*, **290**, 197–210.

Maier, M.S., Araya, E., Seldes, A.M. (2000). Sulphated polyhydroxysteroids from the Antarctic ophiuroid *Gorgonocephalus chilensis*. *Molecules*, **5**, 348–349.

Maier, M.S., Roccatagliata, A.J., Kuriss, A., et al. (2001). Two new cytotoxic and virucidal trisulphated triterpene glycosides from the Antarctic sea cucumber *Staurocucumis liouvillei*. *Journal of Natural Products*, **64**, 732–736.

Manzo, E., Ciavatta, M.L., Nuzzo, G., Gavagnin, M. (2009). Terpenoid content of the Antarctic soft coral *Alcyonium antarcticum*. *Natural Product Communications*, **4**, 1615–1619.

Maschek, J.A., Baker, B.J. (2008). The chemistry of algal secondary metabolism. In: C. D. Amsler (ed.) *Algal Chemical Ecology*. Berlin, Springer, pp. 1–20.

Maschek, J.A., Mevers, E., Diyabalanage, T., et al. (2012). Palmadorine chemodiversity from the Antarctic nudibranch Austrodoris kerguelenensis and inhibition of Jak2-STAT5-dependent HEL leukemia cells. *Tetrahedron*, **68**, 9095–9104.

McClintock, J.B. (1987). Investigation of the relationship between invertebrate predation and biochemical composition, energy content, spicule armament and toxicity of benthic sponges at McMurdo Sound, Antarctica. *Marine Biology*, **94**, 479–487.

McClintock, J.B., Baker, B.J. (1997). Palatability and chemical defense of eggs, embryos and larvae of shallow water Antarctic marine invertebrates. *Marine Ecology Progress Series*, **154**, 121–131.

McClintock, J.B., Karentz, D. (1997). Mycosporine-like amino acids in 38 species of subtidal marine organisms from McMurdo Sound. Antarctica. *Antarctic Science*, **9**, 392–398.

McClintock, J.B., Heine, J., Slattery, M., Weston, J. (1991). Biochemical and energetic composition, population biology, and chemical defense of the Antarctic ascidian Cnemidocarpa verrucosa lesson. *Journal of Experimental Marine Biology and Ecology*, **147**, 163–175.

McClintock, J.B., Slattery, M., Thayer, C.W. (1993). Energy content and chemical defense of the articulate Brachiopod Liothyrella uva (Jackson, 1912) from the Antarctic Peninsula. *Journal of Experimental Marine Biology and Ecology*, **169**, 103–116.

McClintock, J.B., Amsler, M.O., Amsler, C.D., et al. (2004). Biochemical composition, energy content and chemical antifeedant and antifoulant defenses of the colonial Antarctic ascidian *Distaplia cylindrica*. *Marine Biology*, **145**, 885–894.

McClintock, J.B., Amsler, C.D., Baker, B.J. (2010). Overview of the chemical ecology of benthic marine invertebrates along the western Antarctic Peninsula. *Integrative and Comparative Biology*, **50**, 967–980.

McGovern, T.M., Hellberg, M.E. (2003). Cryptic species, cryptic endosymbionts, and geographic variation in chemical defenses in the bryozoan *Bugula neritina*. *Molecular Ecology*, **12**, 1207–1215.

Mellado, G.G., Zubía, E., Ortega, M.J., López-González, P.J. (2005). Steroids from the Antarctic octocoral *Anthomastus bathyproctus*. *Journal of Natural Products*, **68**, 1111–1115.

Miyata, Y., Diyabalanage, T., Amsler, C.D., et al. (2007). Ecdysteroids from the Antarctic tunicate *Synoicum adareanum*. *Journal of Natural Products*, **70**, 1859–1864.

Moles, J., Núñez-Pons, L., Taboada, S., et al. (2015). Anti-predatory chemical defences in Antarctic benthic fauna. *Marine Biology*, **162**, 1813–1821.

Moles, J., Wägele, H., Cutignano, A., Fontana, A., Avila, C. (2016). Distribution of granuloside in the Antarctic nudibranch *Charcotia granulosa* (Gastropoda: Heterobranchia: Charcotiidae). *Marine Biology*, **163**, 54–65.

Moles, J., Wägele, H., Cutignano, A., et al. (2017). Giant embryos and hatchlings of Antarctic nudibranchs (Mollusca: Gastropoda: Heterobranchia). *Marine Biology*, **164**, 114–126.

Moon, B., Park, Y.C., McClintock, J.B., Baker, B.J. (2000). Structure and bioactivity of erebusinone, a pigment from the Antarctic sponge *Isodictya erinacea*. *Tetrahedron*, **56**, 9057–9062.

Morris, B.D., Prinsep, M.R. (1999). Amathaspiramides A-F, novel brominated alkaloids from the marine bryozoan *Amathia wilsoni*. *Journal of Natural Products*, **62**, 688–693.

Noguez, J.H., Diyabalanage, T.K.K., Miyata, Y., et al. (2011). Palmerolide Macrolides from the Antarctic tunicate Synoicum adareanum. *Bioorganic & Medicinal Chemistry*, **19**, 6608–6614.

Núñez-Pons, L., Avila, C. (2014a). Defensive metabolites from Antarctic invertebrates: does energetic content interfere with feeding repellence? *Marine Drugs*, **12**, 3770–3791.

Núñez-Pons, L., Avila, C. (2014b). Deterrent activities in the crude lipophilic fractions of Antarctic benthic organisms: chemical defences against keystone predators. *Polar Research*, **33**, 21624.

Núñez-Pons, L., Avila, C. (2015). Natural products mediating ecological interactions in Antarctic benthic communities: a mini-review of the known molecules. *Natural Product Report*, **32**, 1114–1130.

Núñez-Pons, L., Forestieri, R., Nieto, R.M., et al. (2010). Chemical defenses of tunicates of the genus Aplidium from the Weddell Sea (Antarctica). *Polar Biology*, **33**, 1319–1329.

Núñez-Pons, L., Carbone, M., Paris, D., et al. (2012a). Chemoecological studies on hexactinellid sponges from the Southern Ocean. *Naturwissenschaften*, **99**, 353–368.

Núñez-Pons, L., Carbone, M., Vázquez, J., et al. (2012b). Natural products from Antarctic colonial ascidians of the genera Aplidium and Synoicum: variability and defensive role. *Marine Drugs*, **10**, 1741–1764.

Núñez-Pons, L., Rodríguez-Arias, M., Gómez-Garreta, A., Ribera-Siguán, A., Avila, C. (2012c). Feeding deterrency in Antarctic marine organisms: bioassays with the omnivore amphipod *Cheirimedon femoratus*. *Marine Ecology Progress Series*, **462**, 163–174.

Núñez-Pons, L., Carbone, M., Vázquez, J., Gavagnin, M., Avila, C. (2013). Lipophilic defenses from Alcyonium soft corals of Antarctica. *Journal of Chemical Ecology*, **39**, 675–685.

Núñez-Pons, L., Nieto, R.M., Avila, C., Jiménez, C., Rodríguez, J. (2015). Mass spectrometry detection of minor new meridianins from the Antarctic colonial ascidians *Aplidium falklandicum* and *Aplidium meridianum*. *Journal of Mass Spectrometry*, **50**, 103–111.

Núñez-Pons, L., Avila, C., Romano, G., Verde, C., Giordano, D. (2018). UV-protective compounds in marine organisms from the Southern Ocean. *Marine Drugs*, **16**, 336.

Palermo, J.A., Rodrı, M.F., Spagnuolo, C., Seldes, A.M. (2000). Illudalane sesquiterpenoids from the soft coral Alcyonium paessleri : the first natural nitrate esters. *Journal of Organic Chemistry*, **65**, 4482–4486.

Papaleo, M.C., Fondi, M., Maida, I., et al. (2012). Sponge-associated microbial Antarctic communities exhibiting antimicrobial activity against Burkholderia cepacia complex bacteria. *Biotechnology Advances*, **30**, 272–293.

Pasotti, F., Manini, E., Giovannelli, D., et al. (2015). Antarctic shallow water benthos in an area of recent rapid glacier retreat. *Marine Ecology*, **36**, 716–733.

Patiño Cano, L.P., Manfredi, R.Q., Pérez, M., et al. (2018). Isolation and antifouling activity of azulene derivatives from the Antarctic gorgonian *Acanthogorgia laxa*. *Chemistry & Biodiversity*, **15**, e1700425.

Paul, V.J. (1992). *Ecological Roles of Marine Natural Products*. Cornell University Press, Ithaca, NY.

Pawlik, J.R. (2012). Antipredatory defensive roles of natural products from marine invertebrates. In: E. Fattorusso, W. H. Gerwick, O. Taglialatela-Scafati (eds) *Handbook of Marine Natural Products*. Springer Netherlands, Dordrecht, pp. 677–710.

Peck, L.S. (2018). Antarctic marine biodiversity: adaptations, environments and responses to change. *Oceanography and Marine Biology*, **56**, 105–236.

Peters, K.J., Amsler, C.D., McClintock, J.B., van Soest, R.W.M., Baker, B.J. (2009). Palatability and chemical defenses of sponges from the western Antarctic Peninsula. *Marine Ecology Progress Series*, **385**, 77–85.

Peters, L., Wright, A.D., Krick, A., König, G.M. (2004). Variation of brominated indoles and terpenoids within single and different colonies of the marine bryozoan *Flustra foliacea*. *Journal of Chemical Ecology*, **30**, 1165–1182.

Poloczanska, E.S., Burrows, M.T., Brown, C.J., et al. (2016). Responses of marine organisms to climate change across oceans. *Frontiers in Marine Science*, **3**, 1–21.

Principe, P.P., Fisher, W.S. (2018). Spatial distribution of collections yielding marine natural products. *Journal of Natural Products*, **81**, 2307–2320.

Puglisi, M.P., Becerro, M.A. (2018). *Life in Extreme Environments: Insights in Biological Capability*. CRC Press, Boca Raton, FL.

Puglisi, M.P., Sneed, J.M., Ritson-Williams, R., Young, R. (2019). Marine chemical ecology in benthic environments. *Natural Product Report*, **36**(3), 410–429.

Reyes, F., Fernandez, R., Rodriguez, A., et al. (2008). Aplicyanins A-F, new cytotoxic bromoindole derivatives from the marine tunicate *Aplidium cyaneum*. *Tetrahedron*, **64**, 5119–5123.

Rhimou, B., Hassane, R., Nathalie, B., Coppens, Y., Vannes, U.D.B. (2010). Antiviral activity of the extracts of Rhodophyceae from Morocco. *African Journal of Biotechnology*, **9**, 7968–7975.

Rivera, P. (1996). Plastoquinones and a chromene isolated from the Antarctic brown alga Desmarestia menziesii. *Boletín de la Sociedad Chilena de Química*, **41**, 103–105.

Rodríguez Brasco, M.F., Seldes, A.M., Palermo, J. A. (2001). Paesslerins A and B: novel tricyclic sesquiterpenoids from the soft coral *Alcyonium paessleri*. *Organic Letters*, **3**, 1415–1417.

Schnitzler, I., Pohnert, G., Hay, M., Boland, W. (2001). Chemical defense of brown algae (Dictyopteris spp.) against the herbivorous amphipod Ampithoe longimana. *Oecologia*, **126**, 515–521.

Schoenrock, K.M., Schram, J.B., Amsler, C.D., McClintock, J.B., Angus, R.A. (2015). Climate change impacts on overstory Desmarestia spp. from the western Antarctic Peninsula. *Marine Biology*, **162**, 377–389.

Seldes, A.M., Brasco, M.F.R., Franco, L.H., et al. (2007). Identification of two meridianins from the crude extract of the tunicate Aplidium meridianum by tandem mass spectrometry. *Natural Product Research*, **21**, 555–563.

Sharp, J.H., Winson, M.K., Porter, J.S. (2007). Bryozoan metabolites: an ecological perspective. *Natural Product Report*, **24**, 659–673.

Silchenko, A.S., Kalinovsky, A.I., Avilov, S.A., et al. (2013). Triterpene glycosides from Antarctic sea cucumbers IV. Turquetoside A, a 3-O-methylquinovose containing disulfated tetraoside from the sea cucumber Staurocucumis turqueti (Vaney, 1906) (= Cucumaria spatha). *Biochemical Systematics and Ecology*, **51**, 45–49.

Slattery, M., McClintock, J.B. (1995). Population structure and feeding deterrence in three shallow-water Antarctic soft corals. *Marine Biology*, **122**, 461–470.

Solanki, H., Angulo-Preckler, C., Calabro, K., et al. (2018). Suberitane sesterterpenoids from the Antarctic sponge Phorbas areolatus (Thiele, 1905). *Tetrahedron Letters*, **59**, 3353–3356.

Soldatou, S., Baker, B.J. (2017). Cold-water marine natural products, 2006 to 2016. *Natural Product Report*, **34**, 585–626.

Stoecker, D. (1980). Chemical defenses of ascidians against predators. *Ecology*, **61**, 1327–1334.

Swearingen III, D.C., Pawlik, J.R. (1998). Variability in the chemical defense of the sponge Chondrilla nucula against predatory reef fishes. *Marine Biology*, **131**, 619–627.

Taboada, S., Núñez-Pons, L., Avila, C. (2013). Feeding repellence of Antarctic and sub-Antarctic benthic invertebrates against the omnivorous sea star *Odontaster validus*. *Polar Biology*, **36**, 13–25.

Tian, Y., Li, Y., Zhao, F. (2017). Secondary metabolites from polar organisms. *Marine Drugs*, **15**, 28.

Torssel, K.B.G. (1983). *Natural Product Chemistry. A Mechanistic and Biosynthetic Approach to Secondary Metabolism*. John Wiley, New York.

Tremblay, N., Abele, D. (2016). Response of three krill species to hypoxia and warming: an experimental approach to oxygen minimum zones expansion in coastal ecosystems. *Marine Ecology*, **37**, 179–199.

Turner, J., Bindschadler, R., Convey, P., et al. (2009). *Antarctic Climate Change and the Environment: A Contribution to the International Polar Year 2007–2008*. Scientific Committee on Antarctic Research, Cambridge, UK.

Turner, J., Barrand, N.E., Bracegirdle, T.J., et al. (2014). Antarctic climate change and the environment: an update. *Polar Record*, **50**, 237–259.

Turon, X., Becerro, M.A., Uriz, M.J. (1996). Seasonal patterns of toxicity in benthic invertebrates: the encrusting sponge Crambe crambe (Poecilosclerida). *Oikos*, **75**, 33–40.

Vankayala, S.L., Kearns, F.L., Baker, B.J., Larkin, J.D., Woodcock, H.L. (2017). Elucidating a chemical defense mechanism of Antarctic sponges: a computational study. *Journal of Molecular Graphics and Modelling*, **71**, 104–115.

Vetter, W., Janussen, D. (2005). Halogenated natural products in five species of Antarctic sponges: compounds with POP-like properties? *Environmental Science and Technology*, **39**, 3889–3895.

von Salm, J.L., Wilson, N.G., Vesely, B.A., et al. (2014). Shagenes A and B, new tricyclic sesquiterpenes produced by an undescribed Antarctic octocoral. *Organic Letters*, **16**, 2630–2633.

von Salm, J.L., Witowski, C.G., Fleeman, R.M., et al. (2016). Darwinolide, a new diterpene scaffold that inhibits methicillin-resistant *Staphylococcus aureus* biofilm from the Antarctic sponge *Dendrilla membranosa*. *Organic Letters*, **18**, 2596–2599.

von Salm, J.L., Schoenrock, K.M., McClintock, J. B., Amsler, C.D., Baker, B.J. (2018). The status of marine chemical ecology in Antarctica. Form and function of unique high-latitude chemistry. In: M.P. Puglisi, M. A. Becerro (eds) *Life in Extreme Environments: Insights in Biological Capability*. CRC Press, Boca Raton, FL, pp. 27–69.

Wang, M., Tietjen, I., Chen, M., et al. (2016). Sesterterpenoids isolated from the sponge Phorbas sp. activate latent HIV-1 provirus expression. *Journal of Organic Chemistry*, **81**, 11324–11334.

Wang, Y.J. (2014). The future of marine invertebrates in face of global climate change. *Journal of Coastal Development*, **17**, e105.

Wiencke, C., Clayton, M.N. (2002). *Synopses of the Antarctic Benthos. Antarctic Seaweeds*. A. R.G. Gantner Verlag KG Ruggell, Liechtenstein.

Wiencke, C., Amsler, C.D., Clayton, M.N. (2014). Macroalgae. In: C. De Broyer, P. Koubbi, H. J. Griffiths, et al. (eds) *Biogeographic Atlas of the Southern Ocean*. Scientific Committee on Antarctic Research, Cambridge, UK.

Wilkins, S.P., Blum, A.J., Burkepile, D.E., et al. (2002). Isolation of an antifreeze peptide from the Antarctic sponge *Homaxinella balfourensis*. *Cellular and Molecular Life Sciences*, **59**, 2210–2215.

Wilson, N.G., Maschek, J.A., Baker, B.J. (2013). A species flock driven by predation? Secondary metabolites support diversification of slugs in Antarctica. *PLoS One*, **8**, e80277.

Winston, J.E. (2010). Life in the colonies: learning the alien ways of colonial organisms. *Integrative and Comparative Biology*, **50**, 919–933.

Witowski, C.W. (2015). Investigation of bioactive metabolites from the Antarctic sponge Dendrilla membranosa and marine microorganisms. PhD thesis, University of South Florida.

Young, E.B., Dring, M.J., Savidge, G., Birkett, D. A., Berges, J.A. (2007). Seasonal variations in nitrate reductase activity and internal N pools in intertidal brown algae are correlated with ambient nitrate concentrations. *Plant Cell & Environment*, **30**, 764–774.

Young, R.M., von Salm, J.L., Amsler, M.O., et al. (2013). Site-specific variability in the chemical diversity of the Antarctic red alga *Plocamium cartilagineum*. *Marine Drugs*, **11**, 2126–2139.

Young, R.M., Schoenrock, K.M., von Salm, J.L., Amsler, C.D., Baker, B.J. (2015). Structure and function of macroalgal natural products. In:D. Stengel and S. Connan (eds) *Natural Products from Marine algae. Methods in Molecular Biology*. Humana Press, New York, pp. 39–73.

CHAPTER TWELVE

Metabolic and taxonomic diversity in antarctic subglacial environments

TRISTA J. VICK-MAJORS
Michigan Technological University
AMANDA M. ACHBERGER
Texas A&M University
ALEXANDER B. MICHAUD
Aarhus University

and

JOHN C. PRISCU
Montana State University

12.1 Introduction

Aquatic subglacial habitats occur throughout the cryosphere where basal melting is sufficient to produce aqueous environments (Priscu & Christner, 2004). Heat energy for melting of basal ice is produced by frictional heating due to glacier movement and geothermal heat flux (Fisher et al., 2015). These heat sources in concert with the lowering of the pressure melting point due to the weight and insulating properties of the overlying ice all contribute to basal ice melting. Ultimately, the energy budget of the glacial basal zone determines whether conditions are favourable for liquid water and thus subglacial life. Where conditions are favourable for life, geochemistry and hydrology exert first order control over the microbial metabolisms present (Tranter et al., 2005) by providing the energy required to fuel biosynthetic reactions in the absence of light. Subglacial waters are thought to hold an estimated 10^{21} cells, equivalent to 1600 teragrams of cellular carbon, and may comprise one of the largest freshwater environments on Earth (Priscu et al., 2008). Organisms inhabiting these environments are subject to cold, dark and high-pressure conditions, as well as limited energetic resources, compared to the sunlit biosphere. Here, we review current knowledge on antarctic aquatic environments that lie beneath the ice sheets and the organisms that inhabit them.

12.1.1 History of subglacial exploration in the antarctic

As recently as the 1990s, antarctic subglacial lakes were a mere curiosity in the minds of scientists. Perhaps the earliest evidence of subglacial lakes was provided by aviators who, in the 1960s, noted a relatively flat ice plain near the Russian Vostok Station (Zotikov, 2006). Airborne radio-echo sounding made in the late 1960s and the 1970s to map the bed of the antarctic ice sheet revealed unexpected flat reflectors at the bottom of the ice sheet; these were interpreted as liquid water bodies over which the ice flowed (Oswald & Robin, 1973). Kapitsa and colleagues (1996) summarised satellite data that became available after the flat reflectors were detected, and provided unequivocal evidence that a huge subglacial lake, known as subglacial Lake Vostok, existed beneath Vostok Station. These results prompted scientists to re-examine radio-echo sounding data, resulting in the publication of subglacial lake inventories that initially ranged from 150 lakes (Siegert et al., 2005) to more than 400 lakes (Wright & Siegert, 2012). This rapid expansion in the number of known subglacial lakes resulted from new radar datasets collected in previously unexplored regions of the ice sheet, and from satellite measurements of ice surface elevation change caused by the movement of subglacial water.

Following the seminal publication by Kapitsa et al. (1996), drilling at Vostok Station stopped when the Russian drilling team retrieved cores of what was thought to be Lake Vostok water that had accreted to the bottom of the ice sheet (Jouzel et al., 1999; Karl et al., 1999; Petit et al., 1999; Priscu et al., 1999). This prompted a new, international focus on planning efforts to study antarctic subglacial lake environments. The Scientific Committee on Antarctic Research (SCAR) convened a group of scientists and technologists in July 2000 who were tasked with determining the importance of subglacial lake research and ensuring that the subglacial environment remained pristine. The Subglacial Antarctic Lake Exploration Group of Specialists' (SALEGOS) membership was multinational and interdisciplinary, providing the expertise necessary to advance a detailed science, technology and environmental plan for exploration and research. SALEGOS was tasked with considering all aspects of an exploration and research programme including defining and addressing environmental stewardship and contamination issues related to accessing, observing, entering and sampling subglacial environments. The group actively engaged invited experts to provide additional information and guidance on key issues. Environmental concerns were twofold: (i) the fostering of stewardship and protection of subglacial lake environments; and (ii) the retrieval and processing of samples without compromising their scientific value.

Under the direction of SCAR, SALEGOS reorganised in 2004 to form a SCAR Scientific Research Programme under the title Subglacial Antarctic Lake

Environments (SALE). This group concluded that: (i) the exploration and study of subglacial environments has the potential for great scientific pay-off by advancing our understanding of the evolution of Antarctica, the Earth system, climate and life, which offset the risk involved in accessing these environments; (ii) with proper planning and adoption of precautionary principles, environmental concerns can be adequately addressed and risk minimised; (iii) the investigation of subglacial lake environments can provide a model for maximising scientific return while minimising human caused alteration or degradation of the environment; and (iv) exploring new frontiers requires due diligence to ensure that scientific advances are accomplished while avoiding the loss or harm to unique environments such as those beneath the vast ice sheets of Antarctica.

Owing in large part to the efforts of the SCAR planning groups SALEGOS and SALE, the US National Academy of Sciences convened a National Research Council (NRC) workshop to address the scientific and environmental issues involved in subglacial lake exploration and concluded that transformative science would be obtained from the exploration of these systems, but that it was imperative that strict guidelines for environmental stewardship be followed (National Research Council, 2007). Following the endorsement of the NRC, SALE declared that its terms of reference had been met and the group disbanded in early 2009, concluding that individual nations should consider funding research on subglacial environments.

Following more than a decade of deliberation, antarctic research programmes from Russia (subglacial lake Vostok SLV), the United Kingdom (Ellsworth Subglacial Lake; SLE) and the United States (Whillans Subglacial Lake; SLW) were funded to directly sample the subglacial environment in different regions of Antarctica. Vostok subglacial Lake is a large lake under a deep ice cover (~4000 m) in East Antarctica, which was accessed in 2012. Upon penetrating the ice sheet over Vostok Subglacial Lake, water flooded rapidly into the borehole as hydraulic equilibrium was reached. The water froze during the ensuing winter and was re-cored the following summer season, allowing a sample of frozen lake surface water to be recovered. The analysis of this sample is discussed by Bulat (2016). During December 2012, UK scientists attempted to access Lake Ellsworth, a deep-water lake in central West Antarctica, using a clean access hot-water drill (Siegert et al., 2012). The UK team experienced a series of equipment and operational failures that ended their efforts short of reaching the lake (Siegert et al., 2014). One month later, in January 2013, a US team used a clean hot-water drill (Priscu et al., 2013) to successfully access SLW and retrieve lake water and sediment samples (Tulaczyk et al., 2014) showing for the first time the presence of a thriving microbial ecosystem beneath the antarctic ice sheet (Christner et al., 2014). An interdisciplinary US team has now acquired national funding to drill into Subglacial Lake Mercer in. SLM is a hydraulically active lake that receives a portion of its water from East Antarctica (Siegfried et al., 2014).

12.2 Subglacial habitat diversity

Subglacial aquatic environments can support microbial life as long as the needs for carbon, nutrients (macronutrients nitrogen, phosphorus and sulphur, as well as a suite of micronutrients) and an energy source are met. Much of the progress towards understanding life in subglacial aquatic environments over the past two decades has been through the study of arctic and Alpine glaciers (e.g. Sharp et al., 1999; Skidmore et al., 2000; Gaidos et al., 2009). These systems are reviewed elsewhere (e.g. Skidmore, 2011; Achberger et al., 2017) and will not be the focus here. Sampling of antarctic subglacial aquatic environments has been more limited, although water is widespread beneath the antarctic ice sheets as basal melt water (Llubes et al., 2006; Pattyn, 2010), groundwater (Uemura et al., 2011; Christoffersen & Bougamont, 2014), and in at least one instance, a pocket of marine brine trapped under an antarctic glacier (Blood Falls; Mikucki et al., 2004). However, exploration of antarctic glaciers and ice sheets, in conjunction with remote sensing, has revealed a diversity of subglacial habitats, including debris-rich basal ice and water-saturated subglacial sediments, brines and lakes.

Basal ice, the deepest layer of glacier ice, is in direct contact with the glacier bed. Subglacial sedimentary debris, which is physically deformed by the grinding movement of the glacier above, is entrained in the ice. Impurities, including debris, dissolved solutes, microbial cells and dissolved gases, are concentrated into the liquid water veins that delineate ice crystal boundaries. This concentrated environment contains carbon, nutrient and energy sources, and hosts viable microorganisms that are thought to be active in situ (Doyle et al., 2013).

Ice streams flow over regions of water-saturated unconsolidated sediments (e.g. Fricker et al., 2007), and these wetland-like environments may be the largest subglacial habitat type in Antarctica ($10^4 - 10^5$ km^3 of water; Priscu et al., 2008). Glacial till associated water is predicted to have long residence times (1000–10 000 years; Christoffersen & Bougamont, 2014), which results in high solute concentrations (Skidmore et al., 2010; Michaud et al., 2016) that support, and may be partially maintained by, microbial activity. Few studies have focused on microbial life in subglacial sediments, but viable microorganisms were recovered from beneath the Kamb Ice Stream (Lanoil et al., 2009), and from sediments of SLW (Christner et al., 2014) in West Antarctica. Estimated *in situ* cell concentrations from Kamb and those determined from SLW sediment samples were both on the order of 10^5 cells g sediment^{-1}.

Subglacial lakes are widespread beneath the antarctic ice sheets, with recent inventories tallying >400 (Wright & Siegert, 2012; Siegert et al., 2016). Antarctic subglacial lakes tend to occur in subglacial basins in the interior of the ice sheet, on the flanks of subglacial mountains, and in ice stream regions (Dowdeswell & Siegert, 1999). The lakes range from large, deep lakes to small, shallow lakes. While the lakes located in the interior tend to be hydraulically stable over decadal to millennial time scales, those found beneath fast flowing ice streams are often characterised by their instability in terms of lake volume over time (Gray et al., 2005; Wingham et al., 2006; Fricker et al., 2007, 2015). While there is evidence for microbial life in both active (Christner et al., 2014) and inactive (Jouzel et al., 1999; Karl et al., 1999; Petit et al., 1999; Priscu et al., 1999) lakes, the relative isolation and stability of the latter likely present different challenges than the hydraulic connectivity and fluctuating environment of the former.

Subglacial aquatic environments on the antarctic continent are not only connected to one another, but also to the marine environment. Approximately 75% of the antarctic coast is surrounded by ice shelves, which cover >1.5×10^6 km^2 of the coastal ocean (Rignot et al., 2013), making iceshelf cavities the largest, but perhaps least well studied, aquatic subglacial habitat type. Iceshelf cavities can be considered intermediate environments between continental subglacial aquatic environments and the ice-free or sea-ice-covered Southern Ocean. Water enters iceshelf cavities from adjacent open ocean (e.g. Robinson et al., 2010) and from continental outflows (e.g. Carter & Fricker, 2012) likely leading to a productivity gradient where oligotrophy increases with distance from open water (Lipps et al., 1979; Riddle et al., 2007; Post et al., 2014; Vick-Majors et al., 2015).

12.3 Subglacial geochemistry

Antarctic subglacial environments lack a direct hydrological connection from ice sheet surface to bed. Their relative isolation means that biological processes must be supported by materials that are introduced through melt at the ice sheet base, those liberated from the subglacial sediments or from bedrock through glacial comminution (crushing), and those that can be recycled through biological activity. The observable geochemical environment results from interactions between all of these processes.

12.3.1 Subglacial sediment properties and weathering

Glaciers enhance mineral weathering rates by crushing the bedrock and associated minerals, thereby increasing the surface area to volume ratio of mineral particles, which leads to further enhancement of weathering via water–rock interactions and/or biological activity. Water–rock interactions can dissolve minerals in the absence of biology, but weathering rates are enhanced when life is present (Montross et al., 2013). The freshly exposed mineral surfaces created by glacial comminution are key to supporting microbial metabolism. Mineral-sourced reductants are thermodynamically favourable energy sources for microorganisms in permanently dark sub-glacial environments (Mikucki et al., 2016; Vick-Majors et al., 2016). Microorganisms are capable of utilising reductants from mineral sources to drive carbon fixation and biomass synthesis. Microbial use of mineral-sourced reductants contributes to the overall solute load and the concentrations of metabolic by-products (i.e. CO_2 from respiration and H^+ from sulfide oxidation). For these reasons and despite the permanent cold temperatures, the presence of active microbial ecosystems means that glaciated alpine catchments are characterised by solute generation that is similar to non-glaciated catchments (Prestrud-Anderson et al., 1997; Tranter 2003; Figure 12.1).

The solute composition of subglacial aquatic environments more closely reflects that of non-glaciated catchments than that of glaciated alpine catchments. The solute load derived from biotic and abiotic weathering within subglacial aquatic habitats is determined, in part, by the residence time of water in the subglacial environment, which is a function of the size of the ice mass – water under larger ice masses tends to have longer residence times. Alpine glacier-covered catchments typically have short water residence times compared to polar ice sheet covered catchments, and have solute loads dominated by Ca^{2+} and K^+ (Prestrud-Anderson et al., 1997). Grinding of micas is the most likely source of K^+, while the fast dissolution kinetics of carbonate coupled to glacier-driven bedrock crushing leads to the production of an abundance of Ca^{2+}, even when glaciers override bedrock with only trace amounts of carbonates (Prestrud-Anderson et al., 1997). The indicator ions for silicate weathering, Si and Na^+, are typically not enriched in these shorter residence time glacial catchments (Figure 12.1b), as there is insufficient time to weather silicate minerals, even in the presence of microorganisms. However, in long residence time systems, such as under the antarctic ice sheets, Si and Na^+ are enriched (Figure 12.1) and are similar in quantity to subaerial rivers with equivalent discharge (Michaud et al., 2016; Hawkings et al., 2017). In Lake Whillans, this enrichment of Si and Na^+ holds true for both the crustally derived weathering component, and for the total weathering component, which includes relict solutes of marine origin (Figure 12.1a; Michaud et al., 2016).

12.3.2 Subglacial redox environment

The redox chemistry of an environment impacts the metabolic processes used by microorganisms to generate biomass and conserve energy. Oxygen concentration is a key property determining the redox state. In surface environments

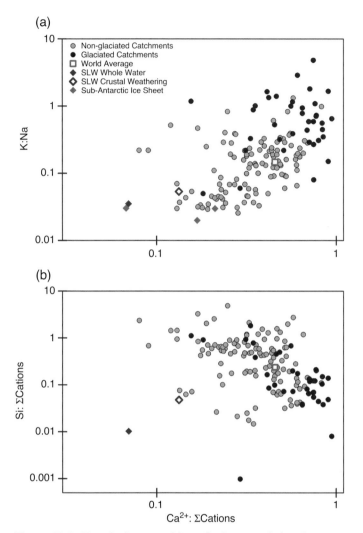

Figure 12.1 Chemical composition of sub-antarctic ice sheet water and meltwater draining glacier-covered and non-glacier-covered catchments. The two panels highlight the typical differences in glaciated versus non-glaciated catchment weathering products: (a) general enrichment of potassium and calcium in glaciated catchments; (b) the typically low concentrations of silica in glacial meltwaters relative to the typically high amounts of calcium. These typical patterns of geochemical composition in glacial meltwater are not seen when water is analysed from beneath the antarctic ice sheet (blue and red diamonds in a and b). Sub-antarctic ice sheet samples include Kamb and Bindschadler Ice Streams and a jökulhaup near Casey Station in Wilkes Land, on the East Antarctic coastline. Data provided courtesy of S. P. Anderson and figure modified from Anderson et al. (1997). (A black and white version of this figure will appear in some formats. For the colour version, please refer to the plate section.)

exposed to sunlight, oxygen is released to the atmosphere as a byproduct of photosynthesis, where it is then used to fuel respiration, often of the organic matter produced by photosynthetic primary producers. Subglacial aquatic environments beneath the antarctic ice sheet receive no solar energy, so primary producers must rely on energy from the catalysis of redox reactions using inorganic minerals (chemolithotrophy). Such metabolisms typically consume oxygen, and antarctic subglacial environments have no direct connections to the surface environment, therefore oxygen is introduced through basal melting of the ice. Oxygen and other atmospheric gases are trapped in glacier ice and, assuming no biological activity within the ice through time, should be introduced to the glacier bed in concentrations that represent the atmospheric composition at the time of ice formation. As ice melts at the base of the ice sheet, these entrained gas bubbles are released to the subglacial environment. Thus, the concentration of oxygen in subglacial environments must depend on the balance between that released by basal melt and that consumed by microbial processes.

Studies of accretion and glacier ice from above the water column of Lake Vostok predicted that the shallow depths of the Lake Vostok water column are supersaturated with atmospheric gases including oxygen (~50 times that of air-equilibrated water), with much of it present as air hydrates (Siegert et al., 2001; McKay et al., 2003). Metabolic activity is thought to drive sub-glacial aquatic environments towards anoxia (Wadham et al., 2010), and anoxic microenvironments may be common where high concentrations of organic matter combine with limited hydrological inputs of oxygen (Tranter et al., 2005). Given the presence of organic carbon and viable microbial cells in the SLV accretion ice, heterotrophic respiration in the lake is hypothesised to cause anoxia in deeper parts of the water column and in the sediments (Karl et al., 1999; Priscu et al., 1999). These predictions are consistent with data collected from SLW. Dissolved oxygen concentrations in SLW were low (~16% of air saturation; Christner et al., 2014), and the concentrations of redox-sensitive but non-bioactive chemical species in the SLW sediments showed that sediment pore waters should be anoxic below a depth of 15 cm (Michaud et al., 2016). A model of dissolved oxygen dynamics in SLW showed that the water column dissolved oxygen concentrations were tightly linked to both biological activity and under-ice hydrology, and suggested that the water column might undergo periods of anoxia (Vick-Majors et al., 2016).

Redox status and pH have important implications for elemental cycling in subglacial environments. For example, sulphur compounds and iron–sulphur minerals have been shown to play key roles in subglacial elemental cycles, and their biotic and abiotic transformations are directly linked to pH and O_2 availability. Pyrite (FeS_2) is commonly found in bedrock minerals, and likely plays a key role in subglacial chemistry (Bottrell & Tranter, 2002; Mitchell et al., 2013). In systems with circumneutral to alkaline pH, FeS_2 can be biologically oxidised to SO_4^{2-} via a thiosulphate ($S_2O_3^{2-}$) intermediate using either O_2 or NO_3^- as an electron acceptor (Harrold et al., 2015), while at acidic pH aerobic oxidation of FeS_2 proceeds abiotically. Under highly reducing conditions, the reduction of SO_4^{2-} to H_2S can be catalysed by microorganisms using organic matter as an electron donor (Jørgensen, 1982), a process that may occur in deep subglacial waters or sediments.

12.4 Diversity of life in subglacial environments

12.4.1 Phylogenetic diversity

Despite the long history of antarctic research, ecological surveys of subglacial environments remain limited. To date, only two subglacial lakes have been sampled (SLM, SLW, SLV) and only a handful of biological measurements exist from subglacial sediments, outflows and sub-iceshelf environments. All of the sites examined thus far have been dominated by bacterial species, primarily from the phyla Proteobacteria (Beta, Delta and Gamma subdivisions), Actinobacteria and Bacteriodetes, and often harboured a less abundant archaeal population (e.g. Rogers et al., 2013; Vick-Majors et al., 2015; Achberger et al., 2016). With the exception of sub-iceshelf ecosystems where marine micro- and macro-eukaryotes have been observed (Lipps et al., 1979; Riddle et al., 2007; Post et al., 2014; Vick-Majors et al., 2015), eukaryotic species are either rare or undetected in antarctic subglacial systems (e.g. Rogers et al., 2013; Christner et al., 2014).

Studies of arctic and mountain glacier systems frequently report the prevalence of iron and sulphur oxidising microorganisms in subglacial environments (e.g. Skidmore, 2011). The ecological importance of such organisms also appears to extend to antarctic subglacial systems. In Lake Whillans, species of the iron and sulphur oxidising genera of *Sideroxydans* and *Thiobacillus* were among the most abundant members of the microbial community (Achberger et al., 2016). These genera were likewise identified in the water-saturated subglacial sediments collected from beneath the Kamb Ice Stream (Lanoil et al., 2009). Additionally, in the marine sub-ice cavity along the southern grounding zone of the Ross Iceshelf (RIS), species of sulphur-oxidising *Thioprofundum* and *Thiohalophilus* were prevalent throughout the water column and sediments (Achberger, 2016). Iron and sulphur compounds may also fuel microbial life at Blood Falls, a subglacial outflow feature at Taylor Glacier in the McMurdo Dry Valleys and in SLV in East Antarctica. Within the iron-rich brine of Blood Falls, several of the dominant organisms were related to genera capable of metabolisms utilising iron and sulphur compounds such as *Thiomicrospira, Desulfocapsa* and *Geopsychrobacter* (Mikucki & Priscu, 2007; Mikucki et al., 2009). Similarly, taxa most closely related to members of the

Acidithiobacillus, Desulfuromonas and *Geopsychrobacter* were identified in accretion ice from SLV (Christner et al., 2006).

In addition to the numerous species associated with iron- or sulphur-based metabolisms, organisms capable of growth using C-1 compounds (e.g. methane and methanol) or nitrogen compounds as energy sources are also common in subglacial antarctic environments. Taxa closely related to species of *Methylophilus*, *Methylobacillus* and *Methylobacterium* were identified in the accretion ice of SLV (Christner et al., 2001; Christner et al., 2006), members of *Methylobacter* were among the most abundant organisms found in the sediments of SLW (Achberger et al., 2016; Michaud et al., 2017). Species of *Polaromonas*, which have been found in numerous cryospheric and high-latitude environments where they have been shown to degrade a wide variety of carbon compounds (Lanoil et al., 2009; Boyd et al., 2011; Hell et al., 2013; Hodson et al., 2015), were also prevalent in SLW. In addition, SLW harboured several species of ammonia (e.g. *Nitrosoarchaeum* and *Nitrosospira*) and nitrite (*Candidatus* Nitrotoga) oxidisers (Achberger et al., 2016), and ammonia-oxidising Archaea (e.g. *Nitrosopumilus*) have been detected in the sub-ice cavity beneath the Ross Iceshelf (Vick-Majors et al., 2015; Achberger, 2016), indicating that reduced nitrogen compounds also help to fuel subglacial ecosystems.

12.4.2 Metabolic diversity

Thick ice covers prevent the penetration of sunlight into subglacial environments. Living in permanent darkness means that microorganisms must derive energy and nutrients from the mineral and organic sources that exist under the ice. Consequently, the hydrological environment, the motion of the glacier, and the type and provenance of the sedimentary and bedrock environments all influence the metabolic diversity and the rates of metabolic activity in subglacial environments (see Section 12.3).

12.4.2.1 *Chemolithotrophy*

Evidence from the antarctic subglacial environments accessed to date suggests that chemolithotrophic metabolisms form the base of the food chain. Rates of dark ^{14}C-bicarbonate incorporation (a proxy for chemosynthetic carbon fixation) have been reported for the SLW water column (32.9 ng C l^{-1} d^{-1}; Christner et al., 2014), the SLW sediment–water interface (100 ng C l^{-1} d^{-1}; Mikucki et al., 2016) and the Blood Falls outflow (14 ng C l^{-1} d^{-1}; Mikucki & Priscu, 2007).

Taxa related to known ammonia and nitrite oxidisers were abundant in the SLW water column, and high proportions of nitrite oxidiser-associated rRNA relative to rDNA suggested that they were active (Achberger et al., 2016). The water column was rich in ammonium, and Δ^{17}O-NO$_3^-$ values of ~ 0‰ showed that the NO_3^- resulted from biological activity, implying that nitrification may be important in fuelling chemosynthetic activity (Christner et al., 2014). However, the attribution of all dark ^{14}C-bicarbonate incorporation in the SLW water column to ammonia oxidation results in an ammonia deficit, implying tight internal cycling of N (Vick-Majors et al., 2016). Nitrification combined with a tight internal N-cycle was also found to be important in a Canadian glacier (Boyd et al., 2011). Detailed analyses of N-cycling in other subglacial environments are needed to determine whether the importance of nitrification as an energy source to fuel chemosynthesis is common across subglacial environments.

Reduced iron and sulphur compounds are also thought to be important chemolithotrophic energy sources in subglacial environments (Sharp et al., 1999; Skidmore et al., 2010). Data from SLV (Christner et al., 2006) showed that the oxidation of iron and sulphide could be foundational food web processes. Functional gene (adenosine-5'-phosphosulphate reductase; APS) and 16S marker gene data from SLW suggest that sulphur-oxidising organisms related to *Sideroxydans* and *Thiobacillus* species comprise a major component of the microbial community present in the surficial sediment layer of the lake (0–2 cm), with abundances decreasing with depth (Purcell et al., 2014; Achberger et al., 2016). That these genes occurred with higher frequency in the upper, oxygenated portions of the sediments (Michaud et al., 2016) implies that the organisms were likely oxidising sulphur, even though APS can also catalyse sulphur reduction. Chemical affinity calculations, which have been shown to correlate with microbial communities and metabolisms (LaRowe & Amend, 2014), showed that pyrite oxidation was the most energetically favourable metabolism in both the surface sediment layer and the SLW water column (Vick-Majors et al., 2016; Michaud et al., 2017), supporting the genetic evidence for the importance of iron–sulphur minerals supporting SLW metabolism.

In Blood Falls, the iron-rich outflow of a marine brine pocket trapped under the Taylor Glacier in East Antarctica (Mikucki et al., 2004), a catalytic sulphur cycle is thought to support microbial life (Mikucki et al., 2009). In this system, SO_4^{2-} is the most abundant electron acceptor; however, isotopic signatures similar to those of marine brines of equivalent ages do not support terminal reduction of SO_4^{2-}. Instead, APS genes were associated with dissimilatory and sulphur-disproportionating organisms. Microorganisms in the Blood Falls brine are thought to oxidise organic matter using Fe(III) as terminal electron acceptor, using sulphur cycling to catalyse the process. This model provides an explanation for the lack of measurable H_2S and persistent concentrations of SO_4^{2-} that characterise the Blood Falls brine.

12.4.2.2 Subglacial carbon metabolisms

All microbial life requires an external carbon source to build biomass or to fuel respiration. A range in concentrations of dissolved organic carbon, a major source of carbon and energy to support heterotrophic microorganisms, has been detected in antarctic subglacial aquatic environments. SLV (17–250 µM estimated from accretion ice; Priscu et al., 1999; Christner et al., 2006) and SLW (220 µM; Christner et al., 2014) contained low concentrations relative to more productive surface environments and to the Blood Falls outflow, which contained 3.5-fold higher concentrations (770 µM) closest to the glacier. Concentrations then decreased to below detection at the most distal point sampled (Mikucki & Priscu, 2007). The ability of subglacial microorganisms to metabolise organic carbon was established using accretion ice from SLV, where 75–99% of organisms were shown to be potentially active and to metabolise glucose and acetate after thawing of ice samples (Karl et al., 1999; Priscu et al., 1999; Christner et al., 2006). Bakermans and Skidmore (2011) and Doyle et al. (2013) showed that heterotrophic microorganisms isolated from antarctic glaciers were capable of respiring while encased in ice at temperatures from ⁻4°C to ⁻33°C. Samples from Blood Falls and SLW also contained active heterotrophs, as shown by incubations with tritiated leucine and thymidine (Mikucki et al., 2004; Christner et al., 2014). The ultimate sources of the organic carbon that supports heterotrophic metabolism must be carbon fixed via chemolithoautotrophy, or relict sedimentary material (Christner et al., 2014; Vick-Majors et al., 2020).

Work in SLW showed that rates of chemoautotrophic carbon fixation surpassed the heterotrophic demand for carbon by a factor of 1.5 (Vick-Majors et al., 2016), while in Blood Falls, heterotrophic rates exceeded autotrophic rates by a factor of ~2 (Mikucki et al., 2004; Mikucki & Priscu, 2007). The depressed rates of heterotrophic activity relative to chemoautotrophic activity in SLW were attributed to calculated low available energy from organic substrates relative to the energy available to support key chemolithotrophic metabolisms (up to 20-fold lower; Vick-Majors et al., 2016). In Blood Falls, the catalytic sulphur cycle predicted to support microbial activity should provide sufficient energy to support microbial growth (Mikucki et al., 2009), but may favour the use of organic substrates over inorganic carbon fixation. A more complete understanding of the energy dynamics controlling subglacial carbon cycling will require further sampling efforts.

The antarctic ice sheets overlay sedimentary material that is thought to be relatively rich in relict organic matter that may support methanogenesis (acetoclastic methanogenesis; Wadham et al., 2012). Work on SLW showed that a methane reservoir exists under the West Antarctic Ice Sheet, with concentrations reaching 300 µM in the deepest sediment layer sampled

(36 cm), and that the methane in that region was most likely produced via the reduction of CO_2 coupled to H_2 oxidation (hydrogenotrophic methanogenesis; Michaud et al., 2017). Based on a stable isotope fractionation model, bacterial oxidation was found to remove >99% of the methane stored in SLW sediment pore waters, likely mitigating the potential release of methane to the atmosphere where subglacial water contacts the ocean at the ice sheet margins (Michaud et al., 2017). This is in contrast to the Greenland Ice Sheet, where methane is transported to the ice sheet margin and released to the atmosphere or oxidised in the proglacial environment (Dieser et al., 2014). Methanotrophy was found to have a high energy yield for microorganisms in SLW and to produce a quantity of biomass in the surficial sediment pore water (26.2 ng C^{-1} d^{-1}) similar to the amount attributed to chemoautotrophic carbon fixation in the water column (32.9 ng C L^{-1} d^{-1}) (Christner et al., 2014; Michaud et al., 2017), implying that methanotrophy is a key metabolic pathway for the production of subglacial microbial biomass. Further sampling will reveal whether this is common to other antarctic subglacial lakes, and whether the conditions to support both hydrogenotrophic and acetoclastic methanogenesis exist under different regions of the ice sheet.

12.5 Conclusions

Antarctic subglacial aquatic habitats are characterised by conditions that seem to typify extreme environments, including permanent darkness, subzero temperatures, energy limitation and isolation from the atmosphere and inputs of freshly derived photosynthetic carbon. While the large ice masses that overlay subglacial habitats are unique to the polar regions, some of the resulting conditions are common to other extreme habitats on Earth. The deep biospheres of marine sediments and continental crust are also permanently dark, disconnected from photosynthetic production, and energy limited (Jørgensen, 2011). Microorganisms in these isolated habitats harness thermodynamic disequilibria through oxidation–reduction reactions, using what is available in the immediate environment in order to maintain metabolism, similar to those of subglacial aquatic environments. At the same time, there are also important differences. In the deep marine biosphere, microorganisms rely on organoclastic sulphate reduction, methanogenesis, or sulphate-mediated anaerobic methane oxidation (Beulig et al., 2018), while in the deep continental biosphere, geogenic H_2 resulting from radioactive decay is an energetically favourable oxidant to support deep microbial life (Lin et al., 2006; Lollar et al., 2014). Microorganisms in such habitats can couple H_2 oxidation with sulphate or iron reduction to drive metabolic processes. This differs from what is known so far about antarctic subglacial environments and the strategies used by life in those environments. Subglacial waters and sediments may be oxygenated, and organic matter, methane, ammonium and iron–sulphur minerals appear to be key energetic substrates. If oxygen is indeed common in antarctic subglacial aquatic environments, its presence may make them unique among deeply buried extreme habitats.

References

Achberger, A.M. (2016). Structure and Functional Potential of Microbial Communities in Subglacial Lake Whillans and at the Ross Ice Shelf Grounding Zone, West Antarctica. PhD thesis, Louisiana State University.

Achberger, A.M., Christner, B.C., Michaud, A.B., et al. (2016). Microbial community structure of subglacial Lake Whillans, West Antarctica. *Frontiers in Microbiology*, **7**, 256–213. doi:10.3389/fmicb.2016.01457

Achberger, A.M., Michaud, A.B., Vick-Majors, T. J., et al. (2017). Microbiology of subglacial environments. In: Margesin, R. (ed.) *Psychrophiles: From Biodiversity to Biotechnology*. Springer International Publishing, Cham, Switzerland, pp. 83–110.

Bakermans, C., Skidmore, M. (2011). Microbial respiration in ice at subzero temperatures (−4°C to −33°C). *Environmental Microbiology Reports*, **3**, 774–782. doi:10.1111/j.1758-2229.2011.00298.x

Beulig, F., Røy, H., Glombitza, C., Jørgensen, B.B. (2018). Control on rate and pathway of anaerobic organic carbon degradation in the seabed. *Proceedings of the National Academy of Sciences of the USA*, **115**, 367–372. doi:10.1073/pnas.1715789115

Bottrell, S.H., Tranter, M. (2002). Sulphide oxidation under partially anoxic conditions at the bed of the Haut Glacier d'Arolla, Switzerland. *Hydrological Processes*, **16**, 2363–2368. doi:10.1002/hyp.1012

Boyd, E.S., Lange, R.K., Mitchell, A.C., et al. (2011). Diversity, abundance, and potential activity of nitrifying and nitrate-reducing microbial assemblages in a subglacial ecosystem. *Applied and Environmental Microbiology*, **77**, 4778–4787. doi:10.1128/AEM.00376-11

Bulat, S.A. (2016). Microbiology of the subglacial Lake Vostok: first results of borehole-frozen lake water analysis and prospects for searching for lake inhabitants. *Philosophical Transactions of the Royal Society A*, **374**, 20140292. doi:10.1098/rsta.2014.0292

Carter, S.P., Fricker, H.A. (2012). The supply of subglacial meltwater to the grounding line of the Siple Coast, West Antarctica. *Annals of Glaciology*, **53**, 267–280. doi:10.3189/2012aog60a119

Christner, B.C., Mosley-Thompson, E., Thompson, L.G., Reeve, J.N. (2001). Isolation of bacteria and 16S rDNAs from Lake Vostok accretion ice. *Environmental Microbiology*, **3**, 570–577. doi:10.1046/j.1462-2920.2001.00226.x

Christner, B.C., Royston-Bishop, G., Foreman, C. M., et al. (2006). Limnological conditions in Subglacial Lake Vostok, Antarctica. *Limnology and Oceanography*, **51**, 2485–2501.

Christner, B.C., Priscu, J.C., Achberger, A.M., et al. (2014). A microbial ecosystem beneath the West Antarctic ice sheet. *Nature*, **512**, 310–313. doi:10.1038/nature13667

Christoffersen, P., Bougamont, M. (2014). Significant groundwater contribution to Antarctic ice streams hydrologic budget. *Geophysical Research Letters*, **41**, 2003–2010. doi:10.1002/2014gl059250

Dieser, M., Hagedorn, B., Christner, B.C., et al. (2014). Molecular and biogeochemical evidence for methane cycling beneath the western margin of the Greenland Ice Sheet. *ISME Journal*, **8**, 2305–2316. doi:10.1038/ismej.2014.59

Dowdeswell, J.A., Siegert, M.J. (1999). The dimensions and topographic setting of Antarctic subglacial lakes and implications for large-scale water storage beneath continental ice sheets. *Geological Society of America Bulletin*, **111**, 254–263.

Doyle, S., Montross, S., Skidmore, M., Christner, B. (2013). Characterizing microbial diversity and the potential for metabolic function at −15°C in the basal ice of Taylor Glacier, Antarctica. *Biology*, **2**, 1034–1053. doi:10.3390/biology2031034

Fisher, A.T., Mankoff, K.D., Tulaczyk, S.M., Tyler,S.W., Foley, N.; WISSARD Science Team (2015). High geothermal heat flux measured below the West Antarctic Ice Sheet. *Science Advances*, **1**, e1500093. doi:10.1126/sciadv.1500093

Fricker, H.A., Scambos, T., Bindschadler, R., Padman, L. (2007). An active subglacial water system in West Antarctica mapped from space. *Science*, **315**, 1544–1548. doi:10.1126/science.1136897

Fricker, H.A., Siegfried, M.R., Carter, S.P., Scambos, T.A. (2015). A decade of progress in observing and modelling Antarctic subglacial water systems. *Philosophical Transactions of the Royal Society A*, **374**, 20140294. doi:10.1098/rsta.2014.0294

Gaidos, E., Marteinsson, V., Thorsteinsson, T., et al. (2009). An oligarchic microbial assemblage in the anoxic bottom waters of a volcanic subglacial lake. *ISME Journal*, **3**, 486–497. doi:10.1038/ismej.2008.124

Gray, L., Joughin, I., Tulaczyk, S., Spikes, V.B. (2005). Evidence for subglacial water transport in the West Antarctic Ice Sheet through three-dimensional satellite radar interferometry. *Geophysical Research Letters*, **32**, L03501.

Harrold, Z.R., Skidmore, M.L., Hamilton, T.L., et al. (2015). Aerobic and anaerobic thiosulfate oxidation by a cold-adapted, subglacial chemoautotroph. *Applied and Environmental Microbiology*, **82**, 1486–1495. doi:10.1128/AEM.03398-15

Hawkings, J.R., Wadham, J.L., Benning, L.G., et al. (2017). Ice sheets as a missing source of silica to the polar oceans. *Nature Communications*, **8**, 14198. doi:10.1038/ncomms14198

Hell, K., Insam, H., Edwards, A., et al. (2013). The dynamic bacterial communities of a melting High Arctic glacier snowpack. *ISME Journal*, **7**, 1814–1826. doi:10.1038/ismej.2013.51

Hodson, A., Brock, B., Pearce, D., Laybourn-Parry, J., Tranter, M. (2015). Cryospheric ecosystems: a synthesis of snowpack and glacial research. *Environmental Research Letters*, **10**, 110201–9. doi:10.1088/1748-9326/10/11/110201

Jouzel, J., Petit, J.R., Souchez, R., et al. (1999). More than 200 meters of lake ice above subglacial Lake Vostok, Antarctica. *Science*, **286**, 2138–2141.

Jørgensen, B.B. (1982). Mineralization of organic matter in the sea bed – the role of sulphate reduction. *Nature*, **296**, 643–645. doi:10.1038/296643a0

Jørgensen, B.B. (2011). Deep subseafloor microbial cells on physiological standby. *Proceedings of the National Academy of Sciences of the USA*, **108**, 18193–18194. doi:10.1073/pnas.1115421108

Kapitsa, A.P., Ridley, J.K., de Q Robin, G., Siegert, M.J., Zotikov, I.A. (1996). A large deep freshwater lake beneath the ice of central East Antarctica. *Nature*, **381**, 684–686. doi:10.1038/381684a0

Karl, D.M., Bird, D.F., Bjorkman, K., et al. (1999). Microorganisms in the accreted ice of Lake Vostok, Antarctica. *Science*, **286**, 2144–2147. doi:10.1126/science.286.5447.2144

Lanoil, B., Skidmore, M., Priscu, J.C., et al. (2009). Bacteria beneath the West Antarctic Ice Sheet. *Environmental Microbiology*, **11**, 609–615. doi:10.1111/j.1462-2920.2008.01831.x

LaRowe, D., Amend, J. (2014). Energetic constraints on life in marine deep sediments. In: Kallmeyer, J., Wagner, D. (eds) *Microbial Life of the Deep Biosphere*. De Gruyter, Berlin, Boston, pp. 279–302.

Lin, L.-H., Wang, P.-L., Rumble, D., et al. (2006). Long-term sustainability of a high-energy, low-diversity crustal biome. *Science*, **314**, 479–482. doi:10.1126/science.1127376

Lipps, J.H., Ronan, T.E., DeLaca, T.E. (1979). Life below the Ross Ice Shelf, Antarctica. *Science*, **203**, 447–449. doi:10.1126/science.203.4379.447

Llubes, M., Lanseau, C., Rémy, F. (2006). Relations between basal condition, subglacial hydrological networks and geothermal flux in Antarctica. *Earth and Planetary Science Letters*, **241**, 655–662. doi:10.1016/j.epsl.2005.10.040

Lollar, B.S., Onstott, T.C., Lacrampe-Couloume, G., Ballentine, C.J. (2014). The contribution of the Precambrian continental lithosphere to global H2 production. *Nature*, **516**, 379–382. doi:10.1038/nature14017

McKay, C.P., Hand, K.P., Doran, P.T., Andersen, D.T., Priscu, J.C. (2003). Clathrate formation and the fate of noble and biologically useful gases in Lake Vostok, Antarctica. *Geophysical Research Letters*, **30**, 124. doi:10.1029/2003GL017490

Michaud, A.B., Skidmore, M.L., Mitchell, A.C., et al. (2016). Solute sources and geochemical processes in Subglacial Lake Whillans. *West Antarctica*, **44**, 347–350. doi:10.1130/G37639.1

Michaud, A.B., Dore, J.E., Achberger, A.M., et al. (2017). Microbial oxidation as a methane sink beneath the West Antarctic Ice Sheet. *Nature Publishing Group*, **10**, 1–8. doi:10.1038/ngeo2992

Mikucki, J.A., Priscu, J.C. (2007). Bacterial diversity associated with Blood Falls, a subglacial outflow from the Taylor Glacier, Antarctica. *Applied and Environmental Microbiology*, **73**, 4029–4039. doi:10.1128/AEM.01396-06

Mikucki, J.A., Foreman, C.M., Sattler, B., Lyons, W.B., Priscu, J.C. (2004). Geomicrobiology of Blood Falls: an iron-rich saline discharge at the terminus of the Taylor Glacier, Antarctica. *Aquatic Geochemistry*, **10**, 199–220. doi:10.1007/s10498-004-2259-x

Mikucki, J.A., Schrag, D.P., Mikucki, J.A., et al. (2009). A contemporary microbially maintained subglacial ferrous 'ocean'. *Science*, **324**, 397–400. doi:10.1126/science.1167350

Mikucki, J.A., Lee, P.A., Ghosh, D., et al. (2016). Subglacial Lake Whillans microbial biogeochemistry: a synthesis of current knowledge. *Philosophical Transactions of the Royal Society A*, **374**, pii: 20140290. doi:10.1098/rsta.2014.0290

Mitchell, A.C., Lafrenière, M.J., Skidmore, M.L., Boyd, E.S. (2013). Influence of bedrock mineral composition on microbial diversity in a subglacial environment. *Geology*, **41**, 855–858. doi:10.1130/G34194.1

Montross, S.N., Skidmore, M., Tranter, M., Kivimaki, A.L., Parkes, R.J. (2013). A microbial driver of chemical weathering in glaciated systems. *Geology*, **41**, 215–218. doi:10.1130/G33572.1

National Research Council (2007). *Exploration of Antarctic Subglacial Aquatic Environments: Environmental and Scientific Stewardship*. The National Academies Press, Washington, DC.

Oswald, G., Robin, G.Q. (1973). Lakes beneath the Antarctic ice sheet. *Nature*, **245**, 251–254. doi:10.1038/245251a0

Pattyn, F. (2010). Antarctic subglacial conditions inferred from a hybrid ice sheet/ice stream model. *Earth and Planetary Science Letters*, **295**, 451–461. doi:10.1016/j.epsl.2010.04.025

Petit, J.R., Jouzel,J., Raynaud, D., et al. (1999). Climate and atmospheric history of the past 420,000 years from the Vostok ice core, Antarctica. *Nature*, **399**, 429–436.

Post, A.L., Galton-Fenzi, B.K., Riddle, M.J., et al. (2014). Modern sedimentation, circulation and life beneath the Amery Ice Shelf, East Antarctica. *Continental Shelf Research*, **74**, 77–87. doi:10.1016/j.csr.2013.10.010

Prestrud-Anderson, S., Drever, J.I., Humphrey, N.F. (1997). Chemical weathering in glacial environments. *Geology*, **25**, 399.

Priscu, J.C., Christner, B.C. (2004). Earth's icy biosphere. *Microbial Diversity and Prospecting*, 130–145.

Priscu, J.C., Adams, E.E., Lyons, W.B., et al. (1999). Geomicrobiology of subglacial ice above Lake Vostok, Antarctica. *Science*, **286**, 2141–2144.

Priscu, J.C., Tulaczyk, S., Studinger, M., et al. (2008). Antarctic subglacial water: origin, evolution, and ecology. In: W.F. Vincent, J. Laybourn-Parry (eds) *Polar Lakes and Rivers*. Oxford University Press Inc., New York, pp. 119–135.

Priscu, J.C., Achberger, A.M., Cahoon, J.E., et al. (2013). A microbiologically clean strategy for access to the Whillans Ice Stream subglacial environment. *Antarctic Science*, **25**, 637–647. doi:10.1017/S0954102013000035

Purcell, A.M., Mikucki, J.A., Achberger, A.M., et al. (2014). Microbial sulfur transformations in sediments from Subglacial Lake Whillans. *Frontiers in Microbiology*, **5**, 594. doi:10.3389/fmicb.2014.00594

Riddle, M.J., Craven, M., Goldsworthy, P.M., Carsey, F. (2007). A diverse benthic assemblage 100 km from open water under the Amery Ice Shelf, Antarctica. *Paleoceanography*, **22**. doi:10.1029/2006pa001327

Rignot, E., Jacobs, S., Mouginot, J., Scheuchl, B. (2013). Ice-shelf melting around Antarctica. *Science*, **341**, 266–270. doi:10.1126/science.1235798

Robinson, N.J., Williams, M.J.M., Barrett, P.J., et al. (2010). Observations of flow and ice-ocean interaction beneath the McMurdo Ice Shelf, Antarctica. *Journal of Geophysical Research*, **115**, C03025. doi:10.1029/2008JC005255

Rogers, S., Shtarkman, Y., Koçer, Z., et al. (2013). Ecology of subglacial lake Vostok (Antarctica), based on metagenomic/metatranscriptomic analyses of accretion ice. *Biology*, **2**, 629–650. doi:10.3390/biology2020629

Sharp, M., Parkes, J., Cragg, B., Fairchild, I.J., Lamb, H. (1999). Widespread bacterial populations at glacier beds and their relationship to rock weathering and carbon cycling. *Geology*, **27**, 107. doi:10.1130/0091-7613(1999)0272.3.co;2

Siegert, M.J., Ellis-Evans, J.C., Tranter, M., et al. (2001). Physical, chemical and biological processes in Lake Vostok and other Antarctic subglacial lakes. *Nature*, **414**, 603–609. doi:10.1038/414603a

Siegert, M.J., Carter, S., Tabacco, I., Popov, S., Blankenship, D.D. (2005). A revised inventory of Antarctic subglacial lakes. *Antarctic Science*, **17**, 453. doi:10.1017/S0954102005002889

Siegert, M.J., Clarke, R.J., Mowlem, M., Ross, N. (2012). Clean access, measurement, and sampling of Ellsworth Subglacial Lake: a method for exploring deep Antarctic subglacial lake environments. *Review of Geophysics*, **50**, RG1003. doi:10.1029/2011rg000361

Siegert, M.J., Makinson, K., Blake, D., et al. (2014). An assessment of deep hot-water drilling as a means to undertake direct measurement and sampling of Antarctic subglacial lakes: experience and lessons learned from the Lake Ellsworth field season 2012/13. *Annals of Glaciology*, **55**, 59–73. doi:10.3189/2014AoG65A008

Siegert, M.J., Ross, N., Le Brocq, A.M. (2016). Recent advances in understanding Antarctic subglacial lakes and hydrology. *Philosophical Transactions of The Royal Society A Mathematical Physical and Engineering Sciences*, **374**, 20140306. doi:10.1098/rsta.2014.0306

Siegfried, M.R., Fricker, H.A., Roberts, M., Scambos, T.A., Tulaczyk, S. (2014). A decade of West Antarctic subglacial lake interactions from combined ICESat and CryoSat-2 altimetry. *Geophysical Research Letters*, **41**, 891–898. doi:10.1002/2013gl058616

Skidmore, M. (2011). Microbial communities in Antarctic subglacial aquatic environments.

In: *Antarctic Subglacial Aquatic Environments.* American Geophysical Union. Oxford University Press Inc., New York, pp. 61–81.

Skidmore, M.L., Foght, J.M., Sharp, M.J. (2000). Microbial life beneath a high Arctic glacier. *Applied and Environmental Microbiology*, **66**, 3214–3220. doi:10.1128/AEM.66.8.3214-3220.2000.Updated

Skidmore, M., Tranter, M., Tulaczyk, S., Lanoil, B. (2010). Hydrochemistry of ice stream beds – evaporitic or microbial effects? *Hydrological Processes*, **24**, 517–523. doi:10.1002/hyp.7580

Tranter, M. (2003). Geochemical weathering in glacial and proglacial environments. In: V. P. Singh, P. Singh, U.K. Haritashay (eds) *Treatise on Geochemistry*. Elsevier, the Netherlands, pp. 189–205.

Tranter, M., Skidmore, M., Wadham, J. (2005). Hydrological controls on microbial communities in subglacial environments. *Hydrological Processes*, **19**, 995–998. doi:10.1002/hyp.5854

Tulaczyk, S., Tulaczyk, S., Mikucki, J.A., et al. (2014). WISSARD at Subglacial Lake Whillans, West Antarctica: scientific operations and initial observations. *Annals of Glaciology*, **55**, 51–58. doi:10.3189/2014aog65a009

Uemura, T., Taniguchi, M., Shibuya, K. (2011). Submarine groundwater discharge in Lützow-Holm Bay, Antarctica. *Geophysical Research Letters*, **38**, L08402. doi:10.1029/2010GL046394

Vick-Majors, T.J., Achberger, A., Santibáñez, P. A., et al. (2015). Biogeochemistry and microbial diversity in the marine cavity beneath the McMurdo Ice Shelf, Antarctica. *Limnology and Oceanography*, **61**, 572–586. doi:10.1002/lno.10234

Vick-Majors, T.J., Mitchell, A.C., Achberger, A.M., et al. (2016). Physiological ecology of microorganisms in subglacial Lake Whillans. *Frontiers in Microbiology*, **7**, 1–16. doi:10.3389/fmicb.2016.01705

Wadham, J.L., Tranter, M., Skidmore, M., et al. (2010). Biogeochemical weathering under ice: Size matters. *Global Biogeochemical Cycles*, **24**, GB3025. doi:10.1029/2009gb003688

Wadham, J.L., Arndt, S., Tulaczyk, S., et al. (2012). Potential methane reservoirs beneath Antarctica. *Nature*, **488**, 633–637. doi:10.1038/nature11374

Wingham, D.J., Siegert, M.J., Shepherd, A., Muir, A.S. (2006). Rapid discharge connects Antarctic subglacial lakes. *Nature*, **440**, 1033–1036. doi:10.1038/nature04660

Wright, A., Siegert, M. (2012). A fourth inventory of Antarctic subglacial lakes. *Antarctic Science*, **24**, 659–664. doi:10.1017/S095410201200048X

Zotikov, I.A. (2006). *The Antarctic Subglacial Lake Vostok.* Springer-Verlag, Berlin/Heidelberg/New York.

PART IV   Life and habitability

HOWELL G.M. EDWARDS

Introduction

The definition of 'habitability' is actually simple in that it means 'capable of being inhabited by living organisms': it does not mean that life already exists there but that once it could have been colonised by an established living entity and that the ambient conditions are such that life could have been sustained there. For the astronomer, a habitable planet needs to be located in the 'Goldilocks zone', which basically requires that in a planetary orbit of its star a temperature range is applied such that liquid water can exist. The astrobiologist has some further requirements and refinements which act as a codicil to this prime factor, namely that once initiated, life can be sustained through the operation of survival strategies which facilitate the colonisation of the planetary surface or subsurface and which protect the organism from hostile external parameters such as intense stellar radiation, desiccation, extreme temperature ramps, chemical toxicity of the planetary regolith, and atmosphere and barometric pressure fluctuation. The planetary geologist, in addition, can impose further restraints based on the availability of a suitable geological niche adaptable for colonisation and a planetary geology that is not critically affected by volcanism or tectonic activity, which can destabilise the living colonies. Life and habitability, therefore, is an important concept that requires a multidisciplinary approach both for its consideration and for the appreciation of the factors which could dictate the emergence and eventual sustenance of life: in this respect the Darwinian idea of the essential criterion of the adaptability of life being the driving force behind life's selection and the survival of the fittest species is paramount – with the proviso that the organism adaptability requires a necessary temporal factor for its successful operation: this means that extreme planetary geological processes or events which can prove to be catastrophic to the survival of the living organism, such as meteorite impact or continuous volcanic activity, are generally deemed to be deleterious to life forms as the changes occur too rapidly for adaptation to be incurred.

It is for this reason that planetary exploration in search for life scenarios has a mantra wherein the question is posed: 'Is life extant or extinct'? In our own Solar System, Mars is the only planetary body which satisfies the 'Goldilocks principle', and then perhaps only marginally so, and on Mars we have a panoply of riders which could be envisaged as rendering life untenable, such as an ambient temperature range between −150 and +25°C wherein liquid water can exist only in prescribed locations subsurface and locked within crystalline rock matrices, an extremely high insolation at the Martian surface of low-wavelength high-energy ultraviolet radiation, an extremely toxic and life-destroying oxidative surface chemistry comprising free radicals such as the hydroxyl OH* radical, peroxides and perchlorates, an oxygen-deficient atmosphere and limited nutrient availability. Despite these reservations, Mars still affords undoubtedly the best chance of our finding evidence for extraterrestrial planetary life in our Solar System as the nearest contender, Venus, has even more extremes such as a strongly acidic atmosphere and a significantly larger temperature range with the upper limit being near +350°C. Astrobiologists, therefore, in addition to probing the Martian regolith surface and subsurface, are looking at other possibilities in our Solar System such as the satellite moons, Europa, Io, Titan and Ganymede, with a consequent broadening of the basal requirements to account for icecap cover, hydrocarbon pools on their rocky surfaces and consistently low temperatures.

Clearly, the only life experiment that we have any experience of is our own and, therefore, we can only base our analytical predictions and experiments on terrestrial analogues of what might be expected extraterrestrially when our 'search for life' instrumentation missions reach their intended targets. The main fulcrum of terrestrial analogue efforts centres on the CHNOPS (carbon, hydrogen, nitrogen, oxygen, phosphorus and sulfur) principle, which basically states that all terrestrial life based on our own carbon chemistry requires the presence of liquid water plus chemical compounds containing carbon, hydrogen, nitrogen, oxygen, phosphorus and sulfur. We just cannot envisage any other possibilities realistically, although some authors and philosophers have considered a silicon-based chemistry with all its implications. Realistically, we are restricted to our carbon-based considerations for extraterrestrial life and the search for life signatures, whether these be extant or extinct, will be based on this concept.

This Part contains two articles which address the broad concept of life and habitability and, in particular, the scenarios which may still harbour life in some of the most extreme environments on planetary surfaces and subsurfaces. Questions that will be addressed in these articles include, what type of extraterrestrial organism might be expected and what signatures would be indicative for its detection and for its unambiguous existence – in this, the role of spectral and geomorphological biomarkers will be a key factor in any analytical determination involved in a life detection mission using remote instrumentation.

CHAPTER THIRTEEN

Analytical astrobiology: the search for life signatures and the remote detection of biomarkers through their Raman spectral interrogation

HOWELL G.M. EDWARDS

University of Bradford

a n d

JAN JEHLIČKA

Charles University

13.1 Introduction

The application of Raman spectroscopic techniques to the characterisation of the protective biochemicals used in the survival strategies of extremophilic organisms in terrestrially stressed environments (Wynn-Williams and Edwards, 2000a, 2000b), coupled with the palaeogeological recognition that early Mars and Earth had maintained similar environments under which Archaean cyanobacteria could have developed (McKay, 1997), has driven the proposal for the adoption of Raman spectroscopy as novel analytical instrumentation for planetary exploration (Edwards & Newton, 1999; Dickensheets et al., 2000; Ellery & Wynn-Williams, 2003). The announcement by the European Space Agency (ESA) that a miniaturised Raman spectrometer would form part of the Pasteur analytical life detection protocol in the ExoMars mission for the search for traces of life on Mars in the AURORA programme, and indeed selecting it for the first-pass analytical interrogation of specimens from the Martian surface and subsurface, has confirmed that Raman spectroscopy will perform a key role for the molecular analytical protocols aboard the ExoMars rover vehicle which was due for launch in May 2020. Unfortunately, a postponement to this launch date has recently been announced by ESA/Roscosmos to 2022. Also, NASA has announced that a Raman spectrometer will be part of the scientific instrumentation aboard its Mars 2020 mission, which will be launched in this year, 2020.

It is undeniable that the most important scientific discovery in a future space mission would be the furnishing of indisputable evidence for life signatures on another planet and whether these have arisen from extant or extinct sources; however, this statement itself generates two very important questions, namely, how do we define life and how would we then recognise it or its residues, which themselves may be significantly degraded due to extreme environmental conditions, in the planetary geological record using remote analytical instrumentation? We must also address the possibility that any extraterrestrial organism identified in space exploration could have possibly originated on Earth and been transported to our planetary neighbours either by our own intervention, reinforcing the need for planetary protection protocols for our spacecraft and landers, or through *panspermia* processes, which could include the transfer through space and deposition of chemical building bricks through delivery by meteorites, comets and asteroids. The precise definition of life is actually rather elusive, and most organised attempts to do so have been eventually deemed unsatisfactory (Shapiro & Schulze-Makuch, 2009; Bedau, 2010; Benner, 2010; Tirard et al., 2010). The NASA definition of life as 'A self-sustaining system capable of Darwinian evolution' incorporates a molecular genesis with replicative procedures and avoids several pitfalls of alternative definitions which have been based upon the ability of the system to reproduce (Cleland & Chyba, 2002).

13.2 Astrobiology *versus* exobiology

There has been much confusion in the literature between the terms astrobiology and exobiology, but as amplified in two elegant articles, astrobiology pre-empted exobiology chronologically by two decades (Cockell, 2001; Martinez-Frias & Hochberg, 2007). Although both are related to the study of extraterrestrial life, the term astrobiology was first used by Lafleur (1941), which was largely ignored and not followed up until the publication of Tikhov's book in 1953 (Tikhov, 1953), itself an enlargement from his use of the term 'astrobotany' in 1949 to describe his belief that vegetation existed on Mars and Venus. Tikhov was the first to suggest that spectral signatures from other planets could be used to assess their biotic potential, which today lies at the heart of remote space exploration. Strughold (1953) also used the term astrobiology in 1953, but thereafter it fell into disuse until Freeman (1983) employed it in 1983, prefacing its revival again in the 1990s with the foundation of the NASA Astrobiology Institute and the worldwide excitement and discourse generated by the discovery of the Allen Hills SNC Martian meteorite, ALH 84001, in Antarctica and the observation of possible biotic morphological signatures therein. As both Cockell (2001) and Martinez-Frias and Hochberg (2007) explain, exobiology was coined first by Lederberg (1960) in 1960 to describe life beyond the Earth; in this context, astrobiology is a much wider term extending beyond the biology and addresses the origins, evolution, distribution and future of life in the universe. This concept has been reinforced in a paper by Soffen (1999) who described astrobiology in an exobiology journal (!) as 'the study of the chemistry, physics and adaptations that influence the origin, evolution and destiny of life'.

Fundamentally, the three basic questions of astrobiology are: how did life begin and evolve, does life exist elsewhere, and what is the future of life on Earth and beyond? It is also clear from the above description that astrobiology has a very different brief from astrochemistry, astrogeology and astrobotany; it is, therefore, the function of analytical astrobiology to apply the principles of chemical, biomolecular, morphological and microbiological analysis to these three baseline questions as outlined above. In this chapter, we shall endeavour to examine the terrestrial criteria for the chemical and biomolecular analyses associated with living organisms and apply these to extraterrestrial exploration envisaged by the inclusion of analytical instrumentation on remote planetary spacecraft and landers. The key question here, of course, is what biochemical species truly define the presence of extinct or extant life, be this terrestrial or extraterrestrial and whether we can recognise such biomarkers in the extreme environments that we expect to find on our neighbouring planets and their satellites?

13.3 The detection of life signatures

It is equally appropriate to consider some of the scientific parameters that will need to be evaluated for the detection of life signatures using remote robotic analytical instrumentation, specifically in the ExoMars 2020 project; hereafter, because of the delayed launch to the ExoMars 2020 mission, we should strictly refer to it as ExoMars 2022. Primarily, the selection criteria for an analytical astrobiological mission such as ExoMars 2022 need to consider the following questions:

- What organisms could have existed and possibly survived the current and past extremes of environment on Mars?
- What type of geological protective niches are to be found which may conceal the traces of relict or extant life on Mars?
- What signatures would these organisms have left in such environments as indicators of their presence and how are we going to recognise them using remote spectroscopic instrumentation?
- What molecules if detected by remote planetary instrumentation, interrogating surface and subsurface specimens, could be considered as constituting a proof of life on Mars?
- Are there terrestrial scenarios or putative Mars analogue sites which could be used as 'models' for the evaluation of these scientific questions and

thereby provide a suitable test-bed for the refinement of spectral data algo-
rithms that can be utilised as forensic evidential confirmation of the presence of biosignatures on Mars?

13.4 The historical Mars

From the birth of our Solar System some 4.6 Gya, the terrestrial geological
record suggests that microbial autotrophic ecosystems already existed on
Earth from 3.5 to 3.8 Gya. There is now much evidence that early Earth and
early Mars were indeed very similar in their physicochemical composition;
since Mars is significantly smaller than Earth, it is therefore very likely that
planetary cooling occurred more rapidly as proposed by McKay (1997). The
planet was probably more temperate and wet and since there is geological
evidence that life had already started on Earth during this period, it seems
reasonable to conclude that life had also started on Mars. By Epoch IV (*ca.* 1.5
Gya to present), however, catastrophic changes on Mars would have compromised the survival of organisms on the Martian surface, and it is possible that
the Martian analogues of terrestrial extremophiles would have been the last
survivors of life on Mars through their environmental adaptation in Martian
geological niche sites.

13.5 Analytical astrobiology of Mars

The detection of biomolecular markers in geological substrates or the subsurface regolith of Mars is a primary goal for astrobiology (Edwards, 2004;
Edwards et al., 2005). However, the evolutionary pressure of environmental
stress on Mars, especially the high levels of low-wavelength ultraviolet radiation insolation, low temperatures, extreme desiccation and hypersalinity,
would have demanded appropriately severe protective strategies adopted by
biological organisms to promote the origin, survival and evolution of microbial life (Cockell & Knowland, 1999). The ultraviolet radiation protection
afforded to subsurface organisms by the iron (III) oxide surface regolith acting
as a low-wavelength filter has been proposed as a key factor for the maintenance of biomolecular activity at the Martian surface (Clark, 1998), but the
same ultraviolet and low-wavelength electromagnetic radiation insolation
generates hydroxyl radicals and peroxides in the surface regolith which
would certainly inhibit the survival of complex biomolecules in the surface
oxidation zone. This process is believed to result in the presence of peroxides
and perchlorates in the Martian regolith which would severely compromise
the survivability of organic molecules there. In this respect the diagenesis,
catagenesis and biodegradation of terrestrial Mars analogues to make the
fossils recognisable in our own geological record would not be transposable
directly to a Martian scenario. Hence, the complex chemical systems comprising terrestrial soils, bitumens and kerogens found in our own planetary

lithology and which provide much valuable information about their sourcing processes would not be expected to occur to the same wide-ranging extent on Mars, although niche environments favourable to their survival may still occur (Pullan et al., 2008; Edwards et al., 2010; Marshall et al., 2010; Jehlička et al., 2010a, 2010b).

However, it is believed that Mars might still preserve a chemical record of early life in rocks from the Noachian era, which overlaps the terrestrial Archaean geological history, from about 3.8 Gya. The search for extinct or extant life on Mars must therefore centre upon the identification and recognition of specially protected niche geological sites, firstly, in regions where they were generated and, secondly, of equal importance, where these biomolecular signatures would be well preserved. The fundamental analytical approach to the astrobiological interrogation of Mars must then consider the detection of key molecular biomarkers, probably within rocks and certainly in subsurface terrain, perhaps even in ancient lacustrine sediments (Doran et al., 1998; Bishop et al., 2004), which will necessitate the deployment of remote analytical sensors with preset protocols and an established database recognition strategy for minerals, biologically modified geological strata and biomolecular residues. Examples from the appropriate and apposite terrestrial analogue sites could therefore include carbonates, carbonated hydroxyfluoroapatite, gypsum, calcium oxalates, porphyrins, carotenoids, scytonemin and anthraquinones (Wynn-Williams & Edwards, 2002; Edwards, 2010).

Clearly, the identification and selection of terrestrial Mars analogue sites (Bishop et al., 2004; Pullan et al., 2008) will be a critical and fundamental step in the development of the analytical astrobiology missions for Mars in two respects: firstly, the understanding of the type of geological formations that have been colonised by extremophilic organisms in terrestrial 'limits of life' situations and habitats; and secondly, the deployment of novel analytical instrumentation which can reveal the presence of the key signatures of extinct and extant life in microniches in the geological record (Wynn-Williams, 1991; Treado & Truman, 1996; Edwards et al., 1997; Wynn-Williams, 1999; Doran et al., 1998; Wynn-Williams & Edwards, 2002; Bishop et al., 2004; Edwards, 2010). The gathering of data from such terrestrial Mars analogue sites is therefore the first step to be taken in the search for extraterrestrial life signatures; in this, the application of Raman spectroscopic techniques has already been demonstrated to be successful through the direct characterisation of the molecular and molecular ionic signatures of biomolecules and their modified structures situated in the terrestrial geological record which does not involve either the physical or the chemical separation of the organic and inorganic components. Some of these terrestrial Mars analogue sites are described below and the data obtained from them using laboratory-based Raman spectroscopic techniques and, more recently, mobile spectrometers, including prototype versions of the flight model destined for the

ExoMars 2022 mission, have advanced our understanding of extremophile behaviour significantly. The detection capability of the Raman spectroscopic instrumentation for biomarker spectral signatures has assisted in the development of a spectral database of recognisable spectral band wavenumbers which can positively identify the presence of biomolecules and associated cyanobacterial colonies in terrestrial geologies of relevance to Mars.

In a special issue of the *Philosophical Transactions of the Royal Society*, in a year that celebrated its 350th anniversary as the longest running scientific journal, several articles highlighted the role of Raman spectroscopy in the characterisation of biosignatures of extremophilic colonisation of geological substrates in a range of stressed terrestrial environments (Edwards et al., 2010; Edwards, 2010; Brier et al., 2010; Carter et al., 2010; Jehlicka et al., 2010a, 2010b; Jorge-Villar & Edwards, 2010; Marshall & Olcott Marshall, 2010; Rull et al., 2010a, 2010b; Sharma et al., 2010; Varnali & Edwards, 2010; Vitek et al., 2010). These articles address the detection of geological and biogeological spectral markers that are relevant to space missions and give a very good appreciation of the Raman spectroscopic requirements that will be essential for the construction of a relevant spectral database (Jorge-Villar & Edwards, 2005) for the ExoMars 2022 and other space missions that have a Raman spectrometer aboard their rover vehicles. Some selected examples of the data which can be provided by the Raman spectroscopic interrogation of terrestrial Mars analogue sites will be highlighted later.

13.6 Biosignatures and biomarkers

A biosignature can be defined as a chemical species or topographical pattern which is **uniquely** derived from living organisms; a biosignature is hence an object, chemical substance and/or distinctive pattern whose origin requires formation by biology alone and includes, for example, microfossil morphology. In terms of the analytical spectral identification of biosignatures relating to the search for life objectives on remote planetary missions, this definition is rather too broad and can be replaced by the following definitive statement: a spectral biosignature is a unique band from a compound which has been synthesised exclusively by biological organisms, henceforth known as a *biomarker*, and which ideally is relatively stable under niche astrogeological conditions. Ambiguous and indefinite biomarkers do exist, however, and unfortunately are frequently incorrectly cited in the literature as genuine, despite the realisation that these can be synthesised both biotically and abiotically under planetary surface and subsurface conditions or in an interstellar medium. Examples of these false biomarkers include n-alkanes, polyaromatic hydrocarbons, N-heterocycles, amino acids, kerogens, urea and carbon. A further subset of biogeomarkers, or biominerals, can be identified from terrestrial geological niches where they are produced as geologically altered material as a result of biological colonisation which has interacted with the prevailing geology. For example, a calcite matrix with oxalic acid produced from the Krebs metabolic cycle of lichen colonies results in the formation of whewellite and weddellite, both of which are hydrated calcium oxalates (mono- and dihydrate, respectively). Evidence of these biominerals in biologically inactive geological strata is deemed terrestrially to be indicative of the presence of extinct biological colonisation. Other potential biomineral biomarkers that can be cited here include idrialite and kladnoite, fused ring metal-aromatic systems and mellite, a compound formed from the degradation of aromatic ring systems in the presence of coordinating and stabilising metal ions such as aluminium (III) as found in aluminosilicate clay substrates.

True biomarkers which have been characterised spectroscopically at the current time are actually rather few in number and comprise: scytonemin (and its family of methylated and methoxylated derivatives), carotenoids, carotanes (degraded and hydrogenated carotenoids), trehalose (a polysaccharide water replacement molecule for biological cells at low temperatures), chlorophyll (the photosynthetic pigment), porphyrins, phycocyanins (radiation protective accessory pigments), DNA and RNA, hopanoids and hopanes, terpenoids and sterols (and their degraded steranes). The thermal diagenesis of complex biomolecules above 200°C results in decomposition products such as polyaromatic hydrocarbons and disordered carbon: although these strictly cannot in themselves be considered as spectroscopic biomarkers for unambiguous diagnostic purposes, nevertheless, their presence in association with other organic molecules can be considered as reliable potential indicators of once extant, or extinct, life. From this brief analysis, it is seen that of these true biomarkers the carotenoids form a highly important family of biomolecules that can afford targets for the Raman spectroscopic interrogation of possible specimens for once-active or still-active biology.

Thus, from the standpoint of analytical spectroscopy it is not sufficient to simply embrace an organic or bioinorganic molecule that is used generically or which can be synthesised by life forms, as several of these can also be synthesised abiotically. Examples of the latter include amino acids and proteins. A further complication arises in that organic molecules degrade under stressed environmental conditions, forming derivatives and eventually carbon. It has already been pointed out that the reactive Martian regolith is highly oxidative and that any water that has migrated to the surface from subsurface aquifers would immediately be rendered liable to attack by low-wavelength ultraviolet radiation to produce hydroxyl radicals which would destructively scavenge organic molecular residues. The realisation that methane, calcium carbonate, carbon and polyaromatic hydrocarbons can be synthesised by geological processes as well as through the degradation of biomolecules means that these molecular carbonaceous materials are not in themselves strictly suitable for classification as biomarkers – even though the detection of methane, carbon or polyaromatic hydrocarbons on Mars would be an exciting and novel discovery in itself, this would not result in the unambiguous conclusion that life did once exist or may even still be extant on Mars!

It is crucial, therefore, that we can identify a suite of spectroscopic molecular biomarkers, whose remote detection on a planetary surface or subsurface would positively and unambiguously indicate the presence of extinct or extant life; here, we need to also narrowly permit the definition of life as essentially cyanobacterial, which represents the earliest identifiable Archaean life forms on the emerging planetary and oceanic Earth, some 3.8 Gya. From spectroscopic and microbiological analytical studies of terrestrial cyanobacteria it has been possible to isolate several biomolecules which truly can be considered as key spectroscopic biomarkers, from which we can construct a Raman spectroscopic database that will act as a true standard of assessment for the presence of life in the biogeological record. Such a definitive list of biomarkers for analytical astrobiology is provided below:

Bioorganic molecules: scytoporphyrinsmin, carotenoids, carotanes, trehalose, phycocyanins, hopanoids
Bioinorganic molecules: whewellite, weddellite, aragonite, mellite, vaterite, chlorophyll

It is interesting to compare this list of potential biomarkers with those already mentioned in two key publications which have also proposed biomarkers: firstly, a paper by Perry et al. (2007), which set out to define biominerals and organominerals as direct and indirect indicators of life. Secondly, in the same year, Parnell et al. (2007) proposed a comprehensive list of biomarkers for study as selective targets for an earlier version of the ExoMars mission, specifically addressing the antibody requirements for the Life Marker Chip instrumentation that was to be carried on that mission. The first comment is that the range of biomarkers selected in these papers is rather large and perhaps needs to be focused on more definitive molecular species. Secondly, a detailed consideration of the organic molecules in these lists reveals that although these are undoubtedly synthesised by living systems and can therefore be truly described as biomolecules, their suitability for unambiguous interpretation as biomarkers of extant or relict life from life detection experiments in analytical astrobiology is another matter. Westall et al. (2015) have recently published an elegant paper that critically describes the type of biosignature which might be expected to be detected on Mars in search for life experiments, concentrating upon chemolithotrophic and anaerobic life development in an extended Martian environment. They recognise that the search for life on Mars will present a particular challenge analytically because of the difficulty in detecting biosignatures from organic molecules in the prevailing hostile environments there. This work extends that of Farmer and des Marais

(1999), Westall and Cavalazzi (2011) and Summons et al. (2011) on the potential for the preservation of organic biomolecules in the Martian regolith. The general conclusion is that organic molecules would not be preserved on the Martian surface due to the extremely intense low-wavelength radiation insolation and especially because of the presence of perchlorates and peroxides. However, Freissinet et al. (2015) have detected low levels in the ppb range of chlorobenzene and C2–C4 chloroalkanes in the mudstone drill holes of Gale Crater using the NASA Mars Curiosity Rover MSL science laboratory GCMS instrument, which they have attributed to the reaction of chlorine produced at the surface of Mars with organic carbon deposits – and it was concluded that the complete destruction of organics at or near the surface of Mars was not a *sine qua non*. In a similar experiment, also at Gale Crater, Webster et al. (2015) have detected the presence of methane and have discussed its possible abiotic or biotic origins. In the only Martian rocks currently available for analysis in terrestrial laboratories, namely Martian meteorites, there have been several reports of the detection of carbon and/or carbonaceous materials which are indicative of the existence in the past in all probability of subsurface organic-bearing materials on Mars (Grady et al., 2004; Steele et al., 2012; Lin et al., 2014).

In the most recent reports this year of the analyses of the Curiosity rover data, Webster et al. (2018) have detected a strong seasonal variation in the concentration of methane by tunable laser spectroscopy in the Martian atmosphere over a 5-year period ranging from 0.24 to 0.65 ppb, with an average value of 0.41 ±0.16 ppb, which they have ascribed to small localised sources of methane release from subsurface reservoirs. In a paper immediately following in the same journal Eigenbrode et al. (2018) have reported further data from the 3.5 billion year lacustrine sedimentary Murray formation at Pahrump Hills in the Gale Crater on Mars using the GCMS instrument in the Sample Analysis at Mars (SAM) suite on Curiosity: they have identified complex aromatic and aliphatic organic molecules in the pyrolysis products released between 500 and 820°C, including thiophenic acid, to which they attribute its sulphurisation as a survival protection.

What also needs to be addressed is the longevity of survival of selected terrestrial biomarkers in the geological record, as this will effectively dictate the usefulness of any selected biomarker target for remote astrobiological analysis, although of course this will necessarily be dependent upon a wide range of environmental effects upon their habitats and the ability of the organisms concerned which produce these biomarkers to adapt rapidly to worsening prevailing environmental conditions.

A very important requirement fulfilled by Raman spectroscopy in the analysis of potential biomarkers is the ability to differentiate between the key molecular species on the basis of their characteristic spectral signatures; this is not only manifest in the discrimination between the relevant organic components of the complex protective biochemicals comprising the stressed biological colonies and the minerals of the geological host matrix but also the identification of the different types of biomarker, such as those exemplified in the list above. Some examples of the Raman spectroscopic detection and identification of biomarkers will now be provided to illustrate the characterisation procedure involved.

13.7 Carotenoids

Because the Raman spectral signature wavenumbers of any molecule or molecular ion are dependent upon the electronic and geometric composition of the particular bonds comprising the target molecule giving rise to the molecular scattering of radiation, the wavenumbers observed in the spectra of a group of closely related compounds, such as the carotenoids of which there are over 400 known, are found to occur over a wavenumber range rather than at a single, precise wavenumber for all molecules. For example, the biogenetic synthesis of carotenes from the parent lycopene structure involves cyclisation, producing alpha- and beta-carotenes, and further hydroxylation in the aliphatic rings gives lutein and zeaxanthin (Weesie et al., 1999). Despite these both being centrosymmetric molecules and having a conjugated 11-ene system with three methyl groups in each ring and a further four pendant methyls situated along the unsaturated –C=C–C– backbone chain, these very similar structures can normally be readily differentiated by analytical Raman spectroscopy. More troublesome, however, is the observation of wavenumber-shifted fundamental bands that occur because of external molecular environmental changes – and these are more difficult to quantify; hence, the importance of constructing reference spectral databases of biomolecules in admixture with other components and particularly with minerals. Finally, it is not merely sufficient to identify the Raman spectral signatures of biomarkers that also have parallels for abiotic organic compounds as this will in itself create an ambiguous interpretation of the spectral data obtained from a remote planetary interrogation. Hence, the observation of the characteristic CH stretching Raman wavenumbers associated with aliphatic organic compounds in the 2800–3000 cm^{-1} region on the ExoMars 2022 mission would not be exclusively indicative *per se* of life signatures on Mars as this functionality also occurs widely in abiotic organic compounds, although this result would be the invaluable direct evidence that organic compounds can survive the hostile Martian environment – itself a major step forward in current knowledge. Figure 13.1 illustrates some typical Raman spectra obtained from a biogeological colonisation of a terrestrial geological niche with the Raman flight-like prototype spectrometer – in this case from a saltern in the Atacama Desert, which shows evidence of three main biological components,

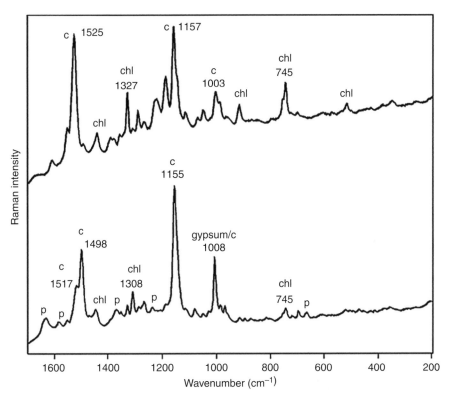

Figure 13.1 Stack-plotted Raman spectra of the biogeological colonisation of gypsum in Salar Grande, Atacama Desert; upper spectrum, surface; lower spectrum, deeper parts. The carotenoid bands can be seen near 25, 157 and 1000 cm^{-1} and the coded labels on the key bands are: c, carotenoid; chl, chlorophyll; p, phycobiliprotein. Raman spectra were excited using green laser excitation at 532 nm. It should be noted that the observed band wavenumbers indicate that different carotenoids are present in the colonised crustal and substratal regions of the specimen (Modified from Vítek et al. 2013.).

carotenoids, chlorophyll and phycobiliprotein in the crust and within a gypsum crystalline substrate.

13.8 Carbon

The degradation of biomolecules in hostile geological environments eventually produces carbon (Edwards et al., 2010) whose signatures in the Raman spectrum have been well described in the laboratory. The idea that it is possible to differentiate between biotic and abiotic carbon formation from the shape of the so-called D and G Raman bands, characteristic of sp^3 and sp^2 hybridised carbon, effectively represented by structures typical of diamond and graphite, has had a long and rather controversial history in the literature, which is still ongoing today. A summary of the extensive literature on this subject is provided by Marshall et al. (2010) who attempt to define a possible methodology for rigorously discriminating between these different carbon sources. However, if we consider the requirements of such finely tuned spectroscopic work that has to be undertaken in the field using portable miniaturised instrumentation with all the sacrifices that have had to be made from comparative laboratory versions, additionally performed under extremely hostile conditions that are mirrored nowhere else terrestrially, then one must pose the question whether it is possible with current remote robotic and miniaturised instrumentation to be able to differentiate between abiotic and biotic carbon? The result then is that the observation of characteristic carbon signatures themselves does not constitute the presence of extinct life, so according to our definition, therefore, carbon cannot be a true biomarker. The same analogy applies to a host of other biochemicals, which have been synthesised abiotically in the laboratory, such as amino acids and sugars. This is the reason that these molecules are not presented as true biomarkers.

13.9 Raman spectroscopy on the ExoMars 2022 mission

The ExoMars RLS instrument science team chose a Raman excitation wavelength of 532 nm in order to optimise the range of accessible biomineral and biological molecular targets in a compromise between a potentially strong fluorescence background emission against an increased Raman spectral intensity with the proviso that commercial laser diodes could be commercially sourced to fit the mass and power requirements of the mission. A laser spot diameter of 50 microns was chosen to balance the requirements of transect sampling of powdered specimens, with grain sizes between 20 and 200 microns, against excessive heating of the sample. It is anticipated that around 400 sample points will be interrogated. The UK flight-like prototype instrument operates in a straight line on the sample surface, with the system autofocusing at each position: the optical elements of the instrument are a 100-mW, 532-nm continuous laser (Spectra-Physics Inc.), a Raman RXN ME probe head with a 25-mm objective and a Holospec f/1.8 spectrograph (Kaiser Optical System Inc.) and a transmission and return optical fibre coupling these elements together. The spectrograph is fitted with a holographic HSG-532 Raman Stokes grating. Details of the specially constructed Peltier-cooled CCD detector are provided in the literature (Edwards et al., 2012). This flight-like prototype and a similar related commercially available, portable system with 532 nm excitation have been used to interrogate a variety of biogeological Mars analogue specimens, as described in the next section.

13.10 Terrestrial extremophilic sites: Mars analogues

Although it is appreciated that the environmental conditions on Mars are probably much more extreme than any found on Earth, the search for life signatures in the Martian regolith inherently depends upon the identification of specific biomarkers that have been produced by the extremophilic colonisation of comparator terrestrial geological niche environments which are to be found in terrestrial 'Mars analogue sites'. A comprehensive list of these sites is now emerging from the literature, and a list of Mars analogue sites which have been investigated by Raman spectroscopy is summarised here. A wide range of instrumentation and protocols has been adopted for spectroscopic site investigation from the sampling of specimens for study in the laboratory using bench-top instrumentation to the adoption of in-field miniaturised spectroscopic instrumentation for the direct interrogation of specimens *in situ* where appropriate.

Cold deserts: Antarctic, Arctic, Spitsbergen, Svalbard, glaciers, snowfields
Hot deserts: Atacama, Negev, Mojave, Death Valley, Arabian, Tenerife
Meteorite craters: Haughton
Meteorites: SNC Martian meteorites
Salterns: Dead Sea, Arabian Desert sabkhas, Atacama Desert
Near-space environment: International Space Station, Biopan satellite missions
Volcanoes: Kilauea, deep-sea vents, grey and black smokers

Several of these sites have been described in detail in a recent special issue (Jehlicka & Edwards, 2014) of the *Philosophical Transactions of the Royal Society* published in 2014 and edited by the present authors in which some 15 papers on the Raman spectroscopy of extremophiles in several Mars analogue sites have been collected and described.

A database of key Raman spectroscopic signatures has been compiled, and although by no means exhaustive, this can afford a reasonable appreciation of the selectivity of discrimination of Raman spectral data for the identification of biogeomarkers and biomarker molecules along with associated geomarkers in real bio-geo systems. A list of the key Raman bands for some key biomarkers and geomarkers is given in Table 13.1, which is taken from Edwards et al. (2014).

13.11 Conclusions

It is clear that a definitive list of biomarkers is now emerging for discussion and eventual acceptance in the scientific analytical spectroscopic community relating to the unambiguous identification of spectroscopic signatures in the search for extinct or extant life in proposed analytical astrobiological missions. The basis of discussion here for potential Raman data identification from extraterrestrial experiments will provide a starting concept that will naturally be enlarged upon and modified by future experiments that will seek to broaden

Table 13.1 Raman bands of biomarkers and associated geomarkers in extremophile exemplars, and their chemical formulae

Name	Formula	Raman wavenumber shifts (cm⁻¹)
Calcite	$CaCO_3$	**1086**, **712**, **282**, 156
Aragonite	$CaCO_3$	**1086**, **704**, **208**, 154
Dolomite	$CaMg(CO_3)_2$	**1098**, **725**, **300**, 177
Magnesite	$MgCO_3$	**1094**, **738**, **330**, **213**, 119
Hydromagnesite	$Mg_5(CO_3)_4(OH)_2 \cdot 4H_2O$	**1119**, **728**, 326, **232**, 202, 184, 147
Gypsum	$CaSO_4 \cdot 2H_2O$	**1133**, **1007**, 669, 628, **492**, **413**
Anhydrite	$CaSO_4$	1015, 674, 628, 500, 416
Quartz	SiO_2	1081, 1064, 808, 796, 696, 500, 542, **463**, 354, 263, **206**, 128
Haematite	Fe_2O_3	**610**, 500, **411**, **293**, 245, **226**
Limonite	$FeO(OH)nH_2O$	693, **555**, 481, **393**, **299**, **203**
Apatite	$Ca_5(PO_4)_3(F,Cl,OH)$	1034, **963**, **586**, **428**
Weddellite	$Ca(C_2O_4)2H_2O$	1630, **1475**, 1411, **910**, 597, **506**, **188**
Whewellite	$Ca(C_2O_4)H_2O$	1629, **1490**, **1463**, 1396, 942, **896**, 865, 596, 521, **504**, 223, 207, **185**, **141**
Chlorophyll	$C_{55}H_{72}O_5N_4Mg$	1438, 1387, **1326**, 1287, 1067, **1048**, **988**, **916**, **744**, **517**, 351
c-phycocyanin	$C_{36}H_{38}O_6N_4$	1655, 1638, 1582, 1463, **1369**, 1338, 1272, 1241, 1109, 1054, **815**, 665, 499
Beta-carotene	$C_{40}H_{56}$	**1515**, **1155**, **1006**
Rhizocarpic acid	$C_{26}H_{23}O_6$	**1665**, 1610, **1595**, **1518**, 1496, 1477, 1347, 1303, 1002, 944, 902, 768, 448
Scytonemin	$C_{36}H_{20}N_2O_4$	1605, **1590**, **1549**, 1444, **1323**, 1283, 1245, **1172**, 1163, 984, 752, 675, 574, 270
Calycin	$C_{18}H_{10}O_5$	1653, **1635**, **1611**, **1595**, **1380**, 1344, 1240, 1155, 1034, **960**, 878, 498, 484
Paretin	$C_{16}H_{12}O_5$	**1671**, 1631, 1613, 1387, 1370, **1277**, 1255, **926**, 571, 519, 467
Usnic acid	$C_{18}H_{16}O_7$	**1694**, 1627, 1607, **1322**, **1289**, **1192**, 1119, **992**, 959, 846, 602, 540
Emodin	$C_{15}H_{10}O_5$	**1659**, 1607, 1577, 1557, **1298**, **1281**, 942, **565**, 467
Atranorin	$C_{19}H_{18}O_8$	**1666**, **1658**, 1632, **1303**, 1294, 1266, **588**
Pulvinic dilactone	$C_{18}H_{10}O_4$	**1672**, **1603**, 1455, **1405**, 1311, **981**, **504**, **458**

From Edwards et al. (2014).
Comparator bands are indicated in bold type face; the numbers are Raman wavenumber shifts in cm⁻¹.

the range of carotenoids, for example, accessed in the spectral database and widen the portfolio to other key biomolecular families such as the scytonemins and hopanoids. This will be seen to be vital for the diagnostic ability of analytical spectroscopy in furtherance of search for life experimental scenarios and for the remote planetary instrumentation envisaged for forthcoming decades particularly in advance of human space missions in our Solar System.

References

Bedau, A. (2010). An Aristotelean account of minimal chemical life. *Astrobiology*, **10**, 1011–1020.

Benner, S.A. (2010). Defining life. *Astrobiology*, **10**, 1021–1030.

Bishop, J.L., Aglen, B.L., Pratt, L.M., et al. (2004). A spectroscopic and isotopic study of sediments from the Antarctic Dry Valleys as analogues for potential palaeolakes on Mars. *International Journal of Astrobiology*, **2**, 273–287.

Brier, J.A., White, S.N., German, C.R. (2010). Mineral-microbe interactions in deep-sea hydrothermal systems: a challenge for Raman spectroscopy. *Philosophical Transactions of the Royal Society A*, **368**, 3067–3086.

Carter, E.A., Hargreaves, M.D., Kee, T.P., Pasek, M. A., Edwards, H.G.M. (2010). A Raman spectroscopic study of a fulgurite. *Philosophical Transactions of the Royal Society A*, **368**, 3087–3098.

Clark, B.C. (1998). Surviving the limits to life at the surface of Mars. *Journal of Geophysical Research Planets*, **103**, 28545–28556.

Cleland,C.E., Chyba, C.F. (2002). Defining life. *Origins of Life and Evolution of the Biosphere*, **32**, 387–393.

Cockell, C.S. (2001). Astrobiology and the ethics of new science. *Interdisciplinary Science Reviews*, **26**, 90–96.

Cockell,C.S., Knowland, J.R. (1999). Ultraviolet screening compounds. *Biological Reviews*, **74**, 311–345.

Dickensheets, D.L., Wynn-Williams, D.D., Edwards, H.G.M., Crowder, C., Newton, E.

M. (2000). A novel miniature confocal Raman spectrometer system for biomarker analysis on future Mars missions after Antarctic trials. *Journal of Raman Spectroscopy*, **31**, 633–635.

Doran, P.T., Wharton, R.A.J., des Marais, D.J., McKay, C.P. (1998). Antarctic palaeolake sediments and the search for extinct life on Mars. *Journal of Geophysical Research Planets*, **103**, 28481–28488.

Edwards, H.G.M. (2004). Raman spectroscopic protocol for the molecular recognition of key biomarkers in astrobiological exploration. *Origins of Life and Evolution of the Biosphere*, **34**, 3–11.

Edwards, H.G.M. (2010). Raman spectroscopic approach to analytical astrobiology: the detection of key geological and biomolecular markers in the search for life. *Philosophical Transactions of the Royal Society A*, **368**, 3059–3066.

Edwards, H.G.M., Newton, E.M. (1999). Applications of Raman spectroscopy for exobiological prospecting. In: J.A. Hiscox (ed.) *The Search for Life on Mars*. British Interplanetary Society, London, pp. 75–83.

Edwards, H.G.M., Russell, N.C., Wynn-Williams, D.D. (1997). Fourier- transform Raman spectroscopic and scanning electron microscopic study of cryptoendolithic lichens from Antarctica. *Journal of Raman Spectroscopy*, **30**, 685–690.

Edwards, H.G.M., Moody, C.D., Jorge Villar, S. E., Wynn-Williams, D.D. (2005). Raman spectroscopic detection of key biomarkers of cyanobacteria and lichen symbiosis in extreme Antarctic habitats:

evaluation for Mars lander missions. *Icarus*, **174**, 560–571.

Edwards, H.G.M., Sadooni, F., Vítek, P., Jehlička, J. (2010). Raman spectroscopy of the Dukhan sabkha: identification of geological and biogeological molecules in an extreme environment. *Philosophical Transactions of the Royal Society A*, **368**, 3099–3108.

Edwards, H.G.M., Hutchinson, I.B., Ingley, R. (2012). The ExoMars Raman spectrometer and the identification of biogeological spectroscopic signatures using a flight-like prototype. *Analytical & Bioanalytical Chemistry*, **404**, 1723–1731.

Edwards, H.G.M., Hutchinson, I.B., Ingley,R., Jehlicka, J. (2014). Biomarkers and their Raman spectroscopic signatures: a spectral challenge for analytical astrobiology. *Philosophical Transactions of the Royal Society A*, **372**, 20140193.

Eigenbrode, J.L., Summons, R.E., Steele, A., et al. (2018). Organic matter preserved in 3-billion-year-old Mudstones at Gale Crater, Mars. *Science*, **360**, 1096–1101.

Ellery, A., Wynn-Williams, D.D. (2003). Why Raman spectroscopy on Mars? A case of the right tool for the right job. *Astrobiology*, **3**, 565–579.

Farmer, J.D., des Marais, D.J. (1999). Exploring for a record of Ancient Martian Life. *Journal of Geophysical Research*, **104**, 26977–26995.

Freeman, K. (1983). Cell astrobiology: shuttle separation of islet cells. *Nature*, **304**, 575.

Freissinet,C., Glavin, D.P., Mahaffy, P.R., et al. and the MSL Science Team (2015). Organic molecules in the Sheepbed mudstone, Gale Crater, Mars. *Journal of Geophysical Research Planets*, **120**, 495–514.

Grady, M.M., Verchovsky, A.B., Wright, I.P. (2004). Magmatic carbon in Martian meteorites: attempts to constrain the carbon cycle on Mars. *International Journal of Astrobiology*, **3**, 117–124.

Jehlička, J., Edwards, H.G.M. (2014). Raman spectroscopy meets extremophiles on Earth and Mars: studies for the successful search for life. *Philosophical Transactions of the Royal Society A*, **372**, 20140207.

Jehlička, J., Vandenabeele, P., Edwards, H.G.M., Culka, A., Capoun, T. (2010a). Raman spectra of pure biomolecules obtained using a hand-held instrument under cold, high-altitude conditions. *Analytical & Bioanalytical Chemistry*, **397**, 2753–2760.

Jehlička, J., Edwards, H.G.M., Culka, A. (2010b). Using portable Raman spectrometers for the identification of organic compounds at low temperatures and high altitudes: exobiological applications. *Philosophical Transactions of the Royal Society A*, **368**, 3109–3126.

Jorge-Villar, S.E., Edwards, H.G.M. (2005). Raman spectroscopy in astrobiology. *Analytical & Bioanalytical Chemistry*, **384**, 100–113.

Jorge-Villar, S.E., Edwards, H.G.M. (2010). Raman spectroscopy of volcanic lavas and inclusions of relevance to astrobiological exploration. *Philosophical Transactions of the Royal Society A*, **368**, 3127–3136.

Lafleur, L.J. (1941). Astrobiology. *Astronomical Society of the Pacific Leaflets*, **143**, 133.

Lederberg, J. (1960). Exobiology: approaches to life beyond the earth. *Science*, **132**, 393.

Lin, Y., Goresy,A.E., Hu, S., et al. (2014). NanoSIMS analysis of organic carbon from the Tissint Martian meteorite: evidence for the past existence of subsurface organic-bearing fluids on Mars. *Meteoritics Planetary Science*, **49**, 2201–2218.

Marshall, C.P., Olcott Marshall, A. (2010). The potential of Raman spectroscopy for the analysis of diagenetically transformed carotenoids. *Philosophical Transactions of the Royal Society A*, **368**, 3137–3144.

Marshall, C.P., Edwards, H.G.M., Jehlicka, J. (2010). Understanding the application of Raman spectroscopy to the detection of traces of life. *Astrobiology*, **10**, 229–243.

Martinez-Frias, J., Hochberg, D. (2007). Classifying science and technology: two problems with the UNESCO system. *Interdisciplinary Science Reviews*, **32**, 315–319.

McKay, C.P. (1997). The search for life on Mars. *Origins of Life and Evolution of the Biosphere*, **27**, 263–289.

Parnell, J., Cullen, D., Sims, M.R., et al. (2007). Searching for life on Mars: selection of molecular targets for ESA's Aurora ExoMars mission. *Astrobiology*, **7**, 578–604.

Perry, R.S., McLoughlin, N., Lynne, B.Y., et al. (2007). Defining biominerals and organominerals: direct and indirect indicators of life. *Sedimentary Geology*, **201**, 157–179.

Pullan, D., Hofmann, B.A., Westall, F., et al. (2008). Identification of morphological biosignatures in Martian analogue field specimens using in situ planetary instrumentation. *Astrobiology*, **8**, 119–156.

Rull, F., Delgado, A., Martinez-Frias, J. (2010a) Micro-Raman spectroscopic study of extremely large atmospheric ice conglomerations (megacryometeors). *Philosophical Transactions of the Royal Society A*, **368**, 3145–3152.

Rull, F., Munoz-Espadas, M.J., Lunar. R., Martinez-Frias, J. (2010b). Raman spectroscopic study of four Spanish shocked ordinary chondrites: Canellas, Olmedilla de Alarcon, Reliegos and Olivenza. *Philosophical Transactions of the Royal Society A*, **368**, 3153–3166.

Shapiro, R., Schulze-Makuch, D. (2009). The search for alien life in our solar system: strategies and priorities. *Astrobiology*, **9**, 1–9.

Sharma, S.K., Misra, A.K., Clegg, S.M., et al. (2010). Time-resolved remote Raman study of minerals under supercritical CO_2 and high temperatures relevant to Venus exploration. *Philosophical Transactions of the Royal Society A*, **368**, 3167–3192.

Soffen, G.A. (1999). Astrobiology. *Life Sciences: ExoBiology*, **23**, 283–288.

Steele, A., McCubbin, F.M., Fries, M.D., et al. (2012). Graphite in the Martian meteorite Allan Hills 84001. *American Mineralogist*, **97**, 1256–1259.

Strughold, H. (1953). *The Red and Green Planet: A Physiological Study of the Possibility of Life on Mars*. Albuquerque: University of New Mexico Press.

Summons, R., Amend, J.P., Bish, D., et al. (2011). Preservation of Martian organic and environmental records. *Astrobiology*, **11**, 157–181.

Tikhov, G.A. (1953). *Astrobiology*. Molodaya Gvardia Publishing House, Moscow.

Tirard, S., Morange, M., Lazcano, A. (2010). The definition of life: a brief history of elusive scientific endeavour. *Astrobiology*, **10**, 1003–1009.

Treado, P.J., Truman, A. (1996). Laser Raman spectroscopy. In: T.J. Wdowiak, D.GAgresti (eds) *Point Clear Exobiology Instrumentation Workshop*. University of Alabama Press, Birmingham, AL, pp. 7–10.

Varnali, T., Edwards, H.G.M. (2010). Ab initio calculations of scytonemin derivatives of relevance to extremophile characterisation by Raman spectroscopy. *Philosophical Transactions of the Royal Society A*, **368**, 3193–3204.

Vítek, P., Edwards, H.G.M., Jehlicka, J., et al. (2010). Microbial colonisation of halite from the hyper-arid Atacama Desert studied by Raman spectroscopy. *Philosophical Transactions of the Royal Society A*, **368**, 3205–3221.

Vítek P., Cámara-Gallego B., Edwards H.G.M. et al. (2013). Phototrophic community in gypsum crust from the Atacama Desert studied by Raman spectroscopy and microscopic imaging. *Geomicrobiology Journal*, **30**, 399–410.

Webster, C.R., Mahaffy, P.R., Atreya, S.K., et al. and the MSL Science Team (2015). Mars methane detection and variability at Gale Crater. *Science*, **347**, 415–417.

Webster, C.R., Mahaffy, P.R., Atreya, S.K., et al. (2018). Background levels of methane in Mars' atmosphere show strong seasonal variation. *Science*, **360**, 1093–1096.

Weesie, R.J., Merlin, J.C., Lugtenburg, J., et al. (1999). Semiempirical and Raman spectroscopic study of carotenoids. *Biospectroscopy*, **5**, 19–33.

Westall, F., Cavalazzi, B. (2011). Biosignatures in rocks. In: Thieland, V., Reitner J. (eds) *Encyclopaedia of Geobiology*. Springer, Berlin, pp. 189–201.

Westall, F., Foucher, F., Bost, N., et al. (2015). Biosignatures on Mars: What, where and how? Implications for the search for Martian Life. *Astrobiology*, **15**, 1–32.

Wynn-Williams, D.D. (1991). Cyanobacteria in deserts –life at the limit? In: B.A. Whitton, M. Potts (eds) *The Ecology of Cyanobacteria: Their Diversity in Time and Space*. Kluwer, Academic Press, Dordrecht, The Netherlands, pp. 341–366.

Wynn-Williams, D.D. (1999). Antarctica as a model for ancient Mars. In: J.A. Hiscox (ed.) *The Search for Life on Mars*. British Interplanetary Society, London, pp. 49–57.

Wynn-Williams, D.D., Edwards, H.G.M. (2000a). Antarctic ecosystems as models for extraterrestrial surface habitats. *Planetary and Space Sciences*, **48**, 1065–1075.

Wynn-Williams, D.D., Edwards, H.G.M. (2000b).Proximal analysis of regolith habitats and protective biomolecules in situ by laser Raman spectroscopy: overview of terrestrial Antarctic habitats and Mars analogs. *Icarus*, **144**, 486–503.

Wynn-Williams, D.D., Edwards, H.G.M. (2002). Environmental UV-radiation: biological strategies for protection and avoidance. In: G.Horneck, C. Baumstarck-Khan (eds) *Astrobiology: The Quest for the Origins of Life*, Springer-Verlag, Berlin, pp. 245–260.

CHAPTER FOURTEEN

Adaptation/acclimatisation mechanisms of oxyphototrophic microorganisms and their relevance to astrobiology

JANA KVÍDEROVÁ
University of South Bohemia

14.1 Introduction

Autotrophic microorganisms, regardless of whether they are using light insolation or chemical reactions to acquire energy for the production of organic compounds, represent the only source of carbon for carbon-based life evolution in a planetary biosphere (Schulze-Makuch & Irwin, 2018). The incorporation of inorganic carbon into organic carbon compounds requires large amounts of energy: light is one of the most available sources of energy for life on the surface of planets orbiting in the habitable zone (Schulze-Makuch & Irwin, 2018) and as an additional energy source for the moons of the Jovian planets in the Solar System. The importance of light as a primary energy source is increased as the mineral resources for the redox (reduction–oxidation) reactions utilised by chemoautotrophic microorganisms are predictably becoming depleted (Leigh et al., 2007).

Anoxygenic photosynthesis, which is found for example in green sulfur bacteria, purple sulfur bacteria and purple non-sulfur bacteria utilising hydrogen, Fe^{2+} ions and H_2S as electron donors, developed shortly after the emergence of life on Earth, ca 3.4 Ga ago (Olson, 2006). Oxygenic photosynthesis appeared later, around 2.5 Ga ago (Olson, 2006; Buick, 2008), resulting in significant atmospheric chemical composition changes from neutral or slightly reducing to oxidative between 2.45 and 2.32 Ga ago (Bekker et al., 2004; Lyons et al., 2014), which facilitated the evolution of complex life forms (Thannickal, 2009). Oxyphototrophic microorganisms, at first prokaryotic cyanobacteria and later eukaryotic algae, became important oxygen and organic carbon sources for heterotrophic life, and this has been maintained to the present time. During their evolutionary history, these had to adapt to changing climatic conditions and they had to develop various adaptation–acclimatisation mechanisms to survive in extreme conditions (Seckbach et al., 2007); these conditions are similar to those observed on other planets and moons in our Solar System and also terrestrially in the Río Tinto (Amils et al., 2007) and Antarctic deserts (Doran et al., 2010).

Therefore, a study of the survival mechanisms of adaptation and acclimatisation of terrestrial oxyphototrophic microorganisms, i.e. cyanobacteria and algae, could contribute to astrobiology research and an understanding of evolutionary processes. A detailed knowledge of the survival and physiological performance of cyanobacteria and algae, especially photosynthesis, under extreme environmental conditions will contribute to an assessment of the parameters of limits of life and habitability. Also, the cell structures and compounds produced in the survival strategies will produce identifiable biosignatures of present or past life.

14.2 Stress and adaptation–acclimatisation mechanisms in oxyphototrophic microorganisms

In the life sciences, there are no uniformly acceptable definitions of adaptation and acclimatisation because in microbiology adaptation is a genetically fixed response while acclimatisation is not (Schulze et al., 2005). Due to the complexity of stress responses upon living organisms, it is difficult to distinguish between strict adaptation and strict acclimatisation, therefore the term 'adaptation–acclimatisation mechanisms' (AAM) has been used to describe the response of organisms in stress-related and ecophysiological studies.

There are many environmental stress factors which can affect the survival of cyanobacteria and algae (Figure 14.1). Stress may be caused either by experiencing the extremes of various operational parameters, e.g. temperatures in hot springs and glacial lake salterns or by the often unpredictable variability of the availability of liquid water in the marginal environments of cold deserts (Elster, 1999) and the poikilo-environments of Gorbushina and Krumbein (1999), which describe the drastic variation of environments which exist outside a normal narrow life-supporting range. Usually, a combination of several stress factors is encountered in the field when studying the sustainability of life in extreme environments. For instance, drought can also be connected with increased irradiance, temperature and salt concentration. Some of the environmental stress factors, such as gravity or magnetic field flux intensity, had not been considered as potential stress factors before space exploration began. At present, the importance of these two factors is increasing with respect to manned spaceflight and space colonisation, especially since cyanobacteria and algae are being considered as a component of regenerative life-support systems (Gòdia et al., 2002). Cyanobacteria and algae are also seen as key factors in astrobiology research for establishing the evaluation of the limits of habitable life in the planets and moons of our Solar System (Nealson, 1997) and more widely in the newly discovered exoplanets within the habitable zone of a host star (refer to e.g. Christiansen (2018) for

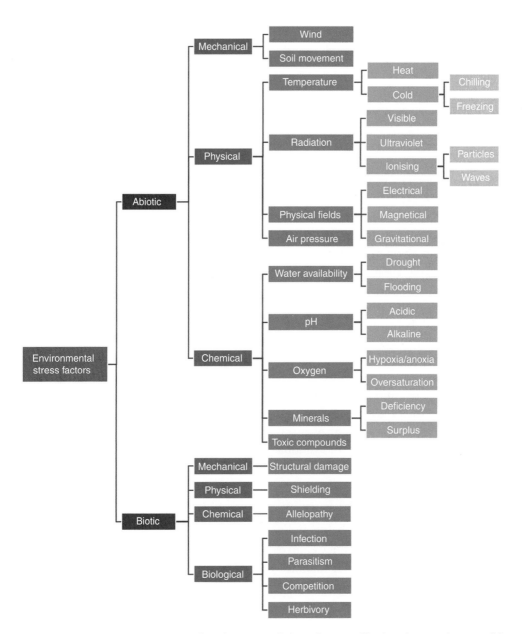

Figure 14.1 Summary of environmental stress factors affecting the oxyphototrophic microorganisms that are important relevant to astrobiology. Modified and extended from Schulze et al. (2005).

exoplanets catalogues), which could be potential harbours of extraterrestrial life (Ballesteros et al., 2019). They are also seen as key indicators and target specimens for the development of life detection techniques (Des Marais & Jahnke, 2019).

General symptoms of stress include the increased expression of stress-related genes such as heat shock proteins (Whitley et al., 1999; Kregel, 2002), the synthesis of stress-related compounds such as carotenoids in response to high irradiance (Bidigare et al., 1993), a growth rate decrease or a growth cessation and a change in metabolic rates (Schulze et al., 2005). The response of the organism to stress exposure depends on the stress intensity and stress duration (Figure 14.2). Severe stress results in a decrease in metabolism and growth rate due to incipient cell damage combined with cell exhaustion. If the microorganism cannot compensate for the effects of even mild and intermediate stress levels, then exhaustion and death are inevitable; the milder the stress experienced, then the more prolonged resistance can be offered and the more chance there is afforded for the organism to develop protective mechanisms against the stress. Mild or intermediate stresses may slightly increase the metabolic rates, and even the growth rates, due to the activation of AAM, resulting in a significantly higher hardiness and resistance, and hence in the acclimatisation and/or adaptation (Schulze et al., 2005) to the environmental stress level.

In general, there are two ways that the organism can cope with an applied stress, namely, to escape (termed a stress avoidance or passive stress response) or to fight (termed a stress tolerance or active stress response). These strategies have been described in detail for cyanobacteria in Kvíderová et al. (2019) and can be summarised for both cyanobacteria and algae, as follows.

- The passive stress response of an organism includes:
  - The escape from adverse conditions to more suitable ones such as:
    - an active migration – using active movement (Hoiczyk, 2000; Clegg et al., 2003) or by buoyancy control (Oliver, 1994);
    - the selection of a suitable habitat – especially with respect to temperature stability, increased humidity or reduced irradiance (predominantly UV insolation, but also extremely high visible radiation levels), usually accomplished through the colonisation of a shielded habitat such as beneath a stone (McKay et al., 1993);
    - the development of complex life cycles with resting stages – this is typical for species that live in an environment in which the suitable growth conditions occur only for short periods time, for example, snow algae (Hoham & Duval, 2001; Komárek & Nedbalová, 2007) or arctic seepage cyanobacteria (Tashyreva & Elster, 2012; Tashyreva & Elster, 2016).

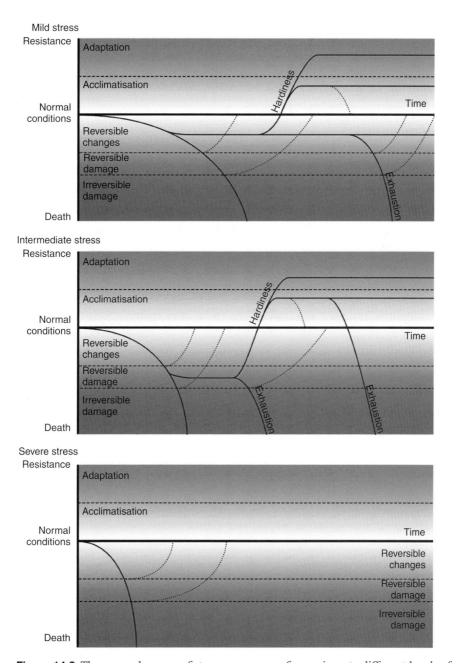

Figure 14.2 The general course of stress responses of organisms to different levels of stress. The full lines indicate possible scenarios of the stress. Dotted lines indicate the predicted response if the stress is released or alleviated.

- The insulation of an organism from its surroundings includes:
  - the generation of additional morphological features, for example, extracellular mucilaginous envelopes and sheaths (e.g. Tashyreva & Elster, 2016), sometimes accompanied by some shielding compounds (Cockell & Knowland, 1999), thick cell walls and supportive colony formation (Xiong et al., 1996);
  - the symbiotic association of different microorganisms, such as microbial biofilms (de los Ríos et al., 2004) and lichens (Schneider et al., 2011).

- The active stress response of an organism includes:
  - The adoption of well-established evolutionary responses, such as:
    - fatty acid compositional changes wherein unsaturation increases the stress resistance (Gombos et al., 1994a, 1994b; Moon et al., 1995);
    - molecular chaperones such as the expression of HSP (Whitley et al., 1999; Kregel, 2002);
    - the generation of osmotically active compounds for the protection of proteins and membranes (Potts, 1999; Oren, 2007);
    - ROS scavenging, wherein enzyme systems are used to combat the stress, consisting of superoxide dismutases, catalases and peroxidases (Ledford & Niyogi, 2005) and by non-enzymatic molecules like carotenoids (Sachindra et al., 2007); ROS may also serve as a molecular signalling device (Baxter et al., 2013);
    - DNA repair which involves genome multiplication as well as formal repair mechanisms, i.e. photoreactivation, excision repair and post-replication repair (Bray & West, 2005), cellular component repairs and de novo synthesis such as D1 protein (Melis, 1999; Takahashi & Badger, 2011).

- Stress-specific responses adopted by the organism will include reaction towards:
  - Temperature, including freezing – cold shock proteins and ice nucleation proteins have not been found in algae but antifreeze proteins have been documented (Li et al., 2009) as well as the identified production of dimethylsulfoniopropionate (Kirst et al., 1991);
  - Radiation (Visible and UV wavelengths) – the synthesis of screening compounds such as mycosporine-like amino acids (Bhatia et al., 2011), scytonemin (Ehling-Schulz & Scherer, 1999; Matsui et al., 2012), phenolic ice nucleation proteins, the chromatic adaptation of (Fujita et al., 1985; Kováčik et al., 2011), several short-term and long-term photoacclimation mechanisms (MacIntyre et al., 2002; Allen,

2003) and mixotrophy used during low-light periods of activity (Moorthi et al., 2009);

- Desiccation – the synthesis of proteins to combat water deficiency (Potts, 1994);
- Salinity – involving ion extrusion or uptake (Avron and Ben-Amotz, 1992; Joset et al., 1996; Apte, 2001);
- pH – changes in cellular proton balance (Gimmler, 2001; Gimmler & Degenhardt, 2001);
- Nutrients – organism uptake mechanisms (Axelsson & Beer, 2001; Shehawy & Kleiner, 2001; Wagner & Falkner, 2001; Raven, 2003).

Cellular 'lines of defence' include, at first, the passive stress response if possible and at the same time or later on the active stress response. These 'lines of defence' include:

1. an avoidance of the stress, i.e. escaping from or hiding from the stress conditions;
2. morphological structures – such as thick cell walls and colony formation to prevent superficial damage from entering the cell interior;
3. the synthesis of protective compounds, e.g. screening pigments, HSP and osmotic compounds to minimise the potential damage either outside the cell (e.g. screening pigments in the extracellular compounds) or inside the cell (e.g. osmotic compound synthesis during desiccation);
4. ROS scavenging – even if the previously mentioned mechanisms can prevent the ROS formation, damage through ROS can occur and compromise important cellular structures such as membranes or proteins directly *in situ*, e.g. the occurrence of high radiation insolation;
5. DNA repair to apply for the case that all previously mentioned responses have failed and the stress damage reaches the cell DNA, which can happen when the organism is exposed to significant UVB low-wavelength radiation, which can cause helical unzipping of the DNA strands.

Usually, both active and passive response strategies are adopted. For instance, the zygospores of snow algae, in which the algae survive periods with reduced water availability, develop thick extracellular envelopes to insulate the cells from their surroundings but also produce large amounts of carotenoids as a protection against high levels of UV and Visible radiation. Soil particles are often attached to these extracellular envelopes and provide additional protection against excessive UV and Visible radiation (Hoham & Duval, 2001; Komárek & Nedbalová, 2007).

14.3 AAM in astrobiology

14.3.1 Limits of life

For an estimation of the habitability of the other planets and moons of our Solar System, and also for exoplanets and exomoons, a knowledge of the ranges for survival of organisms under different environmental gradients is crucial. So far, our knowledge is limited to the observation of terrestrial life forms. However, even these data indicate a broad tolerance of life even in the most extreme terrestrial environments. In very extreme conditions, specialist organisms prevail, and generalists survive in only less severely stressed environments. Although oxygenic photoautotrophic microorganisms are more sensitive to extreme conditions than heterotrophs (Oren & Seckbach, 2001), they are still able to survive a broad range of conditions, some of these very similar to predicted extraterrestrial conditions (Figure 14.3). Evidence of an oxygen-rich environment in the historical Mars itself implies the question – was the high atmospheric oxygen concentration that once occurred on Mars caused by inorganic reactions or by oxygenic photosynthesis (Stamenković et al., 2018)?

For the determination of the growth limits and growth optima of organisms, three types of study are applied based on different degrees of control of the environmental variables. Terrestrial field measurements do not include any control factor for these environmental variables and their effects are, therefore, predicted and determined from multivariate analysis. The response is usually averaged for each community (Kvíderová et al., 2018). The field predictive experiments usually manipulate one or several environmental factors such as temperature or nutrient content, and the effect of these manipulations is considered for the whole community (Elster & Svoboda, 1995; Elster et al., 2001, 2012). For both the field measurements and field manipulations, several environmental factors that may influence the study need to be considered and, in particular, the interactions of individual environmental factors that could influence the metabolic activity but which may remain concealed need to be identified. In laboratory experiments, these environmental factors are strictly defined and their number is limited by the cultivation conditions and exposure criteria. The laboratory experiments are performed either on the natural communities or on the individual pure cultures of the organisms. Before detailed experiments can be commenced, screening studies should be applied such as organism cultivation in crossed gradients (Kvíderová & Lukavský, 2005; Hindák et al., 2013; Kumar et al., 2017). When pure cultures are used, some important interactions may be omitted that become apparent in natural communities. A combination of all three approaches is usually advocated to determine the limits of organism survival and the AAM.

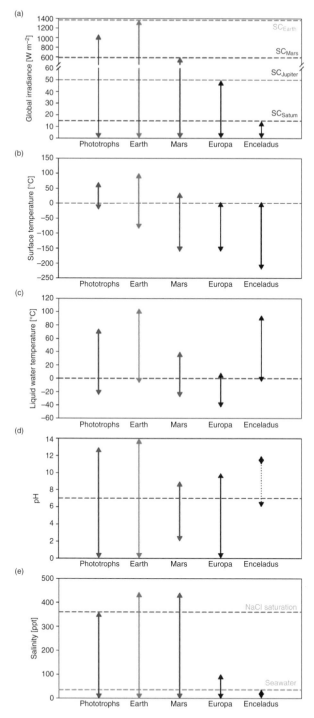

Figure 14.3 The range of occurrence of oxyphototrophic microorganisms on Earth in comparison with the range of environmental variables on Earth and the most astrobiologically important bodies in the Solar System – Mars, Europa and Enceladus. (a) Light, (b) surface temperature, (c) liquid water temperature, (d) pH, (e) salinity. Abbreviations: LW, liquid water availability, subscript indicates the planet/moon; SC, solar constant, subscript indicates the planet/moon.

14.3.2 Light

As light is the most important source for life, being particularly crucial for the survival of phototrophic microorganisms, and is present on the surfaces and in the atmospheres of the inner planets (Schulze-Makuch & Irwin, 2018), for Jupiter and the moons of Saturn the presence of photoautotrophic microorganisms could provide an important carbon input to the local ecosystems. The range of irradiation encountered could provide enough energy for photosynthesis even in the orbit of Saturn (Figure 14.3), since the solar constant calculated in the orbit of Saturn (which has a mean distance of 9.6 AU from the Sun) is approximately one hundredth of the Earth solar constant, i.e. ca. 12 W m^{-2}. Even if the attenuation of the incident light is considered, there still could be enough light energy to support photoautotrophic life. For instance, the cryptoendolithic microorganisms in the Antarctic Dry Valleys are able to survive with irradiances of 1% of the incoming PAR, but their growth rate remains very low (Nienow et al., 1988).

If other stellar systems in the universe are considered, the spectral class of the host star must be taken into account (Kiang et al., 2007a, 2007b). In stars that are more massive than our Sun (G2), which belongs to the spectral class F, the light energy is emitted predominantly in the high-energy UV and blue regions, and the total energy flux at a specified distance of 1 AU is comparatively higher. On the other hand, in stars that are less massive than our Sun, i.e. in stars of spectral classes K and M (Figure 14.4), the maximum amount of light

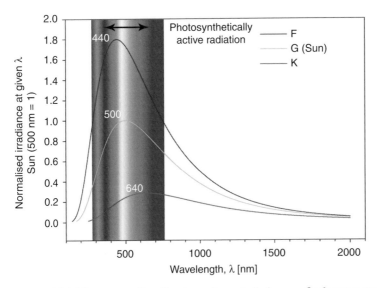

Figure 14.4 The energy flux for stars of spectral classes of relevance to astrobiology. (A black and white version of this figure will appear in some formats. For the colour version, please refer to the plate section.)

energy is emitted in the red and NIR regions and the total energy flux at a specified distance of 1 AU is comparatively lower. Therefore, the photoacclimation and photoprotection mechanisms in receiving organisms may include a novel response to light quality and quantity with respect to the central stars (Kiang et al., 2007a). The recent discovery of photoadaptation to near infrared radiation using chlorophyll *f*, where the maximum absorption occurs at 720–730 nm (Schmitt et al., 2019), indicates that there could be an AAM of photosynthesis to be revealed even in terrestrial organisms that could play an important role in the AAM on exoplanets and exomoons.

14.3.3 Temperature

In general, temperatures on the astrobiologically interesting planets and moons in the Solar Systems are expected to be rather low (Figure 14.3). However, local moderate or high temperature environments cannot the excluded completely from consideration. For instance, hot springs were found at Mt. Erebus, Antarctica, to be inhabited by specific cyanobacteria (Broady, 1984; Soo et al., 2009). The oxyphotosynthetic microorganisms terrestrially can perform photosynthesis at a temperature range from −20° C in the photobionts of antarctic lichens (Kappen et al., 1986; Kappen et al., 1996) to +70°C in thermophilic unicellular cyanobacteria (Allewalt et al., 2006). Polar microorganism especially adapt well to low-temperature environments and are predicted to be able to survive the Martian environment, as has been demonstrated by terrestrial ground simulations as well as by exposure experiments to a near space environment in low Earth orbit on the International Space Station (de Vera et al., 2004; Olsson-Francis et al., 2009; Cockell et al., 2011; Onofri et al., 2012; de Vera et al., 2014; Brandt et al., 2015).

14.3.4 Water availability

Along with temperature, liquid water availability is one of the major determining factors for the habitability of a planet or a moon to organisms. Cyanobacteria and algae as poikilohydric organisms are resistant to desiccation (Potts, 1999; Henley, 2001). For example, even if the polar cyanobacterium *Nostoc* lost 70% of its weight through desiccation compared with the fully hydrated state, it remained photosynthetically active (Kvíderová et al., 2011). In the desiccated state, these organisms are even more resistant to other key stressors such as UVR (Fukuda et al., 2008). In a water-deficient environment, as presented on Mars, it has been predicted that these putative microorganisms could spend the majority of their lifetime in a dehydrated state with only short periods of rehydration. This sparse biosphere and related problems of life detection on Mars are discussed in detail in Cockell and McMahon (2019).

14.3.5 pH and salinity

The salt concentration and pH are of lower importance for considerations of habitability. Nevertheless, photosynthetic microorganisms can survive in extremely acidic, alkaline and/or saline conditions.

A low pH (pH < 4) environment has been found in Planum Meridiani, Mars (Squyres & Knoll, 2005). In its terrestrial geochemical analogue, Río Tinto, Spain (Amils et al., 2007), a simple but complete ecosystem has been discovered (Aguilera et al., 2007a, 2007b) wherein the dominant red alga *Cyanidium* sp. is well adapted to an acidic pH between 1 and 3 (Aguilera et al., 2006; Kvíderová, 2012). In contrast, a high pH (pH > 10) and highly saline environment is proposed for the subsurface oceans of Enceladus (Glein et al., 2015), which can be correlated with the microbial communities found in extremely saline and alkaline lakes such as the Kenyan Rift Valley lakes (Fazi et al., 2018).

High salt concentration as a limiting factor for organism survival is discussed with respect to the present occurrence of liquid water on Mars (McEwen et al., 2011; Dundas et al., 2012; Kereszturi, 2012; Ojha et al., 2015; Dundas et al., 2018). On Earth, hypersaline permanently frozen lakes were found in the Dry Valleys, Antarctica, where the lake bottoms are covered by thick microbial cyanobacterial mats (Mackey et al., 2018).

14.3.6 Space conditions – radiation and altered gravity

The effect of other possible limiting factors to the survival of organisms such as radiation from space, gravity or magnetic field intensity cannot be evaluated in natural conditions on the Earth. Therefore, ground simulations (Angelini et al., 2001; de Vera et al., 2014) and flight experiments (Cockell et al., 2011; Brandt et al., 2015) have been performed, but only for a relatively limited number of photosynthetic microorganisms that were considered important for astrobiology or for potential life-support systems.

The astrobiological flight experiments were designed to consider the effects of exposure to a near space environment (Rettberg et al., 2004; Horneck et al., 2007; Rabbow et al., 2009) and the resulting physiological and morphological flight data are available for a few species, summarised as follows:

- Changes in morphology and ultrastructure. In *Chlorella*, changes in the ultrastructure of energetic organelles, the reduction of starch in the plastid stroma and the increase of condensed chromatin in the nuclei were observed during space flight (Popova et al., 1989; Sytnik et al., 1992; Popova and Sytnik, 1996; Popova, 2003).
- Changes in growth and physiology. There was no significant influence of microgravity on the growth, life cycles and cell size of *Chlorella* observed during the flight experiments (Šetlík et al., 1979; Meleshko et al., 1986).

Increased growth was detected in the cyanobacterium *Nostoc sphaeroides* (Cyanophyta) (Wang et al., 2004), but in contrast, *Anabaena siamensis* (Cyanophyta) showed a slower growth in space (Wang et al., 2006). Slightly reduced photosynthetic activity was found to occur in *Anabaena siamensis* immediately upon landing after exposure in space (Wang et al., 2006).

14.4 Biomarkers for astrobiology

Photosynthetic microorganisms can provide a broad spectrum of biomarkers that can be used for life detection. The main groups of biomarkers are based on morphological structures, isotopic composition, molecular fossils and spectroscopic signatures (Javaux, 2015).

14.4.1 Morphological biomarkers

Morphological biomarkers that can be connected to photosynthetic microorganisms include microfossils, bioweathering, biominerals and sedimentary structures. The morphological biomarkers are easily observed, but their interpretation requires careful consideration as they may be of inorganic origin. The oldest confirmed fossils of cyanobacteria, aged about 2 Ga, were found by Rasmussen et al. (2008). Bioweathering, the erosion of inorganic material due to the activity of microorganisms, is typical for endolithic communities and can be observed for a selection of suitable rocks (Figure 14.5). Biominerals are minerals that have been produced by living organisms. In the case of phototrophic microorganisms, the most commonly encountered biominerals are the silicate shells of diatoms, silica scales of Chrysophyta and calcium scales of coccolithophorids (van den Hoek et al., 1995). These sedimentary structures can be divided further into stromatolites (layered carbonate structures), thromboliths (non-layered carbonated structures), oncolites (small balls with a concentric layered structure) and microbial-induced sedimentary structures (MISS), which are structures connected to microbial mats. Although living stromatolites are known from warm terrestrial environments such as Shark Bay, Australia (Burns et al., 2004), they have also been recorded in the Antarctic (Parker et al., 1981). The occurrence of living thrombolites is rare, e.g. in Lake Clifton, Australia (Moore & Burne, 1994). Living oncolites have been reported from only two Bavarian creeks (Hägele et al., 2006). On the other hand, MISS have been reported in several widespread locations on Earth (Noffke et al., 2001), and morphologically similar structures have recently been observed on Mars (Noffke, 2015).

14.4.2 Isotopic biomarkers

Isotopic biomarkers are based on the discrimination of heavy isotopes for living systems. In the case of cyanobacterial algae, the most important signals

Figure 14.5 (a) Bioweathering of sandstone rock in Svalbard, (b) detail of the endolithic community, (c) image of samples and (d) detection of endolithic community using variable chlorophyll fluorescence where the value of F_0 (minimum fluorescence in the dark) serves as a proxy of the chlorophyll *a* concentration. (A black and white version of this figure will appear in some formats. For the colour version, please refer to the plate section.)

are obtained from $\delta^{13}C$ and $\delta^{15}N$. While the $\delta^{13}C$ content of a living system is relatively stable in marine environments, the freshwater values are found to be more variable (Finlay, 2004). The $\delta^{15}N$ value has been suggested as a potential proof of nitrogen biological fixation (Minagawa & Wada, 1986).

14.4.3 Molecular fossils
Molecular fossils are derived from membrane lipids and other aliphatic hydrocarbons. For cyanobacteria, the presence of 2-methyl hopanoids is typical and is considered as a proof of oxygenic photosynthesis occurring in the geological record (Lyons et al., 2014). The presence of hydrocarbons in fossils as evidence of microscopic algae was suggested by Gelpi et al. (1970).

14.4.4 Spectroscopic biomarkers
Finding all these previously mentioned types of biomarkers requires *in situ* analyses. On the other hand, spectroscopic biomarkers can be used for the

remote observation of atmospheres of exoplanets or exomoons as well as for *in situ* analyses. For the detection of life on an exoplanet or an exomoon, the presence of reasonable amounts of the reduced and oxidised forms of carbon, such as CH_4/CO_2, or combinations of CH_4+O_2, or CH_4+O_3 (or N_2O) should indicate the potential for the presence life (Vázques et al., 2010). On Earth, the relatively high amount of O_2 in the atmosphere is caused by oxygenic photosynthesis, to which algae and cyanobacteria contribute significantly (Chapman, 2013).

For *in situ* detection terrestrially, the presence of photosynthetic pigments, such as chlorophylls and carotenoids, is used for estimation of the biomass of oxyphototrophic microorganisms and their assignment to major taxonomical classes in the field. In spectroscopic field studies, laser-induced fluorescence emission (LIFE) (Storrie-Lombardi et al., 2009; Storrie-Lombardi & Sattler, 2009) and Raman spectroscopy (Jorge Villar & Edwards, 2006; Marshall et al., 2010; Gómez et al., 2011) have been evaluated for the detection of key molecular biomarkers. Chapter 13 gives more details specifically of the Raman spectroscopic detection of biomarkers for the forthcoming ExoMars 2022 mission, which has a dedicated Raman spectrometer aboard the rover vehicle. Carotenoids were found to be the most stable biomarkers in a simulated Mars environment and are hence proposed as a signal for life detection using Raman spectroscopy on Mars (Baqué et al., 2018).

14.4.5 Other putative biomarkers based on photosynthetic activity

Photosynthetic activity can provide other putative biomarkers that could be considered for future adoption in the search for life experiments. Detection of these biomarkers must be tested in extreme environments on Earth to assess the feasibility and detection limits for existing instrumentation and future developments.

Since part of the light captured by an organism is dissipated as heat, then small temperature differences could reveal the presence of photosynthetic microorganisms using sensitive infrared cameras (M. Barták, Masaryk University, Brno, Czech Republic, pers. comm.). Since the thermal resolution of present hand-held infrared cameras is of the order of hundredths of a degree K, the feasibility of this approach for life detection must be evaluated. So far, this approach has not been tested either in the laboratory or in the field.

Only minor, but nevertheless detectable, amounts of captured light energy are re-emitted as fluorescence radiation. The approach based on variable chlorophyll fluorescence (VFC) detection has been commonly used in plant physiology since the 1980s to assess the physiological state of an organism and to evaluate photosynthetic processes (Krause & Weis, 1984; Schreiber & Neubauer, 1990). Presently, fluorescence imaging cameras allow the study of the spatial heterogeneity of the VFC emission at a resolution of ca. 0.2 mm/pixel (FluorCam, Photon Systems Instruments; Nedbal et al., 2000; Nedbal & Whitmarsh, 2004). The fluorescence imaging instrument was thereby able to detect photosynthetic microorganisms in rock samples from Río Tinto, Spain (Gómez et al., 2011), and from Svalbard (Figure 14.5). For the detection of extraterrestrial life, suitable sets of excitation and emission wavelengths should be developed to be able to reveal other photosynthetic systems based on pigments that are different from chlorophylls.

14.5 Conclusion

Oxyphototrophic microorganisms are important primary producers, if not the only ones, in extreme environments on the Earth. They are tolerant to a broad range of environmental conditions that correspond to or partially overlap with conditions on astrobiologically interesting bodies in the Solar System, especially Mars, Europa and Enceladus. The methods used to study oxyphototrophic microorganisms on Earth can be used for extraterrestrial life detection and the characterisation of putative extant life. Likewise, the methods used for the detection of life beyond the Earth may be used for *in situ* terrestrial studies in extreme environments.

14.6 Acknowledgement

This work was supported by the following projects of the Ministry of Education, Youth and Sports – CzechPolar2 (LM2015078) and ECOPOLARIS (CZ.02.1.01/0.0/0.0/16_013/0001708).

References

Aguilera, A., Manrubia, S.C., Gómez, F., Rodríguez, N., Amils, R. (2006). Eukaryotic community distribution and its relationship to water physicochemical parameters in an extreme acidic environment, Río Tinto (Southwestern Spain). *Applied and Environmental Microbiology*, **72**, 5325–5330.

Aguilera, A., Souza-Egipsy, V., Gomez, F., Amils, R. (2007a). Development and structure of eukaryotic biofilms in an extreme acidic environment, Rio Tinto (SW, Spain). *Microbial Ecology*, **53**, 294–305.

Aguilera, A., Zettler, E., Gomez, F., et al (2007b). Distribution and seasonal variability in the benthic eukaryotic community of Rio Tinto (SW, Spain), an acidic, high metal extreme environment. *Systematic and Applied Microbiology*, **30**, 531–546.

Allen, J.F. (2003). State transitions – a question of balance. *Science*, **299**, 1530–1532.

Allewalt, J.P., Bateson, M.M., Revsbech, N.P., Slack, K., Ward, D.M. (2006). Effect of temperature and light on growth of and photosynthesis by *Synechococcus* isolates typical of those predominating in the octopus spring microbial mat community of Yellowstone National Park. *Applied and Environmental Microbiology*, **72**, 544–550.

Amils, R., González-Toril, E., Fernández-Remolar, D., et al. (2007). Extreme environments as Mars terrestrial analogs: The Rio Tinto case. *Planetary and Space Science*, **55**, 370–381.

Angelini, G., Ragni, P., Eposito, D., et al. (2001). A device to study the effect of space radiation on photosynthetic organisms. *Physica Medica*, **XVIII**, 267–268.

Apte, S.K. (2001). Copying with salinity/water stress: cyanobacteria show the way. *Proceedings of the Indian National Science Academy*, B**67**, 285–310.

Avron, M., Ben-Amotz, A. (1992). *Dunaliella: Physiology, Biochemistry and Biotechnology*. Boca Raton, FL: CRC Press.

Axelsson, L., Beer, S. (2001). Carbon limitation. In:L.C. Rai,J.P. Gaur (eds) *Algal Adaptation to Environmental Stresses. Physiological, Biochemical and Molecular Mechanisms*. Springer-Verlag, Berlin/Heidelberg/New York, pp. 21–44.

Ballesteros, F., Fernandez-Soto, A., Martínez, V. (2019). Diving into exoplanets: are water seas the most common? *Astrobiology*. May 2019. Published ahead of print; http://doi.org/10.1089/ast.2017.1720.

Baqué, M., Hanke, F., Böttger, U., et al. (2018). Protection of cyanobacterial carotenoids' Raman signatures by Martian mineral analogues after high-dose gamma irradiation. *Journal of Raman Spectroscopy*, **49**, 1617–1627.

Baxter, A., Mittler, R., Suzuki, N. (2013). ROS as key players in plant stress signalling. *Journal of Experimental Botany*, **65**, 1229–1240.

Bekker, A., Holland, H., Wang, P.-L., et al. (2004). Dating the rise of atmospheric oxygen. *Nature*, **427**, 117.

Bhatia, S., Garg, A., Sharma, K., et al. (2011). Mycosporine and mycosporine-like amino acids: A paramount tool against ultra violet irradiation. *Pharmacognosy Review*, **5**, 138–146.

Bidigare, R.R., Ondrusek, M.E., Kennicutt, M.C., et al. (1993). Evidence for a photoprotective function for secondary carotenoids of snow algae. *Journal of Phycology*, **29**, 427–434.

Brandt, A., de Vera, J.-P., Onofri, S., Ott, S. (2015). Viability of the lichen Xanthoria elegans and its symbionts after 18 months of space exposure and simulated Mars conditions on the ISS. *International Journal of Astrobiology*, **14**, 411–425.

Bray, C.M., West, C.E. (2005). DNA repair mechanisms in plants: crucial sensors and effectors for the maintenance of genome integrity. *New Phytologist*, **168**, 511–528.

Broady, P.A. (1984). Taxonomic and ecological investigations of algae on steam-warmed soil on Mt Erebus, Ross Island, Antarctica. *Phycologia*, **23**, 257–271.

Buick, R. (2008). When did oxygenic photosynthesis evolve? *Philosophical Transactions of the Royal Society of London B: Biological Sciences*, **363**, 2731–2743.

Burns, B.P., Goh, F., Allen, M., Neilan, B.A. (2004). Microbial diversity of extant stromatolites in the hypersaline marine environment of Shark Bay, Australia. *Environmental Microbiology*, **6**, 1096–1101.

Chapman, R.L. (2013). Algae: the world's most important 'plants' – an introduction. *Mitigation and Adaptation Strategies for Global Change*, **18**, 5–12.

Christiansen, J. (2018). Exoplanet catalogs. In: H. J. Deeg, J.A. Belmonte (eds) *Handbook of Exoplanets*. Springer International Publishing, Cham, pp. 1933–1947.

Clegg, M.R., Maberly, S.C., Jones, R.I. (2003). Chemosensory behavioural response of freshwater phytoplanktonic flagellates. *Plant Cell and Environment*, **27**, 123–135.

Cockell, C.S., Knowland, J. (1999). Ultraviolet radiation screening compounds. *Biological Reviews*, **79**, 311–345.

Cockell, C.S., McMahon, S. (2019). Lifeless Martian samples and their significance. *Nature Astronomy*, **3**, 468–470.

Cockell, C.S., Rettberg, P., Rabbow, E., Olsson-Francis, K. (2011). Exposure of phototrophs to 548 days in low Earth orbit: microbial selection pressures in outer space and on early earth. *ISME Journal*, **5**, 1671–1682.

de los Ríos, A., Ascaso, C., Wierzchos, J., Fernández-Valiente, E., Quesada, A. (2004). Microstructural characterization of cyanobacterial mats from the McMurdo Ice

Shelf, Antarctica. *Applied and Environmental Microbiology*, **70**, 569–580.

de Vera, J. P., Horneck, G., Rettberg, P., Ott, S. (2004). The potential of the lichen symbiosis to cope with the extreme conditions of outer space II: germination capacity of lichen ascospores in response to simulated space conditions. *Advances in Space Research*, **33**, 1236–1243.

de Vera, J.-P., Schulze-Makuch, D., Khan, A., et al. (2014). Adaptation of an Antarctic lichen to Martian niche conditions can occur within 34 days. *Planetary and Space Science*, **98**, 182–190.

Des Marais, D.J., Jahnke, L.L. (2019). Biosignatures of cellular components and metabolic activity. In: B.Cavalazzi, F. Westall (eds) *Biosignatures for Astrobiology*. Springer International Publishing, Cham, pp. 51–85.

Doran, P.T., Lyons, W.B., McKnight, D.M. (2010). *Life in Antarctic Deserts and Other Cold Dry Environments: Astrobiological Analogs*. Cambridge University Press, Cambridge.

Dundas, C.M., Diniega, S., Hansen, C.J., Byrne, S., McEwen, A.S. (2012). Seasonal activity and morphological changes in martian gullies. *Icarus*, **220**, 124–143.

Dundas, C.M., Bramson, A.M., Ojha, L., et al. (2018). Exposed subsurface ice sheets in the Martian mid-latitudes. *Science*, **359**, 199–201.

Ehling-Schulz, M., Scherer, S. (1999). UV protection in cyanobacteria. *European Journal of Phycology*, **34**, 329–338.

Elster, J. (1999). Algal versatility in various extreme environments. In: J. Seckbach (ed.) *Enigmatic Microorganisms and Life in Extreme Environments*. Kluwer Academic Publishers, Dordrecht/Boston/London, pp. 215–227.

Elster, J., Svoboda, J. (1995). *In situ* simulation and manipulation of a glacial stream ecosystem in the Canadian High Arctic. In: D. Jenkins, R.C. Ferrier, C. Kirby (eds) *Ecosystem Manipulation Experiments: Scientific Approaches, Experimental Design and Relevant Results*. Commission of the European Communities, Institute of Hydrology, UK and Environment Canada, Brussels, pp. 254–263.

Elster, J., Svoboda, J., Kanda, H. (2001). Controlled environmental platform used in temperature manipulation study of a stream periphyton in the Ny-Ålesund, Svalbard. In: J. Elster, J. Seckbach, W. F. Vincent,O. Lhotský (eds) *Algae and Extreme Environments*. Cramer, Stuttgart, pp. 63–75.

Elster, J., Kvíderová, J., Hájek, T., Láska, K., Šimek, M. (2012). Impact of warming on Nostoc colonies (Cyanobacteria) in a wet hummock meadow, Spitzbergen. *Polish Polar Research*, **33**, 395–420.

Fazi, S., Butturini, A., Tassi, F., et al. (2018). Biogeochemistry and biodiversity in a network of saline–alkaline lakes: implications of ecohydrological connectivity in the Kenyan Rift Valley. *Ecohydrology & Hydrobiology*, **18**, 96–106.

Finlay, J.C. (2004). Patterns and controls of lotic algal stable carbon isotope ratios. *Limnology and Oceanography*, **49**, 850–861.

Fujita, Y., Ohki, K., Murakami, A. (1985). Cromatic regulation of photosystem composition in the photosynthetic system of red and blue-green algae. *Plant and Cell Physiology*, **26**, 1541–1548.

Fukuda, S.-y., Yamakawa, R., Hirai, M., et al. (2008). Mechanisms to avoid photoinhibition in a desiccation-tolerant cyanobacterium, Nostoc commune. *Plant and Cell Physiology*, **49**, 488–492.

Gelpi, E., Schneider, H., Mann, J., Oró, J. (1970). Hydrocarbons of geochemical significance in microscopic algae. *Phytochemistry*, **9**, 603–612.

Gimmler, H. (2001). Acidophilic and acidotolerant algae. In: L.C. Rai, J.P.Gaur (eds) *Algal Adaptation to Environmental Stresses. Physiological, Biochemical and Molecular Mechanisms*. Springer-Verlag, Berlin/Heidelberg/New York, pp. 259–290.

Gimmler, H., Degenhardt, B. (2001). Alkaliphilic and alkalitolerant algae. In: L.C. Rai, J. P. Gaur (eds) *Algal Adaptation to*

Environmental Stresses. Physiological, Biochemical and Molecular Mechanisms. Springer-Verlag, Berlin/Heidelberg/New York, pp. 291–322.

Glein, C.R., Baross, J.A., Waite Jr, J.H. (2015). The pH of Enceladus' ocean. Geochimica et Cosmochimica Acta, 162, 202–219.

Gòdia, F., Albiol, J., Montesinos, J.L., et al. (2002). MELISSA: a loop of interconnected bioreactors to develop life support in Space. Journal of Biotechnology, 99, 319–330.

Gombos, Z., Wada, H., Hideg, E., Murata, N. (1994a). The unsaturation of membrane lipids stabilizes photosynthesis against heat stress. Plant Physiology, 104, 563–567.

Gombos, Z., Wada, H., Murata, N. (1994b). The recovery of photosynthesis from low-temperature photoinhibition is accelerated by the unsaturation of membrane lipids: A mechanism of chilling tolerance. Proceedings of the National Academy of Sciences of the USA, 91, 8787–8791.

Gómez, F., Walter, N., Amils, R., et al. (2011). Multidisciplinary integrated field campaign to an acidic Martian Earth analogue with astrobiological interest: Rio Tinto. International Journal of Astrobiology, 10, 291–305.

Gorbushina, A.A., Krumbein, W.E. (1999). The poikilotrophic micro-organism and its environment. In: J. Seckbach (eds.) Enigmatic Microorganisms and Life in Extreme Environments. Kluwer Academic Press, Dordrecht/Boston/London, pp. 177–185.

Hägele, D., Leinfelder, R., Grau, J., Burmeister, E.-G., Struck, U. (2006). Oncoids from the river Alz (southern Germany): tiny ecosystems in a phosphorus-limited environment. Palaeogeography Palaeoclimatology Palaeoecology, 237, 378–395.

Henley, W.J. (2001). Algae under desiccation stress. In: Elster, J., Seckbach, J., Vincent, W.F., Lhotský, O. (eds) Algae and Extreme Environments. J. Cramer in der Gebr. Borntraeger Verlagsbuchhandlung, Stuttgart, pp. 443–452.

Hindák, F., Kvíderová, J., Lukavský, J. (2013). Growth characteristics of selected thermophilic strains of cyanobacteria using crossed gradients of temperature and light. Biologia, 68, 830–837.

Hoham, R.W., Duval, B. (2001). Microbial ecology of snow and freshwater ice with emphasis on snow algae. In: Snow Ecology: An Interdisciplinary Examination of Snow-Covered Ecosystems, eds.H. G. Jones, J. W. Pomeroy, D. A. Walker, R. W. Hoham (eds). Cambridge University Press, Cambridge, UK, pp. 168–228.

Hoiczyk, E. (2000). Gliding motility in cyanobacteria: observations and possible explanations. Archives of Microbiology, 174, 11–17.

Horneck, G., Debus, A., Mani, P., Spry, J.A. (2007). Astrobiology exploratory missions and planetary protection requirements. In: G.Horneck,P. Rettberg (eds) Complete Course in Astrobiology, Wiley VCH,Weinheim, pp. 353–397.

Javaux, E. J. (2015). Biomarkers. In: M.Gargaud, W.M. Irvine,R., Amils (eds) Encyclopedia of Astrobiology. Springer Berlin Heidelberg, Berlin, Heidelberg, pp. 271–293.

Jorge Villar, S.E., Edwards, H.G.M. (2006). Raman spectroscopy in astrobiology. Analytical and Bioanalytical Chemistry, 384, 100–113.

Joset, F., Jeanjean, R., Hagemann, M. (1996). Dynamics of the response of cyanobacteria to salt stress: Deciphering the molecular events. Physiologia Plantarum, 96, 738–744.

Kappen, L., Bölter, M., Kühn, A. (1986). Field measurements of net photosynthesis of lichens in the Antarctic. Polar Biology, 5, 255–258.

Kappen, L., Schroeter, B., Scheidegger, C., Sommerkorn, M., Hestmark, G. (1996). Cold resistance and metabolic activity of lichens below 0°C. Advances in Space Research, 18, 119–128.

Keresztúri, A. (2012). Review of wet environment types on Mars with focus on duration and volumetric issues. *Astrobiology*, **12**, 586–600.

Kiang, N.Y., Segura, A., Tinetti, G., et al. (2007a). Spectral signatures of photosynthesis. II. Coevolution with other stars and atmosphere on extrasolar worlds. *Astrobiology*, **7**, 252–274.

Kiang, N.Y., Siefert, J., Govindjee, Blankenship, R.E. (2007b). Spectral signatures of photosynthesis. I. Review of Earth organisms. *Astrobiology*, **7**, 222–251.

Kirst, G.O., Thiel, C., Wolff, H., et al. (1991). Dimethylsulfoniopropionate (DMSP) in icealgae and its possible biological role. *Marine Chemistry*, **35**, 381–388.

Komárek, J., Nedbalová, L. (2007). Green cryosestic algae. In:J.Seckbach (ed.) *Algae and Cyanobacteria in Extreme Environments*. Springer, Dordrecht, pp. 323–342.

Kováčik, Ľ., Jezberová, J., Komárková, J., Kopecký, J., Komárek, J. (2011). Ecological characteristics and polyphasic taxonomic classification of stable pigment-types of the genus *Chroococcus* (Cyanobacteria). *Preslia*, **83**, 145–166.

Krause, G.H., Weis, E. (1984). Chlorophyll fluorescence as a tool in plant physiology. II. Interpretation of fluorescence signals. *Photosynthesis Research*, **5**, 139–157.

Kregel, K.C. (2002). Invited review: heat shock proteins: modifying factors in physiological stress responses and acquired thermotolerance. *Journal of Applied Physiology*, **92**, 2177–2186.

Kumar, D., Kvíderová, J., Kaštánek, P., Lukavský, J. (2017). The green alga *Dictyosphaerium* chlorelloides biomass and polysaccharides production determined using cultivation in crossed gradients of temperature and light. *Engineering in Life Sciences*, **17**, 1030–1038.

Kvíderová, J. (2012). Photochemical performance of the acidophilic red alga *Cyanidium* sp. in a pH gradient. *Origins of Life and Evolution of Biospheres*, **42**, 223–234.

Kvíderová, J., Lukavský, J. (2005). The comparison of ecological characteristics of *Stichococcus* (Chlorophyta) strains isolated from polar and temperate regions. *Algological Studies*, **118**, 127–140.

Kvíderová, J., Elster, J., Šimek, M. (2011). In situ response of *Nostoc* commune s.l. colonies to desiccation in Central Svalbard, Norwegian High Arctic. *Fottea*, **11**, 87–97.

Kvíderová, J., Souquieres, C.-E., Elster, J. (2018). Ecophysiology of photosynthesis of *Vaucheria* sp. mats in a Svalbard tidal flat. *Polar Science*.

Kvíderová, J., Elster, J., Komárek, J. (2019). Ecophysiology of cyanobacteria in the Polar Regions. In: A.K. Mishra, D.N. Tiwari, A. N. Rai (eds) *Cyanobacteria*. Academic Press, London, San Diego, Cambridge, Oxford, pp. 277–302.

Ledford, H.K., Niyogi, K.K. (2005). Singlet oxygen and photo-oxidative stress management in plants and algae. *Plant, Cell & Environment*, **28**, 1037–1045.

Leigh, J.A., Stahl, D.A., Staley, J.T. (2007). Evolution of metabolism and early microbial communities. In: W.F.I., Sullivan, J.A., Baross (eds) *Planets and Life. The Emerging Science of Astrobiology*. Cambridge University Press, Cambridge, UK, pp. 222–236.

Li, H.Y., Liu, X.X., Wang, Y.L., Hu, H.H., Xu, X.D. (2009). Enhanced expression of antifreeze protein genes drives the development of freeze tolerance in an Antarctica isolate of Chlorella vulgaris. *Progress in Natural Science*, **19**, 1059–1062.

Lyons, T.W., Reinhard, C.T., Planavsky, N.J. (2014). The rise of oxygen in Earth's early ocean and atmosphere. *Nature*, **506**, 307.

MacIntyre, H.L., Kana, T.M., Anning, T., Geider, R.J. (2002). Photoacclimation of photosynthesis irradiance response curves and photosynthetic pigments in microalgae and cyanobacteria. *Journal of Phycology*, **38**, 17–38.

Mackey, T., Sumner, D., Hawes, I., et al. (2018). Stromatolite records of environmental change in perennially ice-covered Lake Joyce, McMurdo Dry Valleys, Antarctica. *Biogeochemistry*, **137**, 73–92.

Marshall, C.P., Edwards, H.G.M., Jehlicka, J. (2010). Understanding the application of Raman spectroscopy to the detection of traces of life. *Astrobiology*, **10**, 229–243.

Matsui, K., Nazifi, E., Hirai, Y., et al. (2012). The cyanobacterial UV-absorbing pigment scytonemin displays radical-scavenging activity. *Journal of General and Applied Microbiology*, **58**, 137–144.

McEwen, A.S., Ojha, L., Dundas, C.M., et al. (2011). Seasonal flows on warm Martian slopes. *Science*, **333**, 740–743.

McKay, C.P., Nienow, J.A., Meyer, M.A., Friedmann, E.I. (1993). Continuous nanoclimate data (1985–1988) from the Ross Desert (McMurdo Dry Valleys) cryptoendolithic microbial ecosystem. In: D.H. Bromwich, C. R. Stearns (eds) *Antarctic Meteorology and Climatology: Studies Based on Automatic Weather Stations*. American Geophysical Union,Washington, DC, pp. 201–207.

Meleshko, G.I., Shepelev, E.Y., Kordyum, E.I., Šetlík, I., Doucha, J. (1986). *Nevesomost i yeyo vliyanie na mikroorganizmy i rasteniya [Weightlessness and its influence on microorganisms and plants]*. In: N. N. Gurovskiy (ed.) *Rezultaty medicinskich issledovaniy vypolnenych na orbitalnom nauchno-issledovatelskom komplekse 'Salyut-6' – 'Soyuz' [Results of medical research carried out onborad scientic-research complex 'Salyut-6' – 'Soyuz']*. Nauka, Moscow, pp. 369–391.

Melis, A. (1999). Photosystem-II damage and repair cycle in chloroplasts: what modulates the rate of photodamage in vivo? *Trends in Plant Sciences*, **4**, 130–135.

Minagawa, M., Wada, E. (1986). Nitrogen isotope ratios of red tide organisms in the East China Sea: a characterization of biological nitrogen fixation. *Marine Chemistry*, **19**, 245–259.

Moon, B.Y., Higashi, S.I., Gombos, Z., Murata, N. (1995). Unsaturation of the membrane lipids of chloroplasts stabilizes the photosynthetic mashinery against low-temperature photoinhibition in transgenic tobacco plants. *Proceedings of the National Academy of Sciences of the USA*, **92**, 6219–6223.

Moore, L.S., Burne, R. (1994). The modern thrombolites of Lake Clifton, western Australia. In: J. Bertrand-Sarfati, C. Monty (eds) *Phanerozoic Stromatolites II*. Springer, Dordrecht, pp. 3–29.

Moorthi, S., Caron, D.A., Gast, R.J., Sanders, R. W. (2009). Mixotrophy: a widespread and important ecological strategy for planktonic and sea-ice nanoflagellates in the Ross Sea, Antarctica. *Aquatic Microbial Ecology*, **54**, 269–277.

Nealson, K.H. (1997). The limits of life on Earth and searching for life on Mars. *Journal of Geophysical Research: Planets*, **102**, 23675–23686.

Nedbal, L., Soukupová, J., Kaftan, D., Whitmarsh, J., Trtílek, M. (2000). Kinetic imaging of chlorophyll fluorescence using modulated light. *Photosynthesis Research*, **66**, 3–12.

Nedbal, L., Whitmarsh, J. (2004). Chlorophyll fluorescence imaging of leaves and fruits. In: G.C. Papageorgiou, Govindjee (eds) *Chlorophyll a Fluorescence. A Signature of Photosynthesis*. Springer, Dordrecht, pp. 389–407.

Nienow, J.A., McKay, C.P., Friedmann, E.I. (1988). Cryptoendolithic microbial environment in the Ross Desert of Antarctica: light in the photosynthetically active region. *Microbial Ecology*, **16**, 271–289.

Noffke, N. (2015). Ancient sedimentary structures in the < 3.7 ga Gillespie Lake Member, Mars, that resemble macroscopic morphology, spatial associations, and temporal succession in terrestrial microbialites. *Astrobiology*, **15**, 169–192.

Noffke, N., Gerdes, G., Klenke, T., Krumbein, W. E. (2001). Microbially induced sedimentary structures: a new category within the classification of primary sedimentary structures. *Journal of Sedimentary Research*, **71**, 649–656.

Ojha, L., Wilhelm, M.B., Murchie, S.L., et al. (2015). Spectral evidence for hydrated salts in recurring slope lineae on Mars. *Nature Geoscience*, **8**, 829–832.

Oliver, R.L. (1994). Floating and sinking in gas-vacuoolate cyanobacteria. *Journal of Phycology*, **30**, 161–173.

Olson, J.M. (2006). Photosynthesis in the Archean Era. *Photosynthesis Research*, **88**, 109–117.

Olsson-Francis, K., de la Torre, R., Towner, M. C., Cockell, C.S. (2009). Survival of akinetes (resting-state cells of cyanobacteria) in low earth orbit and simulated extraterrestrial conditions. *Origins of Life and Evolution of Biosphere*, **39**, 565–579.

Onofri, S., de la Torre, R., de Vera, J.P., et al. (2012). Survival of rock-colonizing organisms after 1.5 years in outer space. *Astrobiology*, **12**, 508–516.

Oren, A. (2007). Diversity of organic osmotic compounds and osmotic adaptation in cyanobacteria and algae. In: J. Seckbach (ed.) *Algae and Cyanobacteria in Extreme Environments*. Springer, Dordrecht, pp. 639–655.

Oren, A., Seckbach, J. (2001). Oxygenic photosynthesis in extreme environments. In: J. Elster, J. Seckbach, W.F. Vincent, O. Lhotský (eds) *Algae and Extreme Environments*. J. Cramer in der Gebr. Borntraeger Verlagsbuchhandlung, Berlin, Stuttgart, pp. 13–31.

Parker, B.C., Simmons Jr, G.M., Love, F.G., Wharton Jr, R.A, Seaburg, K.G. (1981). Modern stromatolites in Antarctic dry valley lakes. *Bioscience*, **31**, 656–661.

Popova, A.F. (2003). Comparative characteristic of mitochondria ultrastructural organisation in *Chlorella* cells under altered gravity conditions. *Advances in Space Research*, **31**, 2253–2259.

Popova, A.F., Sytnik, K.M. (1996). Peculiarities of ultrastructure of *Chlorella* cells growing aboard the Bion-10 during 12 days. *Advances in Space Research*, **17**, 99–102.

Popova, A.F., Sytnik, K.M., Kordyum, E.L., et al. (1989). Ultrastructural and growth indices of *Chlorella* culture in multicomponent aquatic systems under space flight conditions. *Advances in Space Research*, **9**, 79–82.

Potts, M. (1994). Desiccation tolerance of prokaryotes. *Microbiology and Molecular Biology Reviews*, **58**, 755–805.

Potts, M. (1999). Mechanisms of desiccation tolerance in cyanobacteria. *European Journal of Phycology*, **34**, 319–328.

Rabbow, E., Horneck, G., Rettberg, P., et al. (2009). EXPOSE, an astrobiological exposure facility on the International Space Station – from proposal to flight. *Origins of Life and Evolution of Biospheres*, **39**, 581–598.

Rasmussen, B., Fletcher, I.R., Brocks, J.J., Kilburn, M.R. (2008). Reassessing the first appearance of eukaryotes and cyanobacteria. *Nature*, **455**, 1101.

Raven, J.A. (2003). Inorganic carbon concentrating mechanisms in relation to the biology of algae. *Photosynthesis Research*, **77**, 155–171.

Rettberg, P., Rabbow, E., Panitz, C., Horneck, G. (2004). Biological space experiments for the simulation of Martian conditions: UV radiation and Martian soil analogues. *Advances in Space Research*, **33**, 1294–1301.

Sachindra, N.M., Sato, E., Maeda, H., et al. (2007). Radical scavenging and singlet oxygen quenching activity of marine carotenoid fucoxanthin and its metabolites. *Journal of Agricultural and Food Chemistry*, **55**, 8516–8522.

Seckbach, J., Chapman, D., Garbary, D., Oren, A., Reisser, W. (2007). Algae and cyanobacteria under environmental extremes. In: J. Seckbach (ed.) *Algae and Cyanobacteria in Extreme Environments*. Springer, Dordrecht, pp. 781–786.

Schmitt, F.-J., Campbell, Z.Y., Bui, M.V., et al. (2019). Photosynthesis supported by a chlorophyll f-dependent, entropy-driven uphill energy transfer in Halomicronema hongdechloris cells adapted to far-red light. *Photosynthesis Research*, **139**, 185–201.

Schneider, T., Schmid, E., de Castro, J.V., et al. (2011). Structure and function of the symbiosis partners of the lung lichen (*Lobaria pulmonaria* L. Hoffm.) analyzed by metaproteomics. *Proteomics*, **11**, 2752–2756.

Schreiber, U., Neubauer, C. (1990). O2-dependent electron flow, membrane energization and the mechanism of non-photochemical quenching of chlorophyll fluorescence. *Photosynthesis Research*, **25**, 279–293.

Schulze, E.-D., Beck, E., Müller-Hohenstein (2005). *Plant Ecology*. Springer, Berlin, Heidelberg.

Schulze-Makuch, D., Irwin, L.N. (2018). Energy sources and life. In: D. Schulze-Makuch, L. N. Irwin (eds) *Life in the Universe. Expectations and Constraints*. Springer Nature Switzerland AG, Cham, pp. 75–100.

Šetlík, I., Kordyum, V.A., Meleshko, G.I., et al. (1979). *Experiment Chlorella on the board of Salyut 6*. In: 29th Congress International Astronautical Federation IAF'78. Pergamon Press, London.

Shehawy, R.M., Kleiner, D. (2001). Nitrogen limitation. In: L. C. Rai, J. P. Gaur (eds) *Algal Adaptation to Environmental Stresses. Physiological, Biochemical and Molecular Mechanisms*. Berlin, Heidelberg, New York, Springer-Verlag, pp. 45–64.

Soo, R.M., Wood, S.A., Grzymski, J.J., McDonald, I.R., Cary, S.C. (2009). Microbial biodiversity of thermophilic communities in hot mineral soils of Tramway Ridge, Mount Erebus, Antarctica. *Environmental Microbiology*, **11**, 715–728.

Squyres, S.W., Knoll, A.H. (2005). Sedimentary rocks at Meridiani planum: origin, diagenesis, and implications for life on Mars. *Earth and Planetary Science Letters*, **240**, 1–10.

Stamenković, V., Ward, L.M., Mischna, M., Fischer, W.W. (2018). O2 solubility in Martian near-surface environments and implications for aerobic life. *Nature Geoscience*, **11**, 905–909.

Storrie-Lombardi, M.C., Sattler, B. (2009). Laser-induced fluorescence emission (LIFE): in situ nondestructive detection of microbial life in the ice covers of Antarctic lakes. *Astrobiology*, **9**, 659–672.

Storrie-Lombardi, M.C., Muller, J.-P., Fisk, M.R., et al. (2009). Laser-Induced Fluorescence Emission (LIFE): searching for Mars organics with a UV-enhanced PanCam. *Astrobiology*, **9**, 953–964.

Sytnik, K.M., Popova, A.F., Nechitailo, G.S., Mashinsky, A.L. (1992). Peculiarities of the submicroscopic organization of *Chlorella* cells cultivated on a solid medium in microgravity. *Advances in Space Research*, **12**, 103–107.

Takahashi, S., Badger, M.R. (2011). Photoprotection in plants: a new light on photosystem II damage. *Trends in Plant Science*, **16**, 53–60.

Tashyreva, D., Elster, J. (2012). Production of dormant stages and stress resistance of polar cyanobacteria. In: A. Hanslmeier, S. Kempe, J. Seckbach (eds) *Life on Earth and Other Planetary Bodies*. Springer, Dordrecht, pp. 367–386.

Tashyreva, D., Elster, J. (2016). Annual cycles of two cyanobacterial mat communities in hydro-terrestrial habitats of the High Arctic. *Microbial Ecology*, **71**, 887–900.

Thannickal, V.J. (2009). Oxygen in the evolution of complex life and the price we pay. *American Journal of Respiratory Cell and Molecular Biology*, **40**, 507–510.

van den Hoek, C., Mann, D.G., Jahns, H.M. (1995). *Algae: An Introduction to Phycology*. Cambridge University Press, Cambridge, UK.

Vázques, M., Pallé, E., Montanés Rodrigues, P. (2010). *Earth as a Distant Planet*. Springer, New York, Dordrecht, Heidelberg, London.

Wagner, F., Falkner, G. (2001). Phosphate limitation. In: L.C. Rai, J.P. Gaur (eds) *Algal Adaptation to Environmental Stresses. Physiological, Biochemical and Molecular Mechanisms*. Springer-Verlag,Berlin, Heidelberg, New York, pp. 65–110.

Wang, G., Chen, H., Li, G., et al. (2006). Population growth and physiological characteristics of microalgae in a miniaturized bioreactor during space flight. *Acta Astronautica*, **58**, 264–269.

Wang, G.H., Li, G.B., Li, D.H., et al. (2004). Real-time studies on microalgae under microgravity. *Acta Astronautica*, **55**, 131–137.

Whitley, D., Goldberg, S.P., Jordan, W.D. (1999). Heat shock proteins: a review of the molecular chaperones. *Journal of Vascular Surgery*, **29**, 748–751.

Xiong, F., Lederer, F., Lukavský, J., Nedbal, L. (1996). Screening of freshwater algae (Chlorophyta, Chromophyta) for ultraviolet-B sensitivity of the photosynthetic apparatus. *Journal of Plant Physiology*, **148**, 42–48.

CHAPTER FIFTEEN

Life at the extremes

STEVEN L. CHOWN

Monash University

15.1 Introduction

> The chief danger to our philosophy, apart from laziness and wooliness, is scholasticism ... which is treating what is vague as if it were precise ...
>
> F.P. Ramsey

Humans have always been fascinated by extreme environments and our capability not only to endure them, but also to perform great feats despite them. Such fascination has perhaps been best chronicled in Frances Ashcroft's (2000) extraordinary book, from which I have borrowed this concluding chapter's title. We are likewise captivated by species which seemingly make light of challenges we cannot overcome without technology: Pompeii worms at hydrothermal vents (Ravaux et al., 2013), bar-headed geese migrating over the Himalaya (Scott et al., 2015) and ice fish in the Antarctic (Kim et al., 2019). Their apparent ease, and that of other species living in difficult environments, belies a suite of biological responses which range from the dazzling to the initially mystifying. All of which depend on a sequence of events that begins with individuals and shows no end, except the extraordinary diversity of Earth's life. One of biology's main successes has been to unveil this sequence for many different species in many different environments. Another has been to understand that the boundary between individual and environment is not a hard one. Living organisms are adept at manipulating their environments to suit themselves (Turner, 2000). In a deep-time sense, all species live in a constructed world.

Physiology as a discipline has contributed much to modern understanding of the way in which life works and populations succeed in extreme environments. Its success is owed at least in part to the fact that physiological mechanisms impose limits on what is possible in terms of ecological responses

(Ricklefs & Wikelski, 2002). The chapters in this book reflect the value of the physiological approach. In spanning multiple organisational levels, taxa and ecosystems, they also demonstrate the benefits of cross-over between different fields – an advantage made clear in Krogh's famous (1929) address on the progress of physiology.

In a book that has explored so deeply and vividly life's responses to extreme environments, a final chapter should perhaps pause to consider, again, what really is meant by extreme environments. Across the preceding chapters, such environments are variously described as: permanently or periodically close to the limits for life; harsh; characterised by stress; or marked by a particular set of abiotic conditions, such as high acidity, low temperature, ongoing catastrophe – or some combination thereof. Such variety should not be surprising. As the history of defining 'stress' has demonstrated (Levine, 1985; Hoffmann & Parsons, 1991; He & Bertness, 2014; Del Giudice et al., 2018), coming to a conclusion is not straightforward. In this regard, Popper's (1945) advice can be taken. For the purposes of profitable further work, rather than an endless semantic struggle, the idea that definitions amount to any more than shorthand labels should be abandoned. Thus, for a hierarchically organised world, characterised on the one hand by flows of energy and on the other by flows of information (Eldredge, 1986), extreme environments may take a variety of forms and play out in different ways, just as is the case for stress. In consequence, the main requirement is that, at the very least, the level and subject of investigation are clearly set out. For example, the subject of interest may be extreme events in a given setting (Dowd & Denny, 2020), rather than absolute extremes.

While such an approach may itself appear at the very least semantic, this is not the case. Take as an example a suburban garden landscape in the UK. By the standards applied in most of the chapters in this book, British suburbia would not be considered an extreme environment. Yet it includes landscapes of fear, owing to feline predators, that have a tremendous influence on the behaviour, distribution and long-term success of prey populations (Bonnington et al., 2013). The fact that prey individuals avoid some patches because of predator presence indicates an individual assessment that the environment is likely to be too extreme. Certainly it would elicit stress because of its unpredictability and/or uncontrollability (Del Giudice et al., 2018) (for further discussions of landscapes of fear see also Gallagher et al., 2017; Gaynor et al., 2019). Of course, such patches are not inimical to all life, but this is also true of many environments regularly and comfortably considered extreme. Polar oceans abound with life, as do hydrothermal vents (Van Dover et al., 2002; Chown et al., 2015). Should any discussion of extremes then be abandoned? The answer is clearly 'no'.

15.2 Absolute and relative extremes

All abiotic variables have known natural extreme values. At just 1 K, the Boomerang Nebula is the coldest natural place known in the Universe (Sahai & Nyman, 1997). On Earth, the lowest surface temperatures have been recorded from the East Antarctic ice divide (−98°C) (Scambos et al., 2018), while the highest tend to be from the Lut Desert in Iran, although sites elsewhere, including in Australia, are not far off (Mildrexler et al., 2011). Geothermal springs, hydrothermal vents and deep mines are, of course, much hotter. Average annual precipitation ranges from about 1 mm (Arica, in the Atacama) (Houston, 2006) to >11 500 mm on the Meghalaya Plateau in northeastern India (Murata et al., 2007), though parts of the Dry Valleys in Antarctica may have very low precipitation too (Marchant & Head, 2007). Natural extreme values for other variables such as salinity, pH, light and background radiation have all been documented (Aliyu & Ramli, 2015; Merino et al., 2019). Despite these extreme values, life seems to be able to thrive in some form in most of the environments characterised by such extremes, even when these are spatially extensive and temporally consistent. Indeed, such predictability may be one key to success (Southwood, 1988; Del Giudice et al., 2018).

Of the abiotic variables, temperature has been subject to most scrutiny, perhaps because of its ease of measurement and its importance to life. The temperature extremes within which individuals can survive, metabolise and complete a full life cycle have been comprehensively reviewed (Clarke, 2014). At the high temperature end of the continuum, Archaea (maximum of 122°C) exceed Bacteria (maximum of 100°C) in the temperatures at which a life cycle can be completed, and in turn they exceed the Eukarya (maximum of 60°C), especially the multicellular vertebrates and most invertebrates. Very clearly, extremes for one major group of organisms are different from those of another. Such variation is clearly illustrated by the thermal optima for growth and other traits across the spectrum of life (Figure 15.1).

Variation among major phylogenetic groups in thermal tolerance has long been known and several innovative explanations for these differences have been proposed (Clarke, 2014), including the idea of oxygen and capacity-limited thermal tolerance (Pörtner, 2001). The aim here is not to consider these ideas in detail (e.g. see Verberk et al., 2016; Jutfelt et al., 2018; Pörtner et al., 2018), but rather simply to point out how clearly they illustrate that extremes are in a very real sense defined by organisms (Dowd & Denny, 2020). And this should come as no surprise because evolutionary change, phylogenetic constraint and phenotypic plasticity are such fundamental characteristics of life (Lauder, 1982; Endler, 1986; West-Eberhard, 2003).

Moreover, scientific perceptions of extremes should also be expected to change as understanding grows. For example, for some time the highest thermal tolerance for an active terrestrial arthropod was thought to be for the Australian

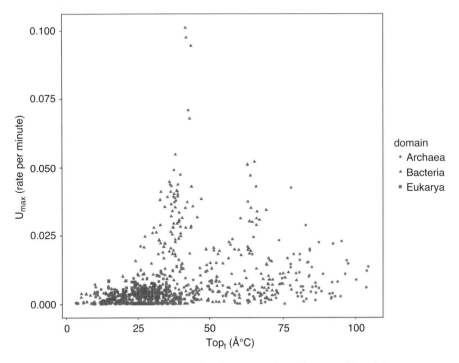

Figure 15.1 Thermal optima (T_{opt}) of various rates (mostly growth) and the rates at those optima (U_{max}) for life's major domains. Data from Corkrey et al. (2016) and Sørensen et al. (2018). Code for extracting this information from these sources is available from the author. (A black and white version of this figure will appear in some formats. For the colour version, please refer to the plate section.)

ant *Melophorus bagoti*, with a survival of 1 hour at 54°C and a critical thermal maximum of 56.7°C (Christian & Morton, 1992), matched perhaps only by the Saharan silver ant *Cataglyphis bomycina* (Wehner et al., 1992) or by two species of pseudoscorpion (Heurtault & Vannier, 1990). In 2015, however, data were published on a mite species with an estimated upper lethal temperature for 50% of the population after 1 hour of exposure to *ca.* 60°C (Wu & Wright, 2015).

Thus, most value is to be gained by understanding how organisms respond through ecological and evolutionary time to the environmental variation they encounter, and what may cause limits to these responses. Environments we perceive as extreme are clearly not so for other organisms, and even in our zeal to use organisms from the planet's abiotic extremes to fast-track our understanding of processes, we may miss the obvious. For example, while Antarctic marine systems are on the lowest side of the ocean temperature continuum, focussing only on this aspect misses the important fact that they are dark for much of the year, covered in sea ice annually, and at shallow depths subject both to anchor ice and glacial scour (Peck, 2018). On the contrary, the benthos has not had to deal with shell-crushing crabs for much of recent evolutionary time (Aronson et al., 2015).

15.3 A focus on the interactions

One useful way to proceed in understanding responses is to adopt an eco-evolutionary approach at the population level, considering the circumstances which lead individuals to have better survival probabilities and likelihood of leaving more offspring which in turn survive to reproduction (Sibly & Calow, 1986), and how these probabilities depend on flows of mass, heat and water (Kearney et al., 2012) (Figure 15.2). The approach can be amended to include the way in which different components of food (lipids, proteins, carbohydrates) are targeted, or to consider specific regulatory pathways (e.g. ion balance).

What such an approach usefully does is draw attention to organism–environment interactions and how they affect population fitness. Extreme

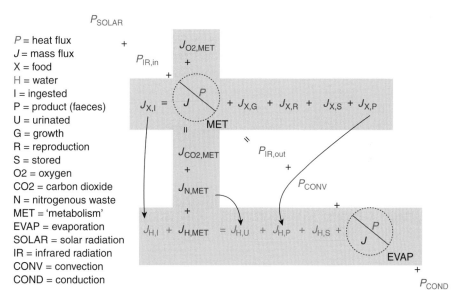

Figure 15.2 Coupled heat/mass balance equations for an organism (figure republished with permission of John Wiley & Sons - Books, from Balancing heat, water and nutrients under environmental change: a thermodynamic niche framework, Kearney et al. 2012 https://doi.org/10.1111/1365-2435.12020; permission conveyed through Copyright Clearance Center, Inc.) The diagonal equation represents the heat balance and, in a biophysical model, are calculated using the physical equations for heat exchange. The horizontal equations, within the shaded grey area, represent connections between the mass balances (food/water/gas exchange/excretion) and the heat balance, and are normally based on empirically determined relationships. (A black and white version of this figure will appear in some formats. For the colour version, please refer to the plate section.)

environments impose uncontrollable or unpredictable effects on these flows in individuals, with deleterious outcomes for survival, reproduction or non-genetic cross-generational influences. If these effects apply across the full distribution of phenotypic variation for the population, the population will eventually become extinct. Whether the species persists in an area will then depend on meta-community dynamics (Leibold & Chase, 2018). Even under absolutely extreme conditions, the differences between control (e.g. minimum metabolic control) and lack thereof (abandoned metabolic control) can be distinguished (Makarieva et al., 2006).

What such an approach also does is dispense with the requirement that extremes need only be abiotic. An extreme environment may be one that has a high density of predators or a pervasive, virulent pathogen. It also draws attention to the fact that abiotic factors interact, and also do so with biotic factors (Hoffmann & Parsons, 1991; Hawlena & Schmitz, 2010; Harley et al., 2017). Although such interactions are not necessarily straightforward to investigate in the field or to replicate in the laboratory, forgetting them is a recipe for overconfident forecasts. Moreover, the approach also suggests what traits might be important to consider when exploring the ways in which environmental factors affect abundance, distribution and meta-community composition (Kearney et al., 2019). It also lends itself to understanding clearly what extreme events mean for particular populations, given that the capabilities of these populations determine the outcome of a given event (Dowd & Denny, 2020), which may itself be defined based on duration, intensity, rate of development and spatial extent (Hobday et al., 2016).

15.4 Unpredictability

When considering extremes, especially from a physiological perspective, much attention is given to the outer limits of variation, whether these limits are defined in an absolute sense (the driest place on Earth), a relative one (the warmest temperature at which an ectothermic vertebrate can complete its life cycle) or in terms of events relative to a given life stage within a population (Bowler & Terblanche, 2008). Yet in the literature on stress, and especially with regard to behavioural and endocrine responses of individuals, unpredictability is frequently emphasised (see Del Giudice et al., 2018).

From an eco-evolutionary perspective, unpredictability is of much importance because it determines population-level outcomes (Simons, 2011; Tufto, 2015). In the absence of predictability, both short-term (phenotypic plasticity) and long-term (adaptation) responses become problematic. In the former case, plasticity is typically low when cues are unreliable (i.e. environmental variation unpredictable) (Tufto, 2000; Chown & Terblanche, 2007). In the latter, bet-hedging is thought to be the main long-term outcome, yet examples of bet-hedging are surprisingly difficult to find, admittedly in part because of the difficulties of demonstrating such a response (Childs et al., 2010; Simons, 2011; Tufto, 2015).

Perhaps one of the clearest illustrations of the importance of predictability in an extreme setting comes from the southwestern part of southern Africa. In the low winter rainfall areas of the Succulent Karoo, plant diversity is exceptionally high for any region globally that is so arid (Figure 15.3). Much higher, for example, than the closely located areas of the Nama Karoo. The exceptional diversity of the Succulent Karoo is attributable in part to the fact that the timing and amount of the low precipitation of the area is reasonably predictable from year to year (Cowling et al., 1999; Desmet & Cowling, 1999). In such settings, population extinction is expected to be far lower than in other circumstances (Tufto, 2015).

Given that unpredictability makes responses to the environment by individuals and populations so difficult, the way in which it may change as the global climate crisis progresses is an exceptionally important concern (Chown et al., 2010; Simons, 2011). While organisms may struggle to keep pace with the velocity of change (Loarie et al., 2009; Gillson et al., 2013), especially given

Figure 15.3 The Succulent Karoo of southern Africa is an exceptionally species-rich arid region, in part owing to the predictability of its rainfall. The image here was taken near Springbok, Northern Cape, South Africa in August 2006. (A black and white version of this figure will appear in some formats. For the colour version, please refer to the plate section.)

the transformations of habitats globally which limit movement and meta-community dispersal, unpredictability of that change, or of circumstances in local settings, will make that struggle even harder. Understanding such unpredictability and how it affects population fitness, especially in the case of extreme events, is pressing.

15.5 Where does that leave extremes?

In Krogh's (1929) essay, he pointed out that by the very nature of the progress of the field, greater specialisation would take place. Certainly, at that time, ecology and physiology were in the process of their now much-discussed divorce (Huey, 1991; Gaston et al., 2009). Shortly thereafter, Ortega Y Gasset (1932) lamented the barbarism of scientific specialisation. Yet for the most part and for many years these divisions have simply grown. That trend continues. But in recent years the value of transdisciplinary research has emerged swiftly, along with useful advice on how to make it work (Brown et al., 2015). These developments are reflected in the fields most occupied with life at the extremes too. Several approaches now seek to integrate biophysics, ecology, evolution and physiology over a range of scales and for both instrumental and non-instrumental approaches (e.g. Gaston et al., 2009; Cooke et al., 2013; Violle et al., 2014). The benefits of so doing are already being reaped from specific investigations of life's responses to extremes (e.g. Clark et al., 2020; Peck, 2020 in this book). Given current momentum, the expectation is that this integration will continue. And it would benefit especially from consideration of the flows of matter and energy in environments ranging across the variability–predictability phase space. To this end, closer integration of what actually happens in the field relative to the laboratory is critical (Dowd & Denny, 2020). So too is a focus on variables other than temperature, and how they interact with the latter (Chown et al., 2011; Peck, 2018).

15.6 Acknowledgements

Grant Duffy produced Figure 15.1. I am grateful to Andrew Clark, Lloyd Peck and Anastassia Makarieva for discussions over many years. Melodie McGeoch read a draft of this work and provided helpful comments. My research is supported by Australian Research Council grant DP170101046.

References

Aliyu, A.S., Ramli, A.T. (2015). The world's high background natural radiation areas (HBNRAs) revisited: a broad overview of the dosimetric, epidemiological and radiobiological issues. *Radiation Measurements*, **73**, 51–59.

Aronson, R.B., Frederich, M., Price, R., Thatje, S., Alistair Crame, J. (2015). Prospects for the return of shell-crushing crabs to Antarctica. *Journal of Biogeography*, **42**, 1–7.

Ashcroft, F. (2000). *Life at the Extremes. The Science of Survival*. Flamingo, London.

Bonnington, C., Gaston, K.J., Evans, K.L., Whittingham, M. (2013). Fearing the feline: domestic cats reduce avian fecundity through trait-mediated indirect effects that increase nest predation by other species. *Journal of Applied Ecology*, **50**, 15–24.

Bowler, K., Terblanche, J.S. (2008). Insect thermal tolerance: what is the role of ontogeny, ageing and senescence? *Biological Reviews*, **83**, 339–355.

Brown, R.R., Deletic, A., Wong, T.H.F. (2015). How to catalyse collaboration. *Nature*, **525**, 315–317.

Childs, D.Z., Metcalf, C.J.E., Rees,M. (2010). Evolutionary bet-hedging in the real world: empirical evidence and challenges revealed by plants. *Proceedings of the Royal Society B*, **277**, 3055–3064.

Chown, S.L., Terblanche, J.S. (2007). Physiological diversity in insects: ecological and evolutionary contexts. *Advances in Insect Physiology*, **33**, 50–152.

Chown, S.L., Hoffmann, A.A., Kristensen, T.N., et al. (2010). Adapting to climate change: a perspective from evolutionary physiology. *Climate Research*, **43**, 3–15.

Chown, S.L., Sørensen, J.G., Terblanche, J.S. (2011). Water loss in insects: an environmental change perspective. *Journal of Insect Physiology*, **57**, 1070–1084.

Chown, S.L., Clarke, A., Fraser, C.I., et al. (2015). The changing form of Antarctic biodiversity. *Nature*, **522**, 431–438.

Christian, K.A., Morton, S.R. (1992). Extreme thermophilia in a Central Australian Ant, Melophorus bagoti. *Physiological Zoology*, **65**, 885–905.

Clark, M.S., Verde, C., Fineschi, S., et al. (2020). Life in the extreme environments of our planet under pressure: climate-induced threats and exploitation opportunities. In: G. di Prisco, A. Huiskes, J. Elster, H. Edwards (eds) *Life in Extreme Environments*. Cambridge University Press, Cambridge.

Clarke, A. (2014). The thermal limits to life on Earth. *International Journal of Astrobiology*, **13**, 141–154.

Cooke, S.J., Sack, L., Franklin, C.E., et al. (2013). What is conservation physiology? Perspectives on an increasingly integrated and essential science. *Conservation Physiology*, **1**, cot001.

Corkrey, R., McMeekin, T.A., Bowman, J.P., et al. (2016). The biokinetic spectrum for temperature. *PLoS ONE*, **11**, e0153343.

Cowling, R.M., Esler, K.J., Rundel, P.W. (1999). Namaqualand, South Africa – an overview of a unique winter-rainfall desert ecosystem. *Plant Ecology*, **142**, 3–21.

Del Giudice, M., Buck, C.L., Chaby, L.E., et al. (2018). What is stress? A systems perspective. *Integrative and Comparative Biology*, **58**, 1019–1032.

Desmet, P.G., Cowling, R.M. (1999). Biodiversity, habitat and range-size aspects of a flora from a winter-rainfall desert in north-western Namaqualand, South Africa. *Plant Ecology*, **142**, 23–33.

Dowd, W.W., Denny, M.W. (2020). A series of unfortunate events: characterizing the contingent nature of physiological extremes using long-term environmental records. *Proceedings of the Royal Society B*, **287**, 20192333.

Eldredge, N. (1986). Information, economics, and evolution. *Annual Review of Ecology and Systematics*, **17**, 351–369.

Endler, J.A. (1986). *Natural Selection in the Wild*. Princeton University Press, Princeton.

Gallagher, A.J., Creel, S., Wilson, R.P., Cooke, S.J. (2017). Energy landscapes and the landscape of fear. *Trends in Ecology and Evolution*, **32**, 88–96.

Gaston, K.J., Chown, S.L., Calosi, P., et al. (2009). Macrophysiology: a conceptual reunification. *American Naturalist*, **174**, 595–612.

Gaynor, K.M., Brown, J.S., Middleton, A.D., Power, M.E., Brashares, J.S. (2019).

Landscapes of fear: spatial patterns of risk perception and response. *Trends in Ecology and Evolution*, **34**, 355–368.

Gillson, L., Dawson, T.P., Jack, S., McGeoch, M. A. (2013). Accommodating climate change contingencies in conservation strategy. *Trends in Ecology and Evolution*, **28**, 135–142.

Harley, C.D.G., Connell, S.D., Doubleday, Z.A., et al. (2017). Conceptualizing ecosystem tipping points within a physiological framework. *Ecology and Evolution*, **7**, 6035–6045.

Hawlena, D., Schmitz, O.J. (2010). Physiological stress as a fundamental mechanism linking predation to ecosystem functioning. *American Naturalist*, **176**, 537–556.

He, Q., Bertness, M.D. (2014). Extreme stresses, niches, and positive species interactions along stress gradients. *Ecology*, **95**, 1437–1443.

Heurtault, J., Vannier, G. (1990). Thermorésistance chez deux Pseudoscorpiones (Garypidae), l'un du désert de Namibie, l'autre de la région de Gênes (Italie). *Acta Zoologica Fennica*, **190**, 165–171.

Hobday, A.J., Alexander, L.V., Perkins, S.E., et al. (2016). A hierarchical approach to defining marine heatwaves. *Progress in Oceanography*, **141**, 227–238.

Hoffmann, A.A., Parsons, E.A. (1991). *Evolutionary Genetics and Environmental Stress*. Oxford University Press, Oxford.

Houston, J. (2006). Variability of precipitation in the Atacama Desert: its causes and hydrological impact. *International Journal of Climatology*, **26**, 2181–2198.

Huey, R.B. (1991). Physiological consequences of habitat selection. *American Naturalist*, **137**, S91–S115.

Jutfelt, F., Norin, T., Ern, R., et al. (2018). Oxygen- and capacity-limited thermal tolerance: blurring ecology and physiology. *Journal of Experimental Biology*, **221**, jeb169615.

Kearney, M.R., Simpson, S.J., Raubenheimer, D., Kooijman, S.A.L.M. (2012). Balancing heat, water and nutrients under environmental change: a thermodynamic niche framework. *Functional Ecology*, **27**, 950–966.

Kearney, M.R., McGeoch, M.A., Chown, S.L. (2019). Where do functional traits come from? The role of theory and models. *Biodiversity Information Science and Standards*; doi:10.3897/biss.3.37500

Kim, B.M., Amores, A., Kang, S., et al. (2019). Antarctic blackfin icefish genome reveals adaptations to extreme environments. *Nature Ecology and Evolution*, **3**, 469–478.

Krogh, A. (1929). The progress of physiology. *American Journal of Physiology*, **90**, 243–251.

Lauder, G.V. (1982). Historical biology and the problem of design. *Journal of Theoretical Biology*, **97**, 57–67.

Leibold, M.A., Chase, J.M. (2018). *Metacommunity Ecology*. Princeton University Press, Princeton, NJ.

Levine, S. (1985). A definition of stress? In:G. P. Moberg (ed.) *Animal Stress*. Springer, Berlin, pp. 51–69.

Loarie, S.R., Duffy, P.B., Hamilton, H., et al. (2009). The velocity of climate change. *Nature*, **462**, 1052–1055.

Makarieva, A.M., Gorshkov, V.G., Li, B.-L., Chown, S.L. (2006). Size- and temperature-independence of minimum life-supporting metabolic rates. *Functional Ecology*, **20**, 83–96.

Marchant, D.R., Head, J.W. (2007). Antarctic dry valleys: microclimate zonation, variable geomorphic processes, and implications for assessing climate change on Mars. *Icarus*, **192**, 187–222.

Merino, N., Aronson, H.S., Bojanova, D.P., et al. (2019). Living at the extremes: extremophiles and the limits of life in a planetary context. *Frontiers in Microbiology*, **10**, 780.

Mildrexler, D.J., Zhao, M., Running, S.W. (2011). Satellite finds highest land skin temperatures on Earth. *Bulletin of the American Meteorological Society*, **92**, 855–860.

Murata, F., Hayashi, T., Matsumoto, J., Asada, H. (2007). Rainfall on the Meghalaya plateau in northeastern India – one of the rainiest places in the world. *Natural Hazards*, **42**, 391–399.

Ortega y Gasset, J. (1932). *The Revolt of the Masses*. W.W. Norton, New York.

Peck, L. (2018). Antarctic marine biodiversity: adaptations, environments and responses to change. *Oceanography and Marine Biology. An Annual Review*, **56**, 105–236.

Peck, L. (2020). The ecophysiology of responding to change in polar marine benthos. In: G. di Prisco, A. Huiskes, J. Elster, H. Edwards (eds.) *Life in Extreme Environments*. Cambridge University Press, Cambridge.

Popper, K. (1945). *The Open Society and Its Enemies. Volume II. The High Tide of Prophecy: Hegel, Marx and the Aftermath*. Routledge & Kegan Paul, London & New York.

Pörtner, H.O. (2001). Climate change and temperature-dependent biogeography: oxygen limitation of thermal tolerance in animals. *Naturwissenschaften*, **88**, 137–146.

Pörtner, H.-O., Bock, C., Mark, F.C. (2018). Connecting to ecology: a challenge for comparative physiologists? Response to 'Oxygen- and capacity-limited thermal tolerance: blurring ecology and physiology'. *Journal of Experimental Biology*, **221**, jeb169615.

Ravaux, J., Hamel, G., Zbinden, M., et al. (2013). Thermal limit for metazoan life in question: in vivo heat tolerance of the Pompeii worm. *PLoS ONE*, **8**, e64074.

Ricklefs, R.E., Wikelski, M. (2002). The physiology/life-history nexus. *Trends in Ecology and Evolution*, **17**, 462–468.

Sahai, R., Nyman, L.-A. (1997). The Boomerang Nebula: the coldest region of the universe? *The Astrophysical Journal*, **487**, L155–L159.

Scambos, T.A., Campbell, G.G., Pope, A., et al. (2018). Ultralow surface temperatures in East Antarctica from satellite thermal infrared mapping: the coldest places on earth. *Geophysical Research Letters*, **45**, 6124–6133.

Scott, G.R., Hawkes, L.A., Frappell, P.B., et al. (2015). How bar-headed geese fly over the Himalayas. *Physiology*, **30**, 107–115.

Sibly, R.M., Calow, P. (1986). *Physiological Ecology of Animals. An Evolutionary Approach*. Blackwell Scientific Publications, Oxford.

Simons, A.M. (2011). Modes of response to environmental change and the elusive empirical evidence for bet hedging. *Proceedings of the Royal Society B*, **278**, 1601–1609.

Sørensen, J.G., White, C.R., Duffy, G.A., Chown, S.L. (2018). A widespread thermodynamic effect, but maintenance of biological rates through space across life's major domains. *Proceedings of the Royal Society B*, **285**, 20181775.

Southwood, T.R.E. (1988). Tactics, strategies and templets. *Oikos*, **52**, 3–18.

Tufto, J. (2000). The evolution of plasticity and nonplastic spatial and temporal adaptations in the presence of imperfect environmental cues. *American Naturalist*, **156**, 121–130.

Tufto, J. (2015). Genetic evolution, plasticity, and bet-hedging as adaptive responses to temporally autocorrelated fluctuating selection: a quantitative genetic model. *Evolution*, **69**, 2034–2049.

Turner, J.S. (2000). *The Extended Organism. The Physiology of Animal-Built Structures*. Harvard University Press, Cambridge, MA.

Van Dover, C.L., German, C.R., Speer, K.G., Parson, L.M., Vrijenhoek, R.C. (2002). Evolution and biogeography of deep-sea vent and seep invertebrates. *Science*, **295**, 1253–1257.

Verberk, W.C., Overgaard, J., Ern, R., et al. (2016). Does oxygen limit thermal tolerance in arthropods? A critical review of current evidence. *Comparative Biochemistry and Physiology A*, **192**, 64–78.

Violle, C., Reich, P.B., Pacala, S.W., Enquist, B.J., Kattge, J. (2014). The emergence and promise of functional biogeography. *Proceedings of the National Academy of Sciences of the USA*, **111**, 13690–13696.

Wehner, R., March, A.C., Wehner, S. (1992). Desert ants on a thermal tightrope. *Nature*, **357**, 586–587.

West-Eberhard, M.J. (2003). *Developmental Plasticity and Evolution*. Oxford University Press, New York.

Wu, G.C., Wright, J.C. (2015). Exceptional thermal tolerance and water resistance in the mite Paratarsotomus macropalpis (Erythracaridae) challenge prevailing explanations of physiological limits. *Journal of Insect Physiology*, **82**, 1–7.

Index